PHYLOGENY RECONSTRUCTION IN PALEONTOLOGY

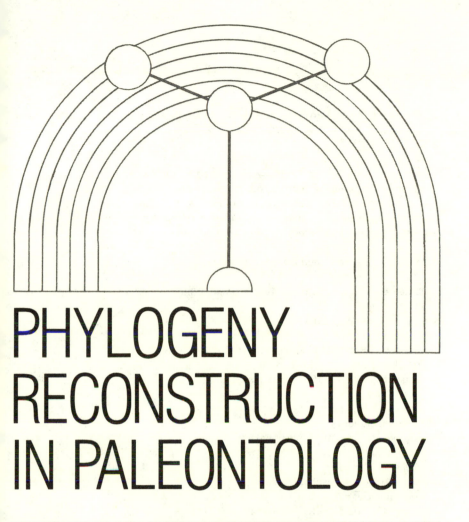

PHYLOGENY RECONSTRUCTION IN PALEONTOLOGY

Robert M. Schoch

Boston University

 VAN NOSTRAND REINHOLD COMPANY
———————————————————————— *New York*

Manufactured in the United States of America.

Published by Van Nostrand Reinhold Company Inc.
115 Fifth Avenue
New York, New York 10003

Van Nostrand Reinhold Company Limited
Molly Millars Lane
Wokingham, Berkshire RG11 2PY, England

Van Nostrand Reinhold
480 Latrobe Street
Melbourne, Victoria 3000, Australia

Macmillan of Canada
Division of Gage Publishing Limited
164 Commander Boulevard
Agincourt, Ontario MIS 3C7, Canada

15 14 13 12 11 10 9 8 7 6 5 4 3 2 1

Library of Congress Cataloging-in-Publication Data
Schoch, Robert M.
    Phylogeny reconstruction in paleontology.
    Bibliography: p.
    Includes index.
    1. Paleontology.   2. Phylogeny.   I. Title.
QE721.S357   1986   575   86-9051
ISBN 0-442-27967-1

Contents

Preface

The construction of phylogenies and evolutionary trees has traditionally been one of the chief activities of many paleontologists. The only direct evidence of past life is within the paleontologist's domain. The reconstruction of phylogeny (this term is discussed and defined in the Introduction) is of interest in its own right and also contributes to the study of evolution and the history of life on earth. Since the general acceptance of evolution, phylogenetic considerations have played a major role in the development of biological classifications.

All too often the reconstruction of phylogenies has been carried out instinctively or intuitively with little clear articulation of methodology or underlying assumptions (Ghiselin, 1972). One commonly held, but naive, view is that we can read phylogeny or evolutionary history directly from the sequence of fossils preserved in the stratigraphic record (Bonde, 1977; Savage, 1977; Simpson, 1975). In many cases elaborate quantitative (phenetic) statistical techniques have been applied to fossil data without any clear concept of how to evaluate the techniques applied or the results. For example, in the context of a detailed study of some Ordovician crinoids, Brower (1973, p. 328) could state: "The method [a morphological distance matrix] has yielded satisfactory results in the two lineages studied quantitatively. Satisfactory should be translated as the lineages and evolutionary trends deduced therefrom largely fit the writer's preconceived notions." To remedy this situation, in the last quarter-century a limited, but often vociferous, number of biologists and paleontologists have rigorously scrutinized certain aspects of phylogeny reconstruction. These debates, although no consensus has yet been reached, are giving rise to the modern foundations of phylogeny reconstruction. The generation of Willi Hennig, Ernst Mayr, and George Gaylord Simpson laid the groundwork, and now a younger generation must continue where they left off. This book is a product of the younger generation. Certain issues that were the subject of intense debate two decades ago are now of primarily historical interest and passed over summarily. On the other hand, considerable attention is focused on particular topics that were not even considered of interest 20 or 30 years ago. It must also be kept in mind that this book represents one paleontologist's viewpoint; although I have attempted to present alternative views (where alternative views were deemed worthy of discussion; I have been selective), it would be intellectually dishonest not to express my own opinions. After all, this book has only one author.

Any discipline must pass through various stages in its development. Referring specifically to historical biogeography, Ball (1975) distinguished three basic phases: a descriptive and empirical phase, a narrative phase, and an analytical phase. The history of paleontological activity, as it relates to phylogeny reconstruction, broadly follows this scheme (Patterson, 1981). Initially fossils were collected and described in a pre- or nonevolutionary and nonphylogenetic context (the modern counterpart of this stage might be viewed as phenetic analysis). Subsequently, with the adoption of evolutionary theories, narrative explanations (scenarios) were introduced to explain the occurrences and putative evolution of fossil organisms. Unfortunately, narrative explanations are not entirely satisfactory in a scientific context. In general, narrative explanations do not establish new empirical facts, do not confirm or discover laws of nature, and do not explicitly make predictions that can be further tested (Ball, 1975; see also Goudge, 1961; Hull, 1974; Patterson, 1981; Ruse, 1973). The final phase has been to take a very analytical and rigorous approach to the reconstruction of phylogeny. As Ball (1975, p. 408) has noted, once we go beyond the descriptive phase of a science, "we leave the world of primary sense data and enter a world of ideas, of theoretical constructs."

With these considerations in mind, the primary aim of this volume is to discuss the theoretical and philosophical bases, foundations, and concepts of phylogeny reconstruction, particularly as practiced by the paleontologist utilizing fossil material. Addressing the latter point specifically, I have included a chapter on the fossil record (chap. 4). Supplementary to the discussions of phylogenetic analysis per se are chapters on historical biogeography and biological classification (with an emphasis on the classification of extinct organisms). The topic of this book is conceptual, philosophical, and theoretical issues (primarily epistemological concerns) in the analysis of the phylogenies of organisms in the strict sense; it does not address the topic of evolution (neither the evolution of organisms, the evolution of communities, nor the evolution of life in general), except in passing. It is also not a volume on paleobiology (for a recent collection of review articles on this subject, see Sepkoski and Crane, 1985) or systematic paleontology. Likewise, it is not a quick and easy "how-to" manual presenting explicit instructions or algorithms on how to reconstruct the phylogenetic relationships for any particular group of organisms, nor is it a compilation of case studies.

This book is not meant as a replacement for, and should not be viewed as directly competing with, the valuable volumes already available that address particular aspects of phylogeny reconstruction or topics of relevance to the paleontologist interested in phylogenetic analysis. In this respect, the following particularly noteworthy books addressing a number of topics relevant to phylogeny reconstruction are recommended: on classification and biology, see Crowson, 1970; on cladistic analysis, see Eldredge and Cracraft (1980), Hennig (1966), Nelson and Platnick (1981), and Wiley (1981); on cladistic and pseudocladistic methods and algorithms, see Funk and Stuessy (1978); on phenetics and numerical taxonomy, see Sneath and Sokal (1973); on traditional and classical evolutionary analysis and classification, see Mayr (1969) and Simpson (1961); on listings of recent symposium volumes on phylogeny reconstruction, see Cracraft and Eldredge (1979), Duncan and Stuessy (1984), Funk and Brooks (1981), Hecht, Goody, and Hecht (1977), Joysey and Friday (1982), Novacek (1982), Patterson (1982), Platnick and Funk (1983), and Sibley (1969); on statistical methods and numerical analysis of biological data, see Jardine and Sibson (1971), Reyment, Blackith, and Campbell (1984), and Simpson, Roe, and Lewontin (1960); on statistical analysis, see Afifi and Azen (1979) and Cheeney (1983); on reviews of the historical background of modern systematics and taxonomy, see Nelson and Platnick (1981), Reyment, Blackith, and Campbell (1984), Simpson (1961), and references cited in these works. The major journal to carry articles devoted to various topics of relevance to phylogeny reconstruction is *Systematic Zoology* (Society of Systematic Zoology); the new journal entitled *Cladistics* (Willi Hennig Society) also promises to be a major contributor to the field.

This volume is the result of my own grapplings with the problems of phylogeny reconstruction in paleontology. As a guiding principle I have adopted Brundin's (1966, p. 21) admonition: "To be able to do well in reconstructing phylogeny, a difficult task under all circumstances, we must use sharp definitions and strict reasoning without any compromise, the best tools ever." I hope I have been able to raise certain topics and issues that are worthy of serious consideration, deal with the differing opinions of various investigators fairly (to this end, I have made liberal use of direct quotations), and present a relatively cogent summary of my own views. I make no claims to have definitive answers. I thank the numerous colleagues who contributed to my studies through discussion, correspondence, and by sending me their reprints. My debt to them is manifested in my citations of their papers. Facilities in which this research was carried out were provided by the Division of Science (Professor Charles P. Fogg, Chairman) of the College of Basic Studies (Dr. Brendan F. Gilbane, Dean), and the Department of Geology (Professor Duncan M. FitzGerald, Chairman) of the College of Liberal Arts (Dr. Geoffrey Bannister, Dean), Boston University. I thank Drs. Fogg, Gilbane, FitzGerald, and Bannister for their support. I also appreciate the interest that Professor Rhodes W. Fairbridge and Mr. Charles S. Hutchinson, Jr. have shown in this project since its inception. A special note of thanks must go to my wife, Cynthia, who has been ever supportive of my endeavors; my little son, Nicholas, remains a continued inspiration. Of course, all errors in fact or interpretation are wholly my own responsibility.

ROBERT M. SCHOCH

References

Afifi, A. A., and S. P. Azen, 1979, *Statistical Analysis: A Computer Oriented Approach*, Academic Press, New York.

Ball, I. R., 1975, Nature and formulation of biogeographical hypotheses, *Syst. Zool.* **24**:407–430.

Bonde, N., 1977, Cladistic classification as applied to vertebrates, in *Major Patterns in Vertebrate Evolution*, M. K. Hecht, P. C. Goody, and B. M. Hecht, eds., Plenum Press, New York, pp. 741–804.

Brower, J. C., 1973, Crinoids from the Girardeau Limestone (Ordovician), *Palaeontographica Americana* **7**:263–499.

Brundin, L., 1966, Transantarctic relationships and their significance, as evidenced by chironomid midges with a monograph of the subfamilies Podonominae and Aphroteniinae and the Austral Heptagyiae, *Kungl. Svenska Vetensksapsakademiens Handlingar, Fjarde Serien* **11**:1–472.

Cheeney, R. F., 1983, *Statistical Methods in Geology*, Allen and Unwin, London.

Cracraft, J., and N. Eldredge, eds., 1979, *Phylogenetic Analysis and Paleontology*, Columbia University Press, New York.

Crowson, R. A., 1970, *Classification and Biology,* Atherton Press, New York.

Duncan, T., and T. F. Stuessy, eds., 1984, *Cladistics: Perspectives on the Reconstruction of Evolutionary History,* Columbia University Press, New York.

Eldredge, N., and J. Cracraft, 1980, *Phylogenetic Patterns and the Evolutionary Process,* Columbia University Press, New York.

Funk, V. A., and D. R. Brooks, eds., 1981, *Advances in Cladistics: Proceedings of the First Meeting of the Willi Hennig Society,* The New York Botanical Garden, Bronx, N.Y.

Funk, V. A., and T. F. Stuessy, 1978, Cladistics for the practicing plant taxonomist, *Syst. Bot.* **3:**159–178.

Ghiselin, M. T., 1972, Models in phylogeny, in *Models in Paleobiology,* T. J. M. Schopf, ed., Freeman, Cooper, San Francisco, pp. 130–145.

Goudge, T. A., 1961, *The Ascent of Life,* University of Toronto Press, Toronto.

Hecht, M. K., P. C. Goody, and B. M. Hecht, eds., 1977, *Major Patterns in Vertebrate Evolution,* Plenum Press, New York.

Hennig, W., 1966, *Phylogenetic Systematics,* University of Illinois, Urbana.

Hull, D. L., 1974, *Philosophy of Biological Science,* Prentice-Hall, Englewood Cliffs, N.J.

Jardine, N., and R. Sibson, 1971, *Mathematical Taxonomy,* Wiley, London.

Joysey, K. A., and A. E. Friday, eds., 1982, Problems of phylogenetic reconstruction, *Syst. Assoc. Spec. Vol.* **21:**1–442.

Mayr, E., 1969, *Principles of Systematic Zoology,* McGraw-Hill, New York.

Nelson, G., and N. I. Platnick, 1981, *Systematics and Biogeography: Cladistics and Vicariance,* Columbia University Press, New York.

Novacek, M. J., convener, 1982, Symposium: Phylogeny and rates of evolution, *Syst. Zool.* **31:**337–412.

Patterson, C., 1981, Methods of paleobiogeography, in *Vicariance Biogeography: A Critique,* G. Nelson and D. E. Rosen, eds., Columbia University Press, New York, pp. 447–490.

Patterson, C., ed., 1982, Methods of phylogenetic reconstruction, *Zool. Jour. Linn. Soc., London* **74:**197–344.

Platnick, N. I., and V. A. Funk, eds., 1983, *Advances in Cladistics, Vol. 2: Proceedings of the Second Meeting of the Willi Hennig Society,* Columbia University Press, New York.

Reyment, R. A., R. E. Blackith, and N. A. Campbell, 1984, *Multivariate Morphometrics,* Academic Press, New York.

Ruse, M., 1973, *The Philosophy of Biology,* Addison-Wesley, Don Mills, Ontario.

Savage, D. E., 1977, Aspects of vertebrate paleontological stratigraphy and geochronology, in *Concepts and Methods of Biostratigraphy,* E. G. Kauffman and J. E. Hazel, eds., Dowden, Hutchinson & Ross, Stroudsburg, Pa., pp. 427–442.

Sepkoski, J. J., Jr., and P. R. Crane, eds., 1985, Tenth anniversary issue, *Paleobiology,* **11:**1–138.

Sibley, C. G., chairman, 1969, *Systematic Biology: Proceedings of an International Congress,* Publication 1692, National Academy of Sciences, Washington, D.C.

Simpson, G. G., 1961, *Principles of Animal Taxonomy,* Columbia University Press, New York.

Simpson, G. G., 1975, Recent advances in methods of phylogenetic inference, in *Phylogeny of the Primates,* W. P. Luckett and F. S. Szalay, eds., Plenum Press, New York, pp. 3–19.

Simpson, G. G., A. Roe, and R. C. Lewontin, 1960, *Quantitative Zoology,* Harcourt, Brace and World, New York.

Sneath, P. H. A., and R. R. Sokal, 1973, *Numerical Taxonomy,* Freeman, San Francisco.

Wiley, E. O., 1981, *Phylogenetics: The Theory and Practice of Phylogenetic Systematics,* Wiley, New York.

PHYLOGENY RECONSTRUCTION IN PALEONTOLOGY

Introduction

This book treats the general subject of the reconstruction of the phylogeny of living and fossil organisms, specifically using paleontological data. Thus, the central tenet or assumption of this book is that evolution of organisms has occurred, and the diversity of life is a result of the change and divergence through time of organisms from a common ancestor (Gaffney, 1979a). As Løvtrup (1984) has aptly pointed out, the very notion of phylogeny implies that evolution has occurred; phylogeny is the product of evolution (Rosen, 1984). Perhaps the most fundamental evidence of evolution is that there is a hierarchical pattern of characteristics among organisms that can be successfully reconstructed and interpreted phylogenetically. Otherwise, it is beyond the scope of this book to set forth the evidence for the reality of evolution; indeed, it is unnecessary, for it has been done many times (see, for instance, Croizat, 1964; C. Darwin, 1859, 1871a, 1871b; Dobzhansky, 1937, 1970; Futuyma, 1979, 1983; Goldschmidt, 1940; V. Grant, 1963, 1977; Haldane and J. Huxley, 1927; J. Huxley, 1942; Jepsen, Simpson, and Mayr, 1949; Lamarck, 1984; Løvtrup, 1982a, 1982b, 1982c; Maynard Smith, 1975, 1978; Mayr, 1963, 1970; H. L. McKinney, 1971; Pianka, 1974; Rench, 1959; Ricklefs, 1979; Riedl, 1978; Scriven, 1959; Simpson, 1953a; Sober, 1984c). For an introduction to the development of and recent advances in evolutionary theory, the reader is referred to Brooks (1983), Buss (1983), Gould (1982), Hitching (1982—a popular account, but worthy of perusal), Ho and Saunders (1979), Mayr (1982), O'Grady (1984), Ridley (1985), Stanley (1979), Vrba (1982a), and Vrba and Eldredge (1984), and to articles contained in the following volumes: Bendall (1983), Grene (1983), Ho and Saunders (1984), Maynard Smith (1982), Milkman (1982), Novacek (1982b), Pollard (1984), Schwartz and Rollins (1979), Scudder and Reveal (1981), and Sober (1984b). In this book I do not intend to advocate a priori any particular theory of an evolutionary mode, process, or mechanism such as Lamarckianism, Darwinism, Neo-Lamarckianism, Neo-Darwinism, or the synthetic theory, or any particular theory as to the overall tempo and pattern of evolution below or above the species level, such as gradualism, saltationism, orthogenesis, or punctuationism. Such considerations arise in this book, but the primary concern is to reconstruct the phylogenetic relationships, *sensu stricto,* of organic beings. The process or mechanism that drives evolution need not be known (indeed, is not known, although many conjectures exist: Rosen, 1984) in order to consider phylogeny, the history and hierarchy of organisms.

What is meant by the term *phylogeny?* The term was originally coined by Haeckel (1866: see Janvier, 1984, p. 41, and Patterson, 1977, p. 584) to refer to the history of the paleontological (based on temporally ordered fossils, in part) development (evolution) of species. Many different concepts of this term are current. In general, phylogeny refers to the pattern of descent, the genealogy, or the evolutionary history of a group of organisms. Some workers, such as Gingerich (1979, p. 42), would define phylogeny broadly to include both the genealogical interrelationships of the members of a group of organisms and the adaptations and relative morphological, physiological, and ecological divergence of the individual organisms. Such a definition might also include hypotheses or information about the evolutionary mechanism, tempo, and mode for a particular group of organisms. Løvtrup (1984, p. 159), with an emphasis on epigenetics and morphogenesis, considered phylogeny to be "the changes to which ontogeny has been subject in the course of time." Janvier (1984, p. 41) states: "Phylogeny expresses the interrelationships of the members of a group, as a result of evolution by descent and transformation of heritable characters." Most workers, however, consider phylogeny to be simply the "evolutionary historical pattern" (Eldredge and Cracraft, 1980, p. 4; cf. Løvtrup's, 1984, "history of evolution") or the "geometric pattern of ancestry and descent among organisms" (Gaffney, 1979a, p. 87)—that is, the phylogenetic trees without any explicit or implicit notion of the evolutionary process per se.

However, even when restricted to the simple historical pattern of evolution, the concept of phylogeny includes two basic components that, though not mutually exclusive, must be separated and made explicit. These are the concepts of relative "phylogenetic relationship" (Hennig, 1965, 1966a) and "direct ancestry." Ideally, for the members of any set of three or more different organisms (such as, for example, members of different species), the relative degrees of relatedness (phylogenetic relationships) of the organisms to one another can be specified. Given any three organisms (or monophyletic taxa as defined in chapter 1) A, B, and C, A and B are said to be more closely related to one another than either is to C if A and B share an ancestor in common that is not an ancestor of C. Sometimes we may feel justified in going beyond a simple hypothesis as to the phylogenetic relationships of several organisms or species to specify that a certain species gave rise to, or is in the direct ancestry of, another species. For example, given our hypothetical example of organisms (or species, but not higher taxa) A, B, and C, where A and B are more closely related to each other than either is to C, we may further hypothesize that either A gave rise to B or B gave rise to A (that is, either A or B itself may be the common ancestor of A and B that is not shared with C). As discussed later in this book, statements concerning the ancestry of organisms are more specific than statements concerning the relative phylogenetic relationships of organisms and accordingly are more heavily assumption laden (Eldredge, 1979a; Eldredge and Tattersall, 1975; Engelmann and Wiley, 1977; Forey, 1982; Gaffney, 1979a, 1979b; Tattersall and Eldredge, 1977). Due to these additional required assumptions, it has been argued that direct ancestor-descendant relationships can seldom, if ever, be objectively recognized (Engelmann and Wiley, 1977; Schoch, 1982). In contrast, many workers (e.g., Fortey and Jefferies, 1982; Gingerich, 1979) consider it an integral part of phylogeny reconstruction to attempt to specify ancestor-descendant relationships. In this book, unless otherwise stated, phylogeny *sensu stricto* is considered minimally to be a simple account, whether conveyed by means of a classification, diagram, prose description, or some other form, of the phylogenetic relationships (relative degrees of relatedness) of certain organisms, species, or monophyletic taxa. Thus, as the term is used here, one need not know nor hypothesize ancestor-descendant relationships in order to specify phylogenetic relationships among organisms. On the other hand, it is agreed that if and when we can specify ancestors and their descendants, this will give us a more complete and specific picture of the pattern of evolution. It must be kept in mind that the concept of phylogenetic relationship, and hence phylogeny as expounded in this volume, applies principally to organisms or taxa that are taxonomically discrete, usually at the species level or higher (cf. Bonde, 1977, p. 791). If we are dealing with three individual organisms (A, B, and C) of a single bisexual species, where A and B mated to produce C, it does not make sense to search for a phylogenetic relationship (i.e., that two are more closely related to each other, share a more recent common ancestor, than either does to the third) between these individuals. Of course we may wish to, and possibly be able to, reconstruct the pattern of mating and descent (genealogy) among these organisms. Likewise, the concept of

phylogenetic relationship may not be strictly applicable in describing relationships between subspecies or varieties that have undergone extensive reticulation or hybridization in the course of their evolution.

From the start it should be made clear to the reader that there are many
and various types of diagrams and charts in the published literature that purport
to illustrate the phylogenetic or other relationships of organisms. The phylogenetic relationships of organisms (terminal taxa) can be depicted using any
type of graphic diagram that groups the terminal taxa hierarchically (e.g., Fig.
1a–e). Branching diagrams that depict exclusively phylogenetic relationships
(sister-group relationships) are usually termed cladograms (Fig. 1a). (Even if
Sokal [in Camin and Sokal, 1965: see Sokal, 1983a, p. 160] originally coined
the term *cladogram* to refer to "a branching . . . network of ancestor-descendant relationships" [Sneath and Sokal, 1973, p. 29], the term is now
generally understood and used with the meaning advocated here [cf. Eldredge
and Cracraft, 1980; Nelson and Platnick, 1981; Wiley, 1981].) Diagrams such
as those illustrated in Figure 1a through e depict only phylogeny *sensu stricto*
as defined in the last section (= "phylogeny diagrams" of Hennig, 1965).
Such diagrams have no temporal, morphological, or geographical axes and
do not specify ancestor-descendant relationships or in themselves imply anything about the evolutionary process. In contrast, phylogenetic or evolutionary
trees (e.g., Fig. 2a) are pictures supposedly depicting the actual evolution of
a group of organisms. On phylogenetic trees ancestral forms may be specified;
the vertical axis usually represents time, and the horizontal axis or axes may
represent morphological and/or geographic divergence. A hypothesized evo-

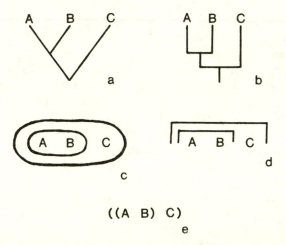

Figure 1. Alternative diagrams that express the hypothesis that taxa A and B are phylogenetically
more closely related to each other than either is to taxon C (taxa A and B share a more recent
common ancestry with each other than either does with C).

lutionary tree composed of ancestors (real or hypothetical) and descendants can be generalized as forming a network of evolutionary relationships connecting taxa without specifying polarities of ancestor-descendant relationships (Farris, 1970; Harper, 1976; Sneath and Sokal, 1973). A network can be "rooted" by singling out a certain taxon (node) as ancestral (primitive). Thus, like a cladogram, a network may yield several trees (several different trees are equivalent to the same network). Some phylogenetic trees purport to show the evolution of major groups of organisms using supraspecific taxa. In such cases each lowest ranking group depicted is often shown by a spindle diagram that varies in width through time (Fig. 2b). The wider the spindle diagram, the more abundant is the group in terms of numbers of individuals or numbers of species. Such diagrams may directly suggest certain evolutionary processes and rates—that is, they may encompass much more than phylo-

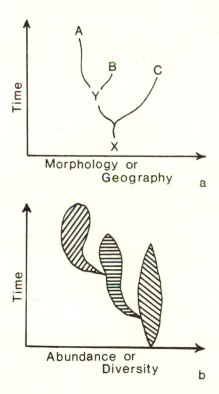

Figure 2. *(a)* A hypothetical evolutionary tree. Time is shown on the vertical axis (progressing from bottom to top), and change in morphology and/or geography is shown on the horizontal axis. *(b)* A hypothetical spindle-diagram evolutionary tree. Time is shown on the vertical axis. Abundance or diversity of a particular taxon is indicated by the horizontal width of the darkened balloon.

genetic relationships and may include depictions of evolutionary stories or scenarios.

In this work reference is made to hierarchical diagrams that are intended to explicitly depict phylogenetic relationships *and* the various types of evolutionary trees and networks as *phylograms* (cf. Mayr's, 1981 [reprinted in Sober, 1984*b*, p. 648] definition of phylogram, which includes only the latter types of diagrams). Other types of commonly published diagrams are dendrograms (branching diagrams based on linking of entities using any criteria—many diagrams, including phylograms, are types of dendrograms in the broad sense: Wiley, 1981, p. 97); phenograms—branching diagrams linking organisms on the basis of estimates of overall similarity or based on cluster analysis of certain correlation coefficients of various data or measurements of organisms, similarity matrices, and principal component and factor analyses; and various kinds of two- and three-dimensional graphs and models illustrating the taxonomic relationships of various groups of organisms (e.g., see Sneath and Sokal, 1973, pp. 259–275). While phenograms, similarity matrices, distance matrices, models of taxonomic relationships, and the like may contain information bearing on the phylogeny of the organisms, they are not in themselves phylograms and should not be confused with phylograms.

PHYLOGENY RECONSTRUCTION FOR FOSSIL VERSUS LIVING ORGANISMS

There has been much recent discussion about the place and value of fossils in phylogeny reconstruction (Kohlberger and Schoch, 1983). On the one have of the history of life on our planet, and in particular of the course of haave of the history of life on our planet, and in particular of the course of evolution" (Hallam, 1977, p. v) and thus provide an "essential time dimension" (Simpson, 1961, p. 67; cf. Gingerich, 1979; Newell, 1959) for phylogeny reconstruction. As S. S. Bretsky (1976, pp. 231–232) states: "Knowledge of the temporal relationships of faunas sets bounds to phylogenetic speculation and frequently permits choice among alternative phylogenetic reconstructions which from morphological evidence alone seem equally tenable." And as Fortey and Jefferies (1982, p. 198) argue: "Since the [phylogenetic] tree includes (indeed consists largely of) ancestors, palaeontological evidence is essential to produce the complete tree." This emphasis on the importance of paleontological data feeds into Gould's (1980*a*, 1980*b*) nomothetic view of paleontology, that paleontology is the sole guardian of "vast times and effects" (Gould, 1980*a*, p. 98) and has much to contribute in its own right above and beyond what can be discovered through analysis of neontological data.

On the other hand, some workers hold an extremely pessimistic view of the value and importance of fossils per se in phylogeny reconstruction (e.g.,

Schaeffer, Hecht, and Eldredge, 1972). Paleontological (fossil) organisms may be viewed as extremely incomplete and imperfect specimens of which we can never know the entire biology (normally only hard parts are preserved, and of course we cannot directly observe their behavior, physiology, and so on). According to this extreme actualistic view, paleontology can never overcome its limits relative to, and must always defer to, neontology (Kohlberger and Schoch, 1983; Patterson, 1981a)—that is, "paleontology must always be subservient to neontology and has no fully independent role in phylogenetic work" (Patterson, 1977, p. 634). Furthermore, some workers believe that two of the prime (though extrinsic) attributes of fossils, those of geographic and stratigraphic (temporal) context, should not be used, at least initially, in constructing phylogenetic hypotheses (Engelmann and Wiley, 1979; McKenna, Engelmann, and Barghoorn, 1977; Schaeffer, Hecht, and Eldredge, 1972). As discussed later in this book, such data may be notoriously unreliable, at least for certain groups of organisms, and it is agreed that the initial mixing of various sorts of data prevents "testing of one system [of data] against the others without circularity" (McKenna, Engelmann, and Barghoorn, 1977, p. 238).

How one views the utility and importance of fossils for reconstructing phylogeny depends on the methodology one adopts. However, it cannot be denied that while the average single fossil specimen may not supply as much information (as many characters for analysis) as the average neontological specimen, paleontology does contribute a wealth of organisms that would not be known otherwise (Kohlberger and Schoch, 1983). Fossil organisms may bear character combinations that greatly influence our hypotheses of homology and character evolution. An otherwise unsuspected diversity of organic beings has been recovered by paleontologists. Furthermore, paleontology *does* supply data on the stratigraphic (temporal) and paleogeographic occurrences of the fossil organisms. Fossils can provide minimal ages for monophyletic groups and amplify biogeographic data on extant groups (Janvier, 1984; Patterson, 1982a). Such data should not be ignored even if its reliability, usefulness, and legitimacy for phylogeny reconstruction are open to question (these topics are discussed throughout the remainder of this book).

One of the theses of this book is that the paleontologist, anatomist, stratigrapher, and geologist all have important roles to play in the reconstruction of phylogeny. The paleontologist, by bringing to light through his or her studies otherwise unknown organisms, contributes to the diversity of forms known to neontologists. The anatomist, through the careful observation and description of the structure of organisms—fossil and extant—that then leads to the study of morphology (the interpretation and generalization of specific anatomical observations), provides the basic characters upon which all methods of phylogeny reconstruction are based to at least some extent. The function of the stratigrapher is to place the fossil organisms in a temporal context. The geologist places the fossils in their paleogeographic and paleoecologic contexts and also serves to check the geologic credibility of the stratigraphic limits of the fossil's range (Shaw, 1964, p. 116). Of course, a single researcher often assumes more than one of the above roles.

SCHOOLS OF PHYLOGENETIC RESEARCH

Prior to the middle of this century, phylogenetic research was badly neglected. In the past many phylogenetic schemes for various groups of organisms were proposed and accepted without being subjected to critical or rigorous scrutiny. It was the processes, modes, and mechanisms of evolution that were of prime concern to researchers. All too often the construction of the phylogenies per se was taken for granted (for a typical example of these abuses, see Osborn, 1929). Phylogeny reconstruction often was, and many times still is, carried out instinctively or intuitively without any clear articulation of methodology or underlying assumptions (Ghiselin, 1972). For example, a recent book on macroevolution (Stanley, 1979), relies heavily on phylogeny reconstruction for its primary data, yet fails to discuss the basis of phylogenetic analysis and uncritically accepts phylogenies as published in the literature. Yet many published phylogenies and evolutionary scenarios are little more than subjectively formulated stories (Shaw, 1969).

Since 1950, however, there has been a movement among certain biologists and paleontologists (primarily taxonomists) to scrutinize and formalize methods of phylogeny reconstruction. In particular two groups, the numerical taxonomists (pheneticists) and the cladists (phylogenetic systematists), have both challenged the traditional methods of phylogeny reconstruction. In turn this movement has forced those practicing the traditional methods to codify their position (commonly known as classical evolutionary methods, or evolutionary systematics). These are the major contemporary schools of phylogeny reconstruction and systematics (Eldredge and Cracraft, 1980; Hull, 1981; Mayr, 1981). In this section each school is described briefly; the actual methodologies of phylogeny reconstruction adopted by certain adherents of these schools are reviewed in chapter 1.

The prime concerns of these schools have not always been the reconstruction of phylogeny per se; more often the first concern has been to establish the proper methods, foundations, and goals of biological classification. However, since the publication of C. Darwin's *Origin of Species* (1859), taxonomists have been forced to consider the question of the relationship of the phylogeny and evolution of the organisms being classified to the classification. Can and should classification reflect phylogeny/evolution? If so, to what extent? C. Darwin (1859) strongly argued that biological classification is and should be based on the natural affinities, or genealogy (= phylogeny), of the organisms classified. This concept immediately poses the problem that if we are to classify organisms via such a system (genealogy), we must have a means to discover such genealogical (phylogenetic) relationships. Or (and the two are not mutually exclusive) do we merely form a "natural" classification, as was often done in pre-Darwinian times, and then read phylogeny from the classification? In either case, taxonomy and classification are intimately associated with phylogeny reconstruction.

Hull (1981, p. 143) has pointed out that

If contemporary philosophers of science agree on anything, it is that scientific classification cannot be theoretically neutral. Nor can there be any prescribed order in which theoretical considerations are introduced into a classification. One cannot begin by producing a theoretically neutral classification and then only later add theoretical interpretations.

Hull (1981), quoting Hempel (1965), argued that the basis of the general theory underlying all biological classifications is phylogenetic; most workers would probably agree with this statement. At the least, the basic units of the classification must be useful (functional) in the theories involving the phylogeny/evolution of organisms (cf. Hull, 1981, p. 144). Even pre-Darwinian, nonphylogenetic, biological classifications were based on natural affinities or genealogical ties of organisms (even if the organisms were not seen as descended from one another, but as materially independent parts of the systematic whole: Agassiz, 1833–1844, discussed in Patterson, 1977; see also Patterson, 1981b). Patterson (1977) has demonstrated that the word *classification* can usually be replaced by *phylogeny* in the work of many pre- and nonevolutionary taxonomists without altering the original authors' intended sense. Moreover, only one major contemporary school of taxonomists—pheneticists or numerical taxonomists—has explicitly argued that biological classifications should be freed from all influences and theories (phylogenetic and otherwise). Yet in practice the pheneticists have been notably unsuccessful in developing classifications free of theory. Simple acts of observation involve inferences (Colless, 1967a; 1967b; Hull, 1970). The concept of homology, the same character in different organisms, has its basis in the concept of *natural affinity* (genealogy, phylogeny). The two other contemporary schools of taxonomy (classical evolutionary taxonomy and cladistics) acknowledge that phylogeny is of more or less importance in influencing their classificatory schemes.

Classical Evolutionary Taxonomy

The *traditional* or *evolutionary* school of systematics and phylogeny reconstruction traces its heritage back to C. Darwin (Mayr, 1981), but in its modern form originates from the *synthetic theory* of evolution and the *new systematics* of the late 1930s and 1940s (Eldredge and Cracraft, 1980; J. Huxley, 1940, 1942). This synthesis bridged the schism that had developed during the first third of the twentieth century between the experimental geneticists on the one hand and the more traditional, field-oriented naturalists on the other (Mayr, 1982; Mayr and Provine, 1980). Many early Mendelian geneticists were anti-Darwinian in that they subscribed to some form of saltationism. Evolutionists of the time often recognized or suggested different forms of mutations for within-population evolution and trans-specific evolution (Gould, 1983a).

The two major conclusions of the synthetic theory were (1) that intraspecific evolution is gradual and explicable in terms of small genetic changes operating via Mendelian mechanisms and (2) that through the population concept and analysis of ecological factors the diversity of organisms and the origin of higher taxa can be explained "in a manner that is consistent with known [i.e., known in the 1940s] genetic mechanisms and with the observational evidence of the naturalists" (Mayr, 1982, p. 567). Originally the synthesis did not espouse any particular mechanism of change more restrictive than that change should occur via known Mendelian mechanisms. But in the 1940s there was more and more insistence that neo-Darwinian natural selection and adaptation is the primary mechanism of evolutionary change—what Gould (1983a) has referred to as the "hardening of the modern synthesis." This strong commitment to adaptationist explanations has recently come under fire (Gould and Lewontin, 1979; Gould and Vrba, 1982; Rachootin and Thomson, 1981). No matter what explanations various adherents of the new synthesis and evolutionary classification espoused, it is important to note that this broad school of thought has historically been deeply concerned with the problem of mechanism and process in evolution. As Hanson (1977, p. 112) has aptly stated, the classical evolutionary school and new systematics can be characterized by a "direct and deep conceptual dependence on the theory of natural selection." Quite the opposite is true for the numerical taxonomic and phylogenetic systematic (cladistic) schools.

The new systematics (J. Huxley, 1940), the forerunner of contemporary evolutionary classification, which went hand in hand with the synthetic theory, was characterized by a vigorous attack on typology in classification and an attempt to take intraspecific variation into account when delimiting and diagnosing species-level taxa (Eldredge and Cracraft, 1980). This attack on typology fits the synthetic concept of slow, steady, gradual variation in time and space between different individuals, populations, and species. Species are not fixed, typological entities or classes (groups) of identical individuals; instead, the individuals composing any species show a range of variation that is the material on which natural selection acts and thus powers the evolutionary process. The basic assumption of evolutionary taxonomists is that the fossil record and the diversity of living organisms can be explained by, and interpreted in light of, evolutionary mechanisms and scenarios as generally expounded in the modern synthesis (Gaffney, 1979a, p. 86; Mayr, 1969; Simpson, 1961; Sneath and Sokal, 1973, p. 17). In their classifications the evolutionary taxonomists often group together organisms on the basis of presumed genetic similarity as judged primarily from the phenotype (Bock, 1977; Mayr, 1969, p. 200). Extremely important to evolutionary classification is the concept that evolutionary and adaptational divergence, the reaching of certain adaptive grades or levels of organization (patristic affinity), and the invasion of new adaptive zones or niches are considered just as important as a basis for classification as are strict branching patterns (phylogeny *sensu stricto*).

The evolutionary school includes in the analysis all available attributes of these organisms [those classified], their correlations, ecological stations, and patterns of distributions and attempts to reflect both of the major evolutionary processes, branching and the subsequent diverging of the branches (clades). This school follows Darwin (and agrees in this point with the cladists) that classification must be based on genealogy and also agrees with Darwin (in contrast to the cladists) "that genealogy by itself does not give classification." (F. Darwin, 1887, vol. 2, p. 247) (Mayr, 1981, p. 510; reprinted in Sober, 1984b, pp. 647–648)

One of the major criticisms of evolutionary classification is its heavy reliance on the synthetic theory of evolution. At times the evolutionary taxonomist may utilize a combination of intuitive judgment (perhaps based on years of experience working with a certain group of organisms) and synthetic theory to arrive at what he or she believes was the most likely course of evolution for a group of organisms. Moreover, the evolutionary school can be broadly eclectic or syncretic at times, borrowing or rejecting data and methods at will from the other major schools. Finally, for evolutionary taxonomists there is not a one-to-one correspondence between their reconstructed phylogenies and their classifications. Rather, prominent evolutionary taxonomists generally acknowledge that there is a component of art to their taxonomy (Simpson, 1945, 1961, p. 110; see Hull, 1970). As Panchen (1982, p. 319) notes, "The standard works on [evolutionary taxonomic] procedure (Simpson, 1961; Mayr, 1969) are to some extent rationalizations of a tradition that is too largely intuitive."

Numerical Taxonomy

As a reaction to the evolutionary school, in the late 1950s and 1960s there emerged the school of classification variously known as numerical, phenetic, computer, or neo-Adansonian (after the eighteenth-century botanist Michel Adanson, who based his classification on overall similarity [Sneath and Sokal, 1973, p. 23]; for general reviews of phenetics and numerical taxonomy, see McNeill, 1978, 1982; Sokal, 1983a, 1983b, 1983c, 1983d; and Sokal and

Sneath, 1973; for a definitive criticism of the phenetic program, see Farris, 1982). "The main concern [of the numerical taxonomists] appeared to be a desire to reformulate the process of delimiting life's orderliness in a more standardized, repeatable, rigorous and objective fashion (Sokal and Sneath, 1963, p. 49)" (Eldredge and Cracraft, 1980, p. 8). In order to accomplish this task, numerical taxonomists espoused estimating the taxonomic affinities of organisms on the basis of mathematical clustering procedures (usually based on overall similarity) using all available taxonomic characters, equally weighted, in coded quantitative form. Such a basis was designed to produce a general-purpose classification that is theoretically neutral (Hull, 1981).

Many numerical taxonomists (also commonly known as pheneticists) explicitly excluded phylogeny and evolutionary considerations a priori as a basis for biological classification (Hull, 1970). The major reasons for this stance were (summarized by Hull, 1970): (1) In most cases phylogenies are unknown and thus cannot be used as a basis for classification. (2) Methods used by evolutionary taxonomists to reconstruct phylogenies tend to be vague or even fallacious, lacking rigor, explicitness, and quantitativeness. (3) Even if with advanced methodologies and techniques (such as those developed by the numerical taxonomists) it may be possible to reconstruct phylogenies for certain groups of organisms, reasonably accurate phylogenies may not be reconstructable for all groups and thus phylogeny could not serve as a universal basis for classification. (4) Even if we knew the complete phylogenies for all groups of organisms, classifications based primarily or solely on phylogeny would still be special-purpose classifications that could not adequately serve biologists as a whole.

Of course, as Farris (1977b, p. 839) in particular has pointed out, pheneticists misrepresent phylogenetically based classifications when they criticize such classifications as being difficult or impossible to apply because often phylogeny cannot be known with certainty. Phylogenetic classifications are not based on true phylogenies per se, but rather on inferred phylogenetic relationships as interpreted according to a definite analytical procedure grounded in empirical data. Likewise, phenetically based classifications are inferred according to some analytical procedure operating on empirical observations. In this sense, phylogenetically and phenetically based classifications are alike in that each system begins with a set of empirical observations that is then transformed into a classification utilizing some definite analytical procedure.

Phylogenetic Systematics

Beginning in Germany in the late 1940s (Hennig, 1950), and then spreading to the English-speaking countries in the late 1960s and 1970s (Brundin, 1966; Crowson, 1970; Hennig, 1965, 1966a, 1966b; Kiriakoff, 1959; Nelson, 1969, 1970, 1971, 1972a, 1972b, 1973), the phylogenetic school of systematics was developed (for brief synopses of the history and theory of cladism, see Dupuis, 1979; and Humphries and Funk, 1984). Hennig and his initial followers (Hennigians) believed that biological classification should directly and unambiguously reflect phylogeny *sensu stricto* (i.e., phylogenetic relationships, relative degrees of relatedness, or the branching patterns of a phylogeny). Given a classification, one should be able to reconstruct the phylogeny *sensu stricto* of a group of organisms from the classification. Hennig termed his system phylogenetic because of his emphasis on phylogeny, but opponents (Mayr, 1969, p. 70) have dubbed the school cladistic due to its emphasis on recognizing natural branching points (cladistic events) and the resulting lineages (clades) in phylogeny. Methods of cladistic analysis are discussed in some detail later in this book.

Originally Hennig (1950, 1965, 1966a) formulated his system based on a dichotomous speciation model and a particular view of the evolutionary process (Gaffney, 1979a, 1979b; Hull, 1981), but it has been demonstrated that cladistic methods of phylogeny reconstruction do not depend on any special view of evolutionary processes (Eldredge and Cracraft, 1980; Nelson and Platnick, 1981; Platnick, 1977b; Wiley, 1981). Most cladists would accept the basic assumptions that biological evolution has occurred and that as evolution occurs, change occurs: New features (including the loss or transformation of old features) originate and are passed on to descendant units (individuals, populations, species, taxa). Most cladists are not particularly concerned about the adaptive significance of the characters they use in reconstructing phylogenies or testing cladograms. However, not all workers would agree with Brooks (1983, p. 2639) that "the theoretical implication of this approach is that ancestral structures somehow constrain future evolutionary options in such a way that we can reconstruct phylogeny without regard for adaptive relationships." A cladogram can be interpreted as the order of branching of sister groups, the history of the successive splitting of lineages, or more simply as "the order of emergence of unique derived characters, whether or not the development of these characters happens to coincide with speciation events" (Hull, 1979, p. 418). Ideally, since different stages of the same organisms (for example, larval and adult forms) are products of the same phylogenetic history, separate cladograms based on distinct stages of the same organisms should be compatible with each other (of course, certain stages may contain more information and thus yield a more detailed phylogeny). However, when applying cladistic methods to organisms, it may not even be necessary to make the basic assumption that evolution has occurred; thus has arisen the school of *modern, transformed,* or *pattern* cladists (also known as natural-order systematics; see Beatty, 1982; Brooks and Wiley, 1985; Charig, 1982; Hull, 1984; Humphries and Funk, 1984; Janvier, 1984; Kluge, 1985; Nelson, 1985; Nelson and Platnick, 1981; Patterson, 1982c; Platnick, 1979; Schoch, 1984a). Logically and epistemologically, pattern cladism is prior to phylogenetic cladism. Phylogenetic cladism (= phylogenetic systematics) can be thought of as a part of the more general research program of pattern cladism (Nelson, 1985).

Cladistic analysis aims to discriminate between more general and less general (particular or special) characters (Platnick, 1979); cladistics is about pattern in nature (Patterson, 1980; cf. McNeill, 1982). Cladistics attempts the "hierarchical clustering of synapomorphies [for an explanation of synapomorphy—shared derived character—see section on cladistic methodology] for its own sake, with the assumption that the regularities exposed are inherent in nature" (Panchen, 1982, p. 307; see also Nelson, 1979). Furthermore, cladistic analysis can be applied to nonbiological systems that change or evolve (cf. Platnick and Cameron, 1977, who note that cladistic analysis can be applied to linguistics; Nelson and Platnick, 1981) and perhaps to any patterned data (Nelson, 1979). "Cladograms depict structural elements of knowledge" (Nelson and Platnick, 1981, p. 14). Hull (1983) has distinguished between cladistics and cladograms in the specific sense and cladograms in the generic sense. In the specific sense cladograms are concerned with species, speciation events, and evolutionary homologies. In the generic sense cladograms are branching diagrams that are resolved using a certain methodology.

Ultimately, cladistic analysis is concerned with discovering lawful regularities, or emergent law, in nature (Brady, 1983). This is accomplished by the extension of a local pattern(s) to broader patterns observed in nature; a law of nature is complete when our predictions are always met and complete. Speculations, in the form of explanations of underlying processes or mechanisms, do not play a part in such a program of research. This is not to deny that process is interesting and important, only to say that it is a different problem (a problem of vital concern to many classical evolutionary taxonomists, but shunned by many strict cladists). In analogy, Newton's law of universal gravitation predicts the gravitational interactions between any masses but does nothing to explain what gravity is, why it exists, and how it operates. Likewise, the strict cladist searches for a pattern (in our case, a phylogenetic pattern of relationships between organisms) with which more and more observations agree but is not particularly concerned with explaining the evolutionary mechanisms and processes that ultimately account for the pattern.

PHILOSOPHY AND SCIENCE

There can be no doubt that philosophy and science go hand in hand. The usual conception is that philosophy deals with concepts, while science applies the concepts to the empirical world (Sober, 1984b). Regardless of whether it should be so in principle, in practice there is often no sharp demarcation between the roles of the philosopher and the scientist—if this is not evident to the reader already, it will become so as he or she pursues this book.

It can be said that there are two major areas of philosophical inquiry that are directly applicable to the practice of science: metaphysics and logic. Within metaphysics, "the study of the most general categories within which we think" (Harré 1972, p. 108), are included ontology, cosmology, and epistemology; logic here refers to the theory, principles, and guidelines of reasoning.

Ontology deals with the theory or theories of being, the nature and relations of being, and the nature and kinds of existence; such topics will be of importance to us later in this book when we deal with such matters as the reality and nature of species, higher taxa, and true versus hypothetical ancestors. Cosmology, in its original conception, is a branch of philosophy that deals with the universe as an orderly system. It is only if there is some kind of inherent order and regularity to the world that we can hope to explain the world in a rational and scientific manner. Thus Popper (1968, p. 15) considered "the problem of cosmology" to be "*the problem of understanding the world*" [italics in the original] and stated that "all science is cosmology." The assumption or belief that the world is inherently orderly also relates to the basic working principle of simplicity (parsimony or economy), which is often used to choose between competing scientific hypotheses that explain the same set of data (observations).

Epistemology deals with the theory of the nature and grounds (basis of acquisition) of knowledge, especially with reference to the limits and validity of kinds of knowledge: How we know what we know. In science there are certain standards that genuine and valid knowledge must meet, and particular scientific methods can be applied to gather such knowledge, but all particular scientific methods have limits beyond which they can no longer claim to gather valid knowledge. Furthermore, knowledge that appeared to be (and perhaps was for all intents and purposes) genuine and valid can be replaced by new knowledge that is genuine and valid (Wiley, 1981). One of the prime concerns of this book is how we can obtain genuine and valid knowledge concerning the phylogeny of biological entities.

Harré (1972) and Wiley (1981) have outlined three basic approaches to epistemology, each of which has been applied by at least some workers in their underlying assumptions concerning phylogeny reconstruction and biological classification. Phenomenalism would limit genuine and valid knowledge to observed facts and events (phenomena). Some pheneticists (numerical taxonomists) would seem to espouse a type of phenomenalism when they advocate merely observing characteristics of organisms and then arranging the organisms into a theory-free classification. Fictionism regards scientific statements and hypotheses as statements that give coherence, order, and unity to sets of data, but the hypotheses themselves may not refer to real entities or processes. Fictionism relates in some ways to nominalism, the theory that there is no universal reality, or at least no universal essences in reality. Workers who regard species and other taxonomic groups of organisms more as convenient groupings than as genuine natural entities are allied with fictionism. Realism holds that concepts, abstractions, and theories have a basis and independent existence outside of the theorist's mind in the real world. There may be theoretical terms for hypothetical entities that do not exist in the real world, but some hypothetical entities do have an existence in the real world. Workers who consider species or other taxonomic units "real" entities in nature can be termed realists.

On a more fundamental level, Popper (1984) distinguishes two basic approaches to the theory of knowledge (epistemology). The traditional approach to epistemology Popper (1984, p. 241) calls observationism. Observationism, as a theory of knowledge, assumes that we receive raw data through our senses (sight, sound, smell, taste, and touch). The sense perceptions are then processed or digested, and as a result of repetitions, associations, generalizations, and induction we arrive at *knowledge.* Popper (1984) rejects this approach and rather suggests (in his theory of knowledge, which has been dubbed evolutionary epistemology) that there is no such thing as raw data independent of our previous hypotheses or theories about the world. One must always start from a hypothesis. Knowledge arises when tentative theories are proposed as possible solutions to a problem and submitted to the process of elimination of error; theories are criticized and tested and eliminated. Historically, all knowledge is invented a priori and then undergoes a process of elimination. Even biological organs used in the acquisition of sense data must have occurred prior (as a priori theories) to the actual acquisition of such sense data. On a fundamental level, simple organisms bearing basic errors in their a priori theories, such that the errors will affect their ability to subsist or adapt, will be eliminated along with their theories. Humans, at least, have the potential ability to separate their theories (expressed in language) from their beings and can thus criticize and eliminate their theories without eliminating themselves. It is important to keep in mind that, from this point of view, even the simplest and most fundamental or basic data is theory laden.

Wiley (1981) distinguishes two basic approaches that are commonly applied to scientific reasoning (i.e., logic): inductivism and logical empiricism. Both forms of logic rely on several basic assumptions or principles. Prime among these are what have become known as Mill's canons and the principle of simplicity (parsimony). As quoted by Harré (1972, p. 38), John Stuart Mill's (1879) two canons are:

(1) *The Canon of Agreement or Congruence:* If two or more instances of the phenomenon under investigation have only one circumstance in common, the circumstance in which alone all the instances agree is the cause (or effect) of the given phenomenon.

(2) *The Canon of Difference:* If an instance in which the phenomenon under investigation occurs, and an instance in which it does not occur have every circumstance in common save one, that one occurring only in the former, the circumstance in which alone the two instances differ is the effect, or the cause, or an indispensable part of the cause of the phenomenon.

To pick between a number of competing hypotheses, all of which fit or explain a certain set of observations, most philosophers and scientists rely on some form of the principle of simplicity. The principle of simplicity (also referred to as parsimony, economy, or Occam's razor) suggests that the simplest hypothesis that fully and adequately explains the data (as perceived and interpreted at the time) in the most economical manner is to be preferred. Thus more complicated explanations, or numerous ad hoc hypotheses for various individual cases, are to be rejected in favor of simpler, unifying explanations and hypotheses. It is important to note that the adoption of a parsimony principle as a criterion to choose between competing, otherwise equal scientific hypotheses need not imply that nature is believed to be *parsimonious*. A parsimony criterion does not necessarily increase the probability of the truth of a statement, but can be viewed as a logical necessity in order to choose between otherwise equally valid hypotheses (Bonde, 1977, p. 744). In its particulars the concept of parsimony is an extremely difficult subject concerning which philosophers and scientists continue to debate (see, for example, Farris, 1983; Sober, 1975, 1983a, 1983b, 1984a). The concept of parsimony is discussed in more detail later in this book (see chap. 3).

Inductivism, as opposed to logical empiricism, has expanded on Mill's canons and the principle of simplicity with the following basic working principles (Harré, 1972; Wiley, 1981). Knowledge develops or grows from the accumulation or collection of independent facts; general laws (hypotheses) can be inferred from sets of particular facts; and the general truth of a scientific law (hypothesis) can be gauged by the number of facts or observations that agree or conform to the law. More generally, the inferential content of a hypothesis arrived at inductively exceeds that of the initial premises (Ball, 1975). There have been many objections made to the inductivist approach. Prime among these objections is the suggestion that we do not perceive independent facts in isolation from some theory or hypothesis; thus theory precedes data gathering. Science does not grow by the mere accumulation of facts that eventually spawn theories; rather, theories are created by scientists (either logically or perhaps through other psychological or intuitive processes) and then new facts are gathered, or old facts are reinterpreted, in light of the theory (perhaps to support or refute the theory).

In contrast to strict inductivists, logical empiricists utilize the hypothetico-deductive method of scientific reasoning (Hempel, 1965, 1966; Popper, 1968; Wiley, 1981). All that is important for a hypothesis to be considered scientific is that it be logically formulated and testable. The hypothesis, or deductions from the hypothesis, must make predictions that can either be confirmed or refuted. The hypothesis must be falsifiable. "The criterion of potential falsifiability is very important, for any hypothesis that cannot theoretically be refuted is at best pseudo-scientific" (Ball, 1975, p. 412). A hypothesis may originate in the mind of the scientist in any way—perhaps suggested by data gathering or perhaps through simple speculation or imagination. A hypothesis is never proven using the hypothetico-deductive method—arguably, there is no such thing as proof or truth in science (except perhaps in closed logical systems, such as some mathematical systems). Of course, here universal hypotheses and potential truths are referred to, not particularistic or empirical statements. Thus, the universal hypothesis that all, both known and unknown (such as future individuals), crows are black can only be falsified, but not proven. However, the particularistic statement that some crows are black can be demonstrated empirically by finding a black crow. If we could some day enumerate every crow that ever lived and would ever live and found all to be black, then our original hypothesis that all crows are black would become a particularistic statement or description. Some would argue that any statement about crows is particularistic because the concept of a crow is not a natural, unrestricted, universal entity, but a spaciotemporally restricted concept. In reference to the subject matter of this book, the Popperian (1968) principle of falsifiability was originally meant to apply to scientific hypotheses that refer to unrestricted universals; the question of whether it is strictly applicable to specific phylogenetic hypotheses is problematic (see Hull, 1983; Panchen, 1982). If facts or observations conform to a hypothesis, the hypothesis is said to be corroborated. If valid or genuine (an epistemological question) facts or observations are not in agreement with the hypothesis, they are said to falsify the hypothesis, and the falsified hypothesis may have to be replaced by a different or modified hypothesis.

Two related and often used concepts that pertain to hypothesis formulation and testing are the concepts of *successive approximation* and *reciprocal illumination* (Schoch, 1984a). Successive approximation is usually used to refer to the fact that an initial hypothesis may conform to or explain the data known when the hypothesis is formulated, but as more data (observations) are gathered, some of which refute or falsify the original hypothesis, the hypothesis may be modified to fit both the old and new data. In contrast, reciprocal illumination (Szalay, 1977a, 1977b) or reciprocal clarification (Hennig, 1966a) refers to the practice of initially keeping certain sets or types of data separate from one another such that one system can be tested against the other without circularity (McKenna, Engelmann, and Barghoorn, 1977; Schoch, 1984a). For example, we might compare a phylogenetic hypothesis for a particular group of organisms based primarily on (or corroborated primarily by) the stratigraphic evidence of fossil forms with the anatomical and biogeographic data of extant forms of the group under consideration.

Finally, it should be explicitly pointed out that, somewhat ironically, the methods adopted in scientific inquiry cannot always be rigorously and scientifically evaluated. Methods of scientific inquiry are used to solve a problem, such as the reconstruction of phylogeny. If one knew the solution to the problem, one could evaluate the efficacy of various methodologies. But, if the correct solution were known, then one would have no need for the methods (Platnick, 1979). Consequently, the evaluation of scientific methodology is often a philosophical, rather than a scientific, pursuit.

WHY STUDY PHYLOGENY?

Why do we care about reconstructing the phylogeny of organisms? It may be considered inherently interesting in its own right. Phylogeny reconstruction is ultimately concerned with reconstructing the history of life, and it is the rare person who is not curious as to his or her origins and place in the world. But more than this, phylogeny reconstruction is of both theoretical and practical value to biology and geology.

At this point it should go without saying that phylogeny reconstruction is of importance to evolutionary biology and theoretical paleontology. The very subject matter of evolutionary biology is the evolution of biological entities; when we refer to evolution, we mean both the pattern of connections—the genealogical affinities or the propinquity of descent—of organisms and the processes, modes, and mechanisms that drive the evolutionary processes. It is possible to study certain aspects of the evolutionary process without explicit reference to phylogenies at or above the species level (e.g., studies of microevolutionary processes in ecological time), or using only the vague notions that a certain group of organisms or species being investigated are closely related to one another. As Fisher (1982, p. 175) has pointed out, for much recent research (e.g., Gould et al., 1977; Raup and Gould, 1974; Raup et al., 1973; Sepkoski, 1978, 1981; Stanley, 1979; Stanley et al., 1981; Van Valen, 1973, 1984) on macroevolution (i.e., evolution above the species level), "it is notable that very few of the patterns that have been considered depend, either for their perception or their analysis, on a detailed understanding of the phylogenetic relationship[s] of all members of the groups being considered." But it is only phylogenetic analysis that in the end can supply the detailed patterns of relationships among species that must ultimately be explained by evolutionary biology. Reconstructed phylogenies will then in turn supply the raw data, highly corroborated monophyletic clades at all levels, necessary for many higher-level studies concerning the tempos, modes, and processes of macroevolution. The term *theoretical paleontology* here refers to the discipline that seeks to generalize from, interpret, and explain the diversity and occurrences (distributions) of the raw entities of paleontology—fossil organisms. It is primarily through phylogenetic analysis, coupled with functional, ecological, and biogeographical studies, that theoretical paleontology achieves these goals.

Fossil entities are also of extreme practical value to the stratigrapher, the biostratigrapher, and the historical geologist. Fossils are one of the best means to establish the relative dating and temporal correlation of the strata that surround them. It is relatively straightforward to correlate as temporally equivalent rocks that contain identical fossil forms. But what of rocks that contain similar but nonidentical fossil entities? By knowing the detailed phylogenetic relationships of the organisms embedded in the rocks, the biostratigrapher may still be able to say something about the temporal relationships (sequencing) of the rock bodies themselves. Thus, phylogenetic analysis is not merely an academic, theoretical pursuit, but one with numerous practical applications.

CHAPTER 1

The Basics of Various Methodologies Used to Reconstruct Phylogeny

CLADISTIC METHODOLOGY

Cladistic (= phylogenetic) methodology has undergone many changes in the years since Willi Hennig (1950, 1965, 1966a, 1966b; see also Brundin, 1966) formulated its basic premises; today there are numerous different ideas concerning the philosophy behind, and the methodology of, cladistic analysis. Here some of the fundamental concepts of cladism are presented, although these will not be unanimously agreed upon by all modern practitioners of the discipline.

In a phylogenetic sense, the basic aim of cladistic analysis has been to distinguish monophyletic (*sensu stricto,* = holophyletic) groups that are usually referred to as clades (from the Greek *klados, branch* or *shoot*—that is, a clade is a separate and distinct branch on the metaphorical phylogenetic tree: J. Huxley, 1958). A group of organisms is monophyletic if it contains all (i.e., all known) and only the descendants of a single common ancestor. A group of organisms may consist of only the descendants of a single common organism, but if it does not include, at least theoretically, all of the descendants of that common ancestor, it is not monophyletic. A group descended from a single common ancestor, and including that single common ancestor, but not including all of the descendants of that common ancestor, is said to be paraphyletic. A group composed of members that do not share a single common ancestor that is included within the group is said to be polyphyletic. Only monophyletic groups—that is, distinct branches whether large or small (but not a part of a branch minus the tip, or only the base of the trunk, or several twigs without the connecting base of the branch)—are considered natural and meaningful within a cladistic framework. Hennig (1965) distinguishes three systems of analysis and classification of organisms: (1) phenetic systems that are based purely on morphological resemblance and may include paraphyletic, polyphyletic, and monophyletic groups; (2) pseudophylogenetic (classical evolutionary) systems that do not admit polyphyletic groups, but include both monophyletic and paraphyletic groups; and (3) the phylogenetic (= cladistic) system that excludes polyphyletic and paraphyletic groups and admits only monophyletic groups. A cladistic analysis can be translated directly into a cladistic classification, and to do so all recognized taxa must be strictly monophyletic.

In the cladistic analysis monophyletic groups are recognized by the possession of common evolutionary novelties (these novelties may be strictly morphological characteristics, or they may be physiological, behavioral, and so on). The justification for recognizing monophyletic groups on the basis of common evolutionary novelties arises from the basic principles and axioms upon which cladistics is based. Platnick (1979) recognizes three such basic principles: (1) that nature is orderly such that patterns (or the pattern) in nature can be represented by some hierarchical means (for instance, by a classification or a branching diagram); (2) that patterns (or the pattern) in nature can be elucidated through the analysis of characters that form internested, hierarchically arranged sets; and (3) that knowledge of phylogeny and/or evolutionary history must be inferred from hypothesized or observed hierarchical patterns of characters. Bonde (1977, p. 750) suggests that there are six general statements to which most phylogenetic systematists will agree axiomatically: (1) The observed diversity of biological organisms is the result of evolution. (2) Individual organisms transfer characteristics, either modified or unmodified, from one generation to another. (3) There are objective or natural biological units composed of organisms (i.e., the units are above the level of the individual organism). (4) There are processes such that the supraindividual units can become divided. (5) Supraindividual units may also become extinct. (6) Known life has a unique origin; life is monophyletic.

Other than the basic assumption of evolution itself, the basic axiom of phylogenetic cladistic methodology is that an ancestor will pass on to all of its descendants, through the continuity of the processes of inheritance, all inheritable features that it possesses. These features may be modified or lost by the descendants, but in general they will be retained, and added to, by the descendants. Any ancestral species, in becoming a distinct species (i.e., through speciation), will normally have evolved certain new evolutionary features. These evolutionary novelties, or derived features, will be passed on to the descendants of the particular species just as will the features that this particular species inherited from its ancestral species. All the descendants of a particular ancestral species will be recognizable by the possession of the evolutionary novelties that they inherited from that particular species. Thus the common possession of these particular evolutionary novelties among a group of organisms will serve to distinguish the group as a monophyletic group. Any inheritable feature observed to occur in organisms will have originated as an evolutionary novelty at some time and will serve to distinguish a monophyletic group at some level.

The evolutionary novelties, or shared and derived characteristics, that serve to distinguish a certain monophyletic group are referred to as synapomorphies of the group. Thus, following the logic of the last paragraph, all characteristics of organisms are synapomorphies at some level. When we are dealing with the same organ or characteristic in two organisms—that is, the organ or characteristic can be traced to a feature characteristic of an ancestor shared by the two organisms under consideration—we refer to such an organ or characteristic as homologous. Any homology defines a synapomorphy at some level and thus distinguishes a monophyletic group. Hence the cladistic credo that homology equals synapomorphy. Characteristics that are shared by all members of a certain monophyletic group, but that are also shared by members outside of the particular monophyletic group and thus define a more inclusive monophyletic group at a higher level, are referred to as symplesiomorphies (shared primitive characteristics) for the particular monophyletic group at hand. It should thus be evident that all characteristics of organisms are also symplesiomorphies at some lower level—that is, a level less inclusive—than the level at which they are synapomorphies.

In general, the term *apomorphy* (apomorphous) refers to an evolutionarily advanced or derived characteristic, and the term *plesiomorphy* (plesiomorphous) refers to an evolutionarily primitive characteristic of an organism. *Synapomorphy* and *symplesiomorphy* refer to derived or primitive characteristics that are shared by two or more taxa, and the term *autapomorphy* refers to an advanced or derived characteristic unique to a specific taxon.

The basic goal of cladistic analysis is to discover the hierarchical pattern of synapomorphies (= homologies) that must exist in nature given that all life has a common origin. In its purest form, cladistic methodology can take the form of an analysis of pattern independent of evolutionary assumptions (transformed cladistics: Charig, 1982; Patterson, 1980; Platnick, 1979). This fact has apparently led some critics of the cladistic program to assert that Hennig's system is nothing more than a modern version of idealistic comparative morphology where synapomorphies are in essence simple idealistic homologies (Dullemeijer, 1980, p. 177).

An investigator utilizing cladistic analysis will often express the hypothesized phylogenetic relationships of the organisms under investigation by using some kind of diagram (usually a cladogram) or convention (such as a written classification) that readily depicts hierarchical phylogenetic relationships, as indicated by the distribution of synapomorphies, between taxa. In the cladogram depicted in Figure 1-1a, A and B are united relative to C, A and B share synapomorphies relative to C, A and B share a more recent common ancestor than either does with C, and A and B form a monophyletic group relative to C. In sum, A and B are more closely related to each other than either is to C. It is often convenient to label the node points (e.g., 1 and 2 in Fig. 1-1b) and list the synapomorphies for each node point that distinguish the groups at the particular levels, or to use bars or some other pictorial device to indicate the synapomorphies that unite various taxa (e.g., Fig. 1-1c).

Given the relationships shown in Figure 1-1a, taxa A and B are often referred to as sister groups or sister taxa relative to C. This is basically another way of saying that in the realm of analysis of the three taxa A, B, and C, A and B are more closely related to each other than either is to C. The term *sister group* can also be used in the original (Hennig, 1966a) strict sense to refer to two taxa that share a more recent common ancestor than either does with any third group, whether the third group happens to be under consideration in the particular analysis or not (see Cartmill, 1981, for a discussion of this point). Løvtrup (1977b, p. 808) refers to "a pair of taxa that arise

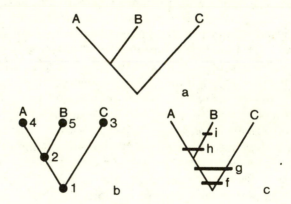

Figure 1-1. *(a)* A basic cladogram depicting the hypothesis that taxa A and B are more closely related to each other than either is to taxon C. *(b)* The same cladogram as in *a* with the node points numbered. In an actual example, synapomorphies that would be listed for node 1 are found in A, B, and C. Synapomorphies listed for node 2 are found in A and B, but not in C. Synapomorphies listed for nodes 3, 4, or 5 are found in C, A, or B, respectively. *(c)* The same cladogram as in *a* with bars indicating synapomorphies that unite the various taxa. For example, synapomorphies "f" and "g" are found in all three taxa, "h" is found in A and B, and "i" is found only in B.

through a process of dichotomy from the taxon in which they are included" as *twin taxa*.

In a cladistic analysis, monophyletic groups are recognized by the possession of shared derived characters (synapomorphies); any claim that a particular group is monophyletic must be substantiated by demonstrated synapomorphies. From this it follows that any two taxa that, within a certain context, are putative sister groups must be united by the joint possession of certain synapomorphies, but in each of the sister groups there must be certain characters that occur in a more primitive state (or at least in a different derived state) than in the other sister group. "This mosaic-like distribution of relatively primitive and derivative characters in related species and species-groups" (Hennig, 1965; reprinted in Sober, 1984b, p. 611) is what is termed the *heterobathmy of characters* (Hennig, 1965; reprinted in Sober, 1984b, p. 613). This concept is fundamental to cladistic analysis. The presence of only primitive characters in a group indicates that the group is probably paraphyletic. "The more complex is the mosaic of heterobathmic characters which we have at our disposal in a chosen group of species, the more surely can their phylogenetic relationships be deduced from it" (Hennig, 1965; reprinted in Sober, 1984b, p. 613).

Some cladists distinguish two different types of groups of organisms (Patterson, 1981c). A *crown-group* (Jefferies, 1979) or simply a *group* (Hennig, 1966a, 1966b) is composed of all descendants (living and extinct) of the latest common ancestor of the living members of a certain monophyletic group. A *stem-group* (possibly paraphyletic) is composed of all those extinct forms that are more closely related to a certain crown-group than to any other group. Together a crown-group and its stem-group form a more inclusive monophyletic group.

Construction of Basic Cladograms

Given two taxa, there is only one possible cladogram (Fig. 1-2a), and thus the problem of the relationships between two taxa may be viewed as inherently uninteresting or trivial. For more than two taxa, Nelson and Platnick (1981) distinguish three types of cladograms: primary cladograms are fully resolved cladograms (fully dichotomous, e.g., Fig. 1-2b), secondary cladograms are partially resolved cladograms (e.g., Fig. 1-2c), and tertiary cladograms are totally unresolved cladograms (e.g., Fig. 1-2d). The aim of basic cladistic analysis is to arrive at fully resolved cladograms. Given three taxa, there are three possible primary cladograms and one tertiary cladogram (Fig. 1-3). Given four taxa, there are 15 primary cladograms, 10 secondary cladograms, and a tertiary cladogram. For five taxa there are 105 possible primary cladograms alone. Obviously, it is impractical in most cases to intuitively evaluate all of the possible cladograms for more than three or four taxa at a time.

The simplest informative cladogram, that involving three taxa where two are more closely related to each other relative to the third (e.g., Fig. 1-3a), can be constructed on the basis of one character uniting two of the taxa relative to the third. We can then analyze the distribution of another, independent character among the three taxa under consideration and determine if the second character yields the same cladogram. Since there are only three informative cladograms, the expectation is that by chance alone, the cladogram yielded by the second character will agree with that of the first character one-third (33%) of the time. Assuming that the first two characters yield the same cladogram, we can then analyze the distribution of a third character among the taxa under consideration. The probability that the cladogram based on a third character will be identical to that based on the first and second characters is 33% times 33%, or approximately 11%. If we add a fourth character, the probability of all four cladograms being in agreement is $33\%^3$, or approximately 4%. Adding a fifth character, the probability of total agreement is $33\%^4$, or approximately 1%. Thus the chance that the same cladogram will be arrived at by chance alone using several independent characters is relatively low.

There are two basic practical approaches to reconstructing the cladistic (phylogenetic) relationships of a set of three or more organisms (or monophyletic taxa) given that we know the distribution of certain characters among the organisms under consideration. One approach is to choose any three of the organisms and on the basis of the distribution of characters among these three organisms unite two of them relative to the third. This approach forms the basic three-taxon problem, its statement, and solution (e.g., ((A, B) C)). The two taxa that are united are united on the basis of putative synapomorphies (for ways to distinguish apomorphy from plesiomorphy see the sec-

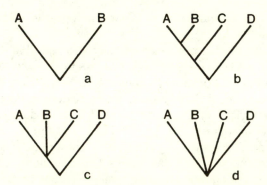

Figure 1-2. (a) The only possible cladogram given two taxa. (b) A primary cladogram: a cladogram that has been fully resolved dichotomously. (c) A secondary cladogram: a cladogram that is only partially resolved. (d) A tertiary cladogram: a totally unresolved cladogram.

tion on determination of morphocline polarity). By dealing with only three taxa at a time, we know all three of the possible cladograms, and the preferred cladogram is deductively selected by excluding the other two possibilities (Patterson, 1981c). (In some cases one may choose to deal directly with the 15 possible primary cladograms given a four-taxon problem, but if the analysis involves more than four taxa, the number of possible cladograms becomes prohibitively large.) Given an initial three-taxon statement, additional taxa can be added one by one on the basis of whether their characters unite them with taxon (A), (B), (AB), (C), or (ABC). In this manner an extremely complex cladogram showing the relationships of numerous terminal taxa can be built up piecemeal by treating the problem as a succession of three-taxon problems.

A second approach is to use the distribution of character states for each character among the taxa under consideration in order to construct a series of preliminary cladograms for all of the taxa involved. (Again, in order to do this one must have an idea as to which character states are apomorphies and which are plesiomorphies.) These preliminary cladograms, based on one character each, will usually be fairly coarse. Given a certain character with only two character states, one derived (e.g., A) and one primitive (e.g., a), a cladogram based on only this character will unite all (and only) taxa (as an unresolved polychotomy if there are more than two such taxa) that bear A relative to the taxa that bear a. The taxa bearing a are not united among themselves (do not form their own natural group) on the basis of the shared primitive a; these taxa merely do not belong to the group characterized by possession of A (of course, there is the possibility that a derived character state of some other character does indeed unite the taxa bearing a). The series of coarse, preliminary, one-character cladograms can then be combined one by one to arrive eventually at a single cladogram for all the taxa under consideration using all characters available.

As a hypothetical, simplified example of the use of the first approach (that

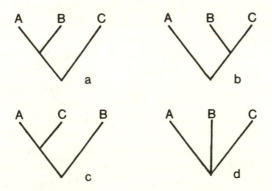

Figure 1-3. The four possible cladograms given three taxa.

of solving a succession of three-taxon problems), suppose that we have six taxa (1 through 6) with a set of character distributions as shown in Table 1-1. Initially, before any analysis is undertaken, the known relationships among the taxa (where, based on an assumption of the unitary origin of life on earth, it is assumed that all of the taxa had a common ancestor at some point, but no relationships between the particular taxa are known) can be depicted by a completely unresolved cladogram (Fig. 1-4a). Choosing any three taxa at random—for instance, 1, 2, and 4—this initial three-taxon problem can be solved on the basis of the known distribution of characters. Taxa 1, 2, and 4 all share A in common, but 2 and 4 are further united by the possession of B (Fig. 1-4b). Thus, the solution is (1 (2, 4)).

A fourth taxon—for instance, 3—can then be selected from the remaining three taxa (3, 5, 6). The three-taxon problem to be solved now is that involving taxa 1, (2, 4), and 3. Again, A is shared in common by all of the taxa, and (2, 4) and 3 are further distinguished by the possession of B. Thus, the solution is (1 (3 (2, 4))) (Fig. 1-4c). However, in solving this last problem, 2 and 4 were grouped together as a single taxon. Once it is established that 3 is more closely related to 2 and/or 4 than it is to 1, the three-taxon problem involving the relationships of 2, 3, and 4 must be solved. In analyzing the distribution of characters among taxa 2, 3, and 4, it is apparent that character B is common to all three taxa and 3 and 4 are further distinguished by the shared possession of character C. Thus the solution to this problem is (2 (3, 4)) (Fig. 1-4d). The resolved relationships of taxa 2, 3, and 4 can then be added to the cladogram depicted in Figure 1-4c to yield the cladogram of Figure 1-4e.

The cladogram depicted in Figure 1-4e is composed of two monophyletic groups, one consisting solely of taxon 1 and another composed of taxa 2, 3, and 4. A fifth taxon—for example, taxon 6—can be selected from the remaining taxa to be analyzed, and the three-taxon problem 1, (2, 3, 4), 6 can be solved. The solution to the problem is readily seen to be (1 ((2, 3, 4) 6)) as (2, 3, 4) and 6 are united by the possession of character B (Fig. 1–4f). The relationships among taxa 2, 3, 4, and 6 must next be resolved. It is already known that 3 and 4 form a group relative to 2 (Fig. 1-4d) and so this problem can first be approached by solving the problem of the relationships between 2, (3, 4), and 6. Taxa (3, 4) and 6 are distinguished from 2 by the possession of character C, and thus the solution is (2 ((3,4) 6)) (Fig. 1-4g). Next the three-taxon problem involving 3, 4, and 6 must be resolved. It is apparent that 4 and 6 are united relative to 3 by the possession of character D (Fig. 1-4h). The cladogram depicted in Figure 1-4h can then be combined with the cladogram depicted in Figure 1-4e to yield the cladogram of Figure 1-4i.

The cladogram depicted in Figure 1-4i is composed of two monophyletic taxa, namely 1 and (2, 3, 4, 6). To these can be added the remaining taxon to be analyzed, 5, and the three-taxon problem involving 1, 5, and (2, 3, 4, 6) is posed. Taxon 5 and taxon (2, 3, 4, 6) are united relative to 1 (Fig.

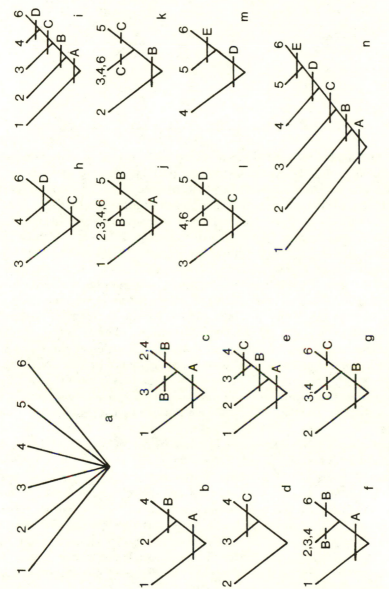

Figure 1-4. Series of cladograms used to resolve the phylogenetic relations among the hypothetical taxa listed in Table 1-1; see text for explanation.

Table 1-1. Hypothetical Data Matrix (+ = Presence,
− = Absence)

			Character			
	A	B	C	D	E	F
Taxon						
1	+	−	−	−	−	−
2	+	+	−	−	−	−
3	+	+	+	−	−	−
4	+	+	+	+	−	−
5	+	+	+	+	+	−
6	+	+	+	+	+	+

1-4*j*) by the possession of character B. Within taxon (2, 3, 4, 6) it is known that taxon (3, 4, 6) is monophyletic relative to 2 (Figs. 1-4*g* and 1-4*i*), so the three-taxon problem involving 2, (3, 4, 6), and 5 must be solved. By the distribution of character C, the solution to this problem is readily seen to be (2 ((3, 4, 6) 5)) (Fig. 1-4*k*). Within taxon (3, 4, 6) it is known that taxa 4 and 6 form a monophyletic group relative to 3 (Fig. 1-4*h*), so the three-taxon problem involving 3, (4, 6), 5 must next be solved. As (4, 6) and 5 are united relative to 3 by the possession of character D, the solution is (3 ((4, 6) 5)) (Fig. 1-4*l*). The relationships of 4, 5, and 6 can then be analyzed. Taxa 5 and 6 are characterized by E relative to 4 (Fig. 1-4*m*). Finally, the cladogram depicted in Figure 1-4*m* can be combined with the cladogram depicted in Figure 1-4*i* to yield the fully resolved cladogram of Figure 1-4*n*.

To give a somewhat more complicated, but also perhaps more realistic, example, we can utilize the hypothetical data matrix of Brooks (1984) for the distributions of the states of 12 characters among seven taxa (Table 1-2). In this example it is assumed that 0 represents the primitive character state, and higher absolute values for character states represent progressively more derived states.

Beginning with any three taxa at random, we might initially choose taxa 1, 2, and 7 for analysis. The first thing we might notice is that all three taxa are united by the joint possession of derived character states of characters C, J, and L; however, this does not help us resolve the relationships among these three taxa. Analyzing the other characters, it is seen that only taxon 1 is apomorphous for character A, only taxon 2 is fully apomorphous for characters D and H, and only taxon 7 is apomorphous for characters E, F, G, I. A derived condition for character H unites taxa 1 and 2 relative to 7; thus the relationship among these three taxa is ((1, 2) 7) as depicted in Figure 1-5*a*.

When we add another taxon—for instance, taxon 5—we immediately see that taxon 5 shares the derived condition of characters E, F, and G with taxon 7, and all of the taxa thus far (taxa 1, 2, 5, and 7) share derived conditions of characters C, J, and L. Furthermore, character J is further derived in taxa

Table 1-2. Hypothetical Data Matrix

						Character						
	A	B	C	D	E	F	G	H	I	J	K	L
Taxon												
1	1	0	1	0	0	0	0	1	0	1	0	1
2	0	0	1	1	0	0	0	3	0	1	0	1
3	0	1	1	1	0	0	0	2	1	1	1	1
4	0	0	1	0	1	0	1	0	0	2	1	0
5	0	0	1	0	1	1	2	0	0	2	0	1
6	0	0	1	0	1	1	3	0	0	2	0	1
7	0	0	1	0	1	1	4	0	-1	2	0	1

Source: After Brooks, 1984, p. 120.

5 and 7 than it is in taxa 1 or 2. This gives the cladogram shown in Figure 1-5*b*.

Next considering taxon 6, it is readily seen that taxon 6 is united with taxa 5 and 7 on the basis of the shared possession of the derived states of characters E, F, and G, and also the further derived condition (character state "2") of character J. The relationships among taxa 5, 6, and 7 can be resolved by noting that taxa 6 and 7 share the possession of further derived conditions (character states "3" and "4," respectively) of character G relative to the state of this character (character state "2") in taxon 5. Thus we arrive at the cladogram shown in Figure 1-5*c*.

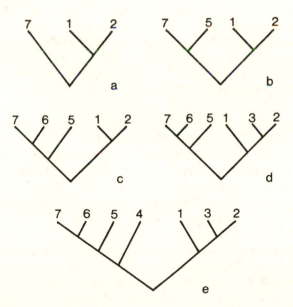

Figure 1-5. Series of cladograms used to resolve the phylogenetic relations among the hypothetical taxa listed in Table 1-2; see text for explanation.

Taking taxon 3 into consideration next, this taxon shares a derived state for character H with taxa 1 and 2. The relationships between taxa 1, 2, and 3 are resolved by noting that only taxa 2 and 3 bear the derived state of character D, and taxa 2 and 3 share relatively more derived states of character H. This gives us the cladogram of Figure 1-5d.

Finally, we must consider taxon 4. The relationships of taxon 4 do not appear to be as unambiguous as are the relationships of the other taxa considered above. Taxon 4 appears to share a derived state of character K with taxon 3, but taxon 4 also appears to be primitive for character L whereas all of the other six taxa share a derived state for this character. Taxon 4, however, is united with taxa 5, 6, and 7 by the possession of the derived state of three characters (states "1" of characters E and G and state "2" of character J); thus, by the criterion of parsimony, taxa 4, 5, 6, and 7 are hypothesized to form a monophyletic group; it is suggested that the joint possession of a putative synapomorphy between taxa 3 and 4 represents homoplasy, and the putative primitive state of character L in taxon 4 represents a reversal. Among taxa 4, 5, 6, and 7, taxa 5, 6, and 7 are united by the further derived state of character G. The final cladogram representing the phylogenetic relationships among taxa 1 through 7 is shown in Figure 1-5e.

Using the basic approach outlined above for resolving complex dichotomous cladograms, Nelson and Platnick (1981, p. 250) distinguish between minimum, greater-than-minimum, and maximum modes of resolution. There will be one unique, maximally efficient (composed of a minimum number of steps) solution for resolving any more-than-three-taxon problem; such a minimum mode of resolution will consist of a series of three-taxon problems solved in a particular order. For the six taxa analyzed in the first example above (Fig. 1-4; Table 1-1), the minimum mode of resolution would involve beginning with taxa 4, 5, and 6; this is resolved as (4 (5, 6)). Next the problem 4, (5, 6), 3 is solved. Subsequently taxon 2 is taken into consideration, and finally taxon 1 is included in the analysis. Considering these six taxa in the above sequence will lead to the resolution of the final cladogram in the minimum number of steps. The maximum mode of resolution is the opposite of the minimum mode of resolution; there are several different sequences in which the taxa under consideration can be incorporated into the analysis such that a maximum number of steps must be taken to arrive at the final, fully resolved cladogram. A greater-than-minimum mode of resolution involves some number of steps between that of the minimum mode of resolution and that of the maximum mode of resolution.

As an example of the second approach, that of using each character separately to construct a coarse cladogram, assume that we have the same taxa (1 through 6) with the same set of character distributions (Table 1-1) as used in the first example above. Character A is common to all six taxa (Fig. 1-6a). Character B unites taxa 2 through 6 (Fig. 1-6b). Character C unites taxa 3 through 6 (Fig. 1-6c). Character D unites taxa 4 through 6 (Fig. 1-6d). Character E is common to taxa 5 and 6 (Fig. 1-6e).

Choosing two characters at random—for instance, B and E—and combining the character cladograms for characters B and E yields the cladogram depicted in Figure 1-6*f*. Adding to this the character cladogram for C yields the cladogram of Figure 1-6*g*. Finally, uniting this cladogram with the character cladogram for D yields the fully resolved cladogram depicted in Figure 1-6*h*.

In many cladistic analyses, conflicts in character state distributions (putative synapomorphies) may be encountered. The basic methodological principle used is that of parsimony (economy, simplicity, Occam's razor). The preferred hypothesis of phylogenetic relationships is that which is based on the congruent distribution of the largest number of independent characters and correspondingly requires the least number of ad hoc hypotheses of homoplasy. If two alternative cladograms for the same set of taxa are based on an equal number of independent characters, it may not be possible to arrive at a fully resolved cladogram; rather, there may be some polychotomous branchings. Any cladistic hypothesis of relationships is subject to refutation or falsification by the discovery of characters whose distribution contradicts the initial cladistic hypothesis. Any cladistic hypothesis predicts that future (yet to be discovered) synapomorphies among the taxa under consideration will be congruent with the proposed cladogram. These topics are dealt with in chapters 2 and 3. Here it should be noted that a set of character distributions can be used to arrive inductively at a preferred cladogram, or using a more hypothetico-

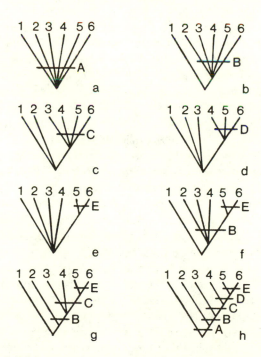

Figure 1-6. Series of cladograms used to resolve the phylogenetic relations among the hypothetical taxa listed in Table 1-1; see text for explanation.

deductive approach, a hypothesis of relationships (or more simply, a hypothesis of synapomorphies in the form of a cladogram) among certain organisms can be posited and then tested by the distribution of putative synapomorphies among the organisms under consideration.

In a more general context, including a nonphylogenetic context, some investigators think of a cladogram in the following terms:

A cladogram is a branching structure joining certain terms (representing taxa) that are related by some unspecified relation. In itself, a cladogram conveys no sense of phylogeny, common ancestry, phenetic resemblance, gradistic resemblance, ecological resemblance, or any other relation that might conceivably join the terms (representing taxa). (Nelson and Platnick, 1981, p. 172)

In this context, general synapomorphy may be viewed as the general restriction of certain subsets of taxa (entities) within other sets—that is, general synapomorphy may not always correspond to phylogenetic synapomorphy (shared and derived characters) as used in this book, but phylogenetic synapomorphy is a category of general synapomorphy.

In constructing general cladograms (those without a specifically phylogenetic inference, although such may be attributed to them later), cladistic workers (following more or less the tradition of Hennigian cladistics) advocate the use of only positive occurrences of characters to form a hierarchy of sets and subsets; phenetic workers (adhering to the numerical taxonomic tradition) utilize both positive (presence) and negative (absence) occurrences of characters; and gradistic workers (in the tradition of classical evolutionary systematics) use a combination of positive and/or negative occurrences in any particular case (Nelson and Platnick, 1981). In general, a cladistic analysis (particularly when utilizing only positive occurrences or shared derived characters) assumes that the characters analyzed are hierarchically ordered, to some degree, into sets and subsets. If there is no such hierarchical order, then the cladogram will show character conflict (different characters will define different, incompatible sets). Performing a cladistic analysis is one way to determine if the characters under consideration do form such a hierarchical pattern (see section on parallelism and reticulation in cladograms).

In a nonphylogenetic context, positive occurrences of characters generally correspond to shared and derived characters in a phylogenetic context, and both correspond to specific characters (characters that distinguish a restricted subset of taxa or entities analyzed) as opposed to general characters that are common to all entities under consideration or a larger (more inclusive) subset of entities. Negative occurrences are not informative in and of themselves in that they merely (redundantly) indicate the presence of positive occurrences in other taxa. Using only the concepts of positive and negative occurrences, false positive occurrences in a nonphylogenetic context refer to nonhomologies or homoplasies in an explicitly phylogenetic context. Nelson and Platnick (1981, pp. 195–196) outline four basic ways to identify false positive occurrences that are thought to be initially shared among a restricted set of taxa under consideration: (1) Further detailed analysis of the actual characters may reveal that the positive characters shared by the organisms (taxa) actually differ in some fundamental way and thus are not the same shared character after all. Rather, they are separate characters that distinguish smaller subsets of taxa. (2) With further analysis, it may be found that the positive characters under consideration actually characterize a more inclusive group—that is, they have a more general distribution among taxa than initially believed; perhaps they are obscure or poorly developed, or lost, in certain taxa. (3) What was initially interpreted as a positive occurrence of a character may actually be a negative occurrence of another character (i.e., the absence of a character). It may be very difficult to determine what a positive or negative occurrence is without an explicitly phylogenetic context. In a phylogenetic context, positive occurrences represent derived traits (evolutionary novelties) and negative occurrences represent primitive traits. Both primitive and derived characters may be represented by real (tangible or observable) structures. Or in some cases the derived character—the positive occurrence—may be the loss of a primitive character. (4) Further analysis may demonstrate that certain characters are not characters at all. From the standpoint of the organism (or other entity) being analyzed they are not characters or traits; rather, they have been imposed upon the organism by the mind of the investigator. In some cases a mixture of characters, with noncongruent distributions among various taxa, may be viewed by the investigator as a single "character" in a particular organism. This, of course, can lead to problems when the investigator searches for the same character in other organisms.

If we can resolve all character conflicts, then all of the positive occurrences (or derived traits in a phylogenetic context) can be combined in a single cladogram. Such a cladogram is parsimonious in accounting for all characters as occurring (or arising) only once (within one restricted subset); the same character does not occur in two or more, otherwise independent subsets. Such a cladogram is an efficient summary of the data at hand. If there are character conflicts, then we search for and prefer the cladogram that takes the most characters (positive occurrences) into account—that is, it is the most efficient summary of the data.

Yet as long as there is conflict among positive occurrences, there is a problem that may be investigated: namely, of the conflicting occurrences, which are real and which not? This residual problem cannot be solved, except perfunctorily, through the use of a clustering procedure. Its solution is possible only through study of organisms and new knowledge of, or new insight into, their real characteristics. (Nelson and Platnick, 1981, p. 199)

Even if various clustering techniques can be applied to a particular data set and yield consistent and unambiguous results, this does not guarantee, or even indicate, that the results (or inferences based on them) are meaningful.

Cladograms can be constructed for terminal taxa at any level, and any cladogram constructed for supraspecific taxa implies a number of less-inclusive cladograms. Thus, if we propose a cladogram (Fig. 1-7) to solve the three-taxon problem of the relationships of reptiles, birds, and mammals with the solution (mammals (reptiles, birds)), then this cladogram implies that given any bird, any reptile, and any mammal, they will be related as follows: (any mammal (any reptile, any bird)). If for some specific case of a bird, a reptile, and a mammal this relationship can be shown to be false, then the more-inclusive cladogram must be false. The falsity of the more-inclusive cladogram may be due to the fact that some of the terminal taxa are unnatural (i.e., nonmonophyletic).

As Nelson and Platnick (1981) have noted, cladistics is often criticized as inadequate because any single cladogram may correspond to several different evolutionary truths (or perhaps none at all), and there are a number of evolutionary events or scenarios that produce character distributions that cannot be depicted as simple dichotomous cladograms. In the case of events that yield multiple branchings on cladograms, one may not be able to distinguish between multiple speciations, actual ancestors, hybridization, or cases where we are simply ignorant of the relationships of the organisms under consideration. It is sometimes suggested that the concept of dichotomous branchings of cladograms is a simplifying assumption that does not reflect how evolution really occurs. These critics perhaps fail to realize that cladistics does not necessarily attempt to reconstruct the actual details of the evolution of organisms (in some cases it need not even deal with evolution); it is not a theory of evolution but a method of systematics and, as used in this book, a method of reconstructing the phylogenetic relationships of organisms. In some instances a cladogram may not correspond to any real evolutionary or historical events, but it may still be considered to have a truth of its own (Nelson and Platnick, 1981, p. 220)—for instance, if the distribution of characters it summarizes is correct. Cladists readily acknowledge that there are limits to the information contained in cladograms. The well-defined boundaries of knowledge within cladistic analyses is an advantage of this methodology. The phylogeneticist and systematist, along with the historian of human events, must always keep

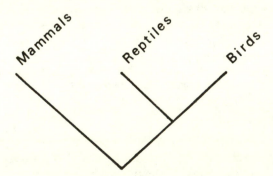

Figure 1-7. One possible solution to the three-taxon problem of resolving the phylogenetic relationships among reptiles, birds, and mammals.

in mind the "uncomfortable fact which is too easy to ignore: that a certain amount of the story has totally and finally disappeared" (M. Grant, 1958, p. 18).

Wagner's Groundplan-Divergence Method of Phylogeny Reconstruction

In the late 1940s and early 1950s, the botanist Dr. W. H. Wagner, Jr. (University of Michigan), independently of Hennig's work, developed a methodology to attempt to construct phylogenetic trees objectively (see Wagner, 1969, 1980, 1984). Wagner's method (now commonly referred to as the Groundplan-divergence Method, but originally called the Visual Groundplan Graph: Wagner, 1984) is based on principles similar to those of Hennig's cladistic methodology; however, there are also subtle but important differences in philosophy and practice between the two. Unfortunately, at times there has been some confusion between cladograms *sensu* Wagner and cladograms *sensu stricto* (as used by Hennigian cladists and as defined in this book).

The goals of Wagner's method are more ambitious than those of Hennigian cladistics.

The Groundplan-divergence Method is not concerned only with branching sequences and recency of common ancestry. It is aimed at finding and describing all of the pathways of genetic change. The method attempts to explain taxonomic diversity in the form of a phylogenetic tree but without respect to time or place. The tree is based upon taxonomy, not the reverse (taxonomy being defined here as phenetic classification, clustering by relative amounts of resemblance and difference). . . . Phylogenetics attempts to record the genetic changes themselves [as manifested in observable characters], not the rate of these changes. A complete phylogeny, then, involves the amount (grades or levels), directions (clades or lines), and sequences (steps or series of shared advancements) of divergences in genotypes, all involving biological changes. It is neither concerned with chronological time (as represented in fossil strata) nor space (as represented in geographical variation). The cladogram of Groundplan-divergence shows the branchings in respect entirely to estimated amount of evolution, i.e., grades, levels, or patristic distances. (Wagner, 1980, p. 175)

While Hennigian cladistics in the pure sense is concerned only with degrees of relatedness, not with ancestors or any quantification of degree of divergence from an ancestral form (only the topology of a strict cladogram is important; the lengths of branches are meaningless), Wagner's cladograms are concerned with both. Wagner's cladograms, or cladistic trees, include ancestors at non-terminal nodes or branch points (the ancestors may be hypothetical or correspond to real, known organisms), and the lengths of the branches are based on the amount of divergence of taxa from ancestral forms. Wagner (1980, p. 176, fig. 1) distinguishes between three types of cladistics and cladograms or cladistic trees. Patriocladistics is the cladistics that Wagner advocates; the lengths of branches on a cladistic tree correspond to degrees of divergence. In chronocladistics the lengths of the branches on a cladistic tree correspond to the times of first appearance in the fossil record of the taxa in question. In typocladistics the cladistic tree is superimposed on a map showing the geographic distribution of the taxa under consideration.

The basic steps of Wagner's method of working out evolutionary relationships of taxa are:

Classify the plants [or other organisms under consideration; here Wagner is referring primarily to a low-level, alpha taxonomy, which will provide the units of analysis], find the character trends, estimate an ancestor on the basis of the most generalized character states, find the relative amount of advancement for each species and variety and form the tree by shared divergent characters. (Wagner, 1980, p. 174)

In more detail, Wagner's procedure is as follows (the seven numbered steps of the procedure follow those proposed by Wagner [1980]).

1. According to Wagner (1980, p. 178) "correct phenetic classification underlies any phylogenetic analysis." The first step in analyzing the phylogeny of a group using the Groundplan-divergence Method is to gain a thorough knowledge of the group and erect an alpha-level (usually subspecies, species, and genera) taxonomy for the entities to be analyzed. This is done according to the similarities and differences of the organisms—that is, on the basis of their phenetics. Species must be correctly identified and allocated to their correct genera before the phylogenetic analysis begins. The species, genera, families, or other low taxonomic groups of the selected taxonomy are then used in the analysis. Wagner does not use the concepts of terminal taxa or monophyly; theoretically, any taxon at any level could be ancestral to any other taxon.

2. Before the phylogenetic analysis is carried out, any hybrid taxa and their direct derivatives must be recognized and removed. Hybrid taxa may obscure the phylogenetic relationships of the nonhybrid taxa. Hybrids, at least among plants, may be produced continually due to repeated interbreeding between the parent species and thus may be polyphyletic. Hybrid taxa are usually recognized as being intermediate in all or most of their characters relative to the parental taxa. Other, extrinsic criteria, such as coexistence with the parental taxa or occurrence in disturbed habitats, may also be used to recognize hybrids. Hybrids among extant plants can often be recognized if they are polyploids.

3. Once the hybrid taxa are eliminated from initial consideration, the next step is to undertake an analysis of character trends among the taxa under consideration (i.e, form morphoclines). Every trend or pathway must be considered separately, even if they originate from identical primitive states. Strictly homologous characters must be compared in order to establish character trends. Wagner (1980, p. 182) suggests using the classical tests of homology (see section on homology). For plants that are closely enough related that they can be made experimentally to hybridize, homology may also be established by the ability of character states to combine in a hybrid individual. If they are not homologous, they will not combine to form intermediates.

Wagner (1980) also advocates recognizing true (and otherwise unrecognizable) parallelisms by the use of other correlated character-state trends (see section on morphocline polarity). In order to determine which of the alternative states of a given character is primitive, Wagner relies heavily on the concept that general = primitive, and he also utilizes out-group comparison. In general, Wagner prefers to deal with two-state characters if possible. The primitive state is scored as 0 and the derived state as 1. If there are more than two character states, Wagner suggests assigning intermediate scores to the states that are intermediate between the most primitive and the most derived states. Wagner stresses that characters whose trends and polarity are ambiguous, dubious, or questionable should be rejected and not used in the analysis. Wagner concludes (1980, p. 184) that in many cases, for a particular group under consideration, one may end up with ten or fewer good, usable characters.

4. The next step is to determine the common ancestor (groundplan) for the group under analysis. The common ancestor is nothing more than an amalgamation of all of the primitive or generalized states of the characters being used in the analysis. The common ancestor may very rarely correspond to an actual, known organism (the common ancestor of Wagner's method can be thought of, in cladistic terms, as the primitive morphotype of the group in question).

5. After determining the common ancestor for the group, one calculates the level of divergence (or patristic distance) of each taxon under consideration from the common ancestor. The patristic distance of any taxon is simply the sum of the specializations of that taxon (advanced or divergent states from that of the ancestor, where each fully advanced state is scored as 1), including any parallelisms, reversals, or convergences. Wagner (1969, 1980) suggests using a target graph where the ancestor is placed at the center and the divergent taxa are plotted around the ancestor; successive concentric levels correspond to greater degrees of divergence (Fig. 1-8).

6. Once divergence levels are calculated for the taxa under consideration, they can be linked together on the basis of shared divergent (i.e., advanced) characters. Common ancestors (groundplans) are also calculated for all pairs and groups of taxa. This procedure is done parsimoniously such that a minimum of parallelism is hypothesized. As an example, if capital letters represent derived states and lower-case letters represent primitive states, the taxa with the divergence formulas of ABCDEFGhij and AbCdeFgHIJ would be connected by the common ancestor AbCdeFghij (Wagner, 1980; see Fig. 1-8). This part of the analysis is extremely similar to that of Hennigian cladistic analysis, except that in standard cladistic analysis ancestral forms are not hypothesized.

7. After a patrocladistic tree has been constructed for the nonhybrid taxa, the hybrid taxa can be inserted, showing their reticulate origin from the parents. The divergence index of a hybrid will usually be the sum of the divergence indexes of both of the parents divided by two (the hybrid being intermediate in character states).

Wagner's Groundplan-divergence Method has been used primarily in analyses of the species of plants within a single genus or closely related group of genera as a part of monographic studies (see citations in Wagner, 1980). It has also found much application in helping to teach concepts in systematic botany (Wagner, 1969, 1980). Computerized Wagner Tree methods (e.g., Farris, 1970; Kluge and Farris, 1969) have been based on Wagner's original procedures as outlined above, and these computer programs have in turn been modified for manual use (C. H. Nelson and Van Horn, 1975; Whiffin and Bierner, 1972).

Løvtrup's Cladistic Analysis

Using Hennigian cladistics as a base, Løvtrup (1977*a*) has elaborated on his philosophy of evolution and his methodology of phylogeny reconstruction. Løvtrup expresses phylogenetic and evolutionary relationships graphically with what he terms dichotomous phylogenetic dendrograms (Løvtrup, 1977*a*, p. 25; see Fig. 1-9). These dendrograms do not correspond to strict Hennigian cladograms, but rather are a cross between cladograms and conventional phylograms (phylogenetic trees). "The horizontal lines in a phylogenetic dendrogram represent the process of dichotomy through which new taxa arise, while the vertical lines represent periods of time during which taxonomic characters have been acquired" (Løvtrup, 1977*a*, p. 25). Each vertical line represents a monophyletic taxon that includes all of its subordinate taxa (in conventional cladograms, such as Figs. 1-1–1-7, the apex occurs at the bottom of the page and progressively lower taxa [less inclusive, or more subordinate

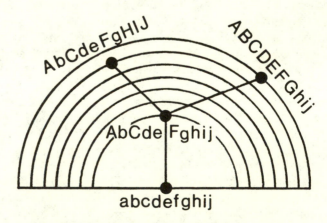

Figure 1-8. A sample Wagner phylogram drawn on a target graph.

taxa] occur toward the top of the page). Løvtrup (1977a) views the course of phylogenetic evolution as progressing from the apex to the base of a phylogenetic dendrogram. Thus the more-inclusive taxon (what Løvtrup [1977a, p. 22] terms superior) exists in nature before it undergoes splitting into its included (inferior) taxa (twin taxa). "Superior taxa have arisen before the inferior ones which they include" (Løvtrup, 1977a, p. 55). In Løvtrup's (1977a) notation, the taxon T_j is divided into taxa T_{j+1} and each taxon T_{j+1} is divided into taxa T_{j+2}—that is, T_j is immediately superior to taxa T_{j+1}, and in turn taxa T_{j+1} are immediately superior to taxa T_{j+2}. Taxon T_j originated (or had its beginnings) before any of the included taxa T_{j+1} came into existence, and likewise each T_{j+1} had its beginnings before any of its included subtaxa came into existence. Løvtrup (1977a, p. 25) states: "The creation of the taxon T_{j+1}, i.e., the origination of all of its taxonomic characters [thereby the taxon becomes fully definable], has occurred during a period of time lasting from the subdivision of the taxon T_j, in which it is included, until it itself became subdivided into two taxa T_{j+2}." Thus, according to Løvtrup, a taxon has not fully originated, is not fully definable, until it itself undergoes splitting into subordinate (inferior) taxa. Accordingly, extant low-ranking (inferior) taxa, such as some species and subspecies, are not fully definable and have not completely originated according to Løvtrup's system. Løvtrup (1977a, 1977b) dates the origination of a taxon from that point in time when it undergoes division into subordinate taxa; thus, twin taxa (sister taxa that originate from, or begin to differentiate due to, the splitting of the immediately superior taxon) may have different ages of final origination (they themselves become fully defined at different times). In contrast, Hennig (1965, 1966a) and most other workers (and as used in this book) date the origination (age of origin) of a particular taxon from the time that it begins to differentiate, along with its sister taxon (or taxa), via the splitting of the immediately superior taxon. Thus, according to Hennig (1965, 1966a), sister taxa in the strict sense have identical origination dates and are of equal age. The time when a taxon (i.e., an initially

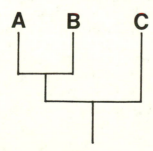

Figure 1-9. A Løvtrupian phylogenetic dendrogram. In Løvtrup's terminology, taxa A and B are "secondary twins," and the taxon including A and B and the taxon C are "primary twins."

homogeneous reproductive community) undergoes division into subordinate taxa is usually considered the "age of division" of the taxon (Hennig, 1965).

Løvtrup (1977*a*) emphasizes the use of extant organisms for phylogeny reconstruction. Indeed, he states that "only extant animals can be classified [here, classification = phylogeny]" and "the discovery of a new fossil has no impact on classification" (Løvtrup, 1977*a*, p. 21). However, Løvtrup (1977*a*, p. 21) does believe that "most fossil groups may be allocated to definite sites in the phylogenetic classification of living animals." In deciding what characters to use for phylogenetic reconstruction, Løvtrup (1977*a*) advocates using as many independent characters as possible, along with a weighting of types of characters. Løvtrup (1977*a*) considers losses to be more common than gains (correspondingly, gains may be more heavily weighted). He also considers morphological form to be relatively more labile than non-morphological characters (molecular-biological data such as "various chemical features, physiological mechanisms and histological differentiation patterns": Løvtrup, 1977*a*, p. 44) and thus places more emphasis on the latter. As for the problem of the independence of characters, Løvtrup (1977*a*, p. 53) notes that "our ignorance about epigenetic mechanisms implies that it is very seldom possible to decide upon the interdependence [and thus possible independence] of characters." In this regard, Løvtrup (1977*a*, p. 54) dismisses relative complexity as a reason to accord high phylogenetic weight to a certain morphological character because "complexity may at times arise through causally connected epigenetic processes, and in such cases the first link in the sequence may be a relatively simple agent whose origin through convergence is not at all implausible."

Løvtrup's basic method of actually reconstructing phylogenetic relationships is based on his theorem that "in a basic classification [i.e., in dealing with three taxa; see Fig. 1-9] the pair of secondary twins [the two of the three terminal taxa under consideration that are more closely related to each other relative to the third] have more characters in common than any of the other pairs of taxa" (Løvtrup, 1977*a*, p. 41). Løvtrup (1977*a*, p. 41) terms this the "rule of the quantitative approach to phylogenetic classification." It is based on three assumptions concerning the characters employed in constructing the phylogeny/classification: "(1) that they should represent not individual taxa, but the phylogenetic lines leading from the bifurcations in the dendrogram, (2) that new characters should arise with a reasonably constant frequency, and (3) that they should be independent" (Løvtrup, 1977*a*, p. 43). Løvtrup is forced to adopt his quantitative approach because he apparently does not believe that one can distinguish between apomorphic and nonapomorphic features in most cases. Given Løvtrup's views, it is easy to understand that he can state:

The basic principles involved in ordinary numerical taxonomy [i.e, phenetics], the use of many characters and the principle of parsimony [note that this is not necessarily parsimony in the sense of minimizing ad hoc hypotheses of homoplasy—see discussion in chap. 3] must, in my opinion, be employed in every attempt to establish an objective phylogenetic classification. I am therefore convinced that the hierarchical numerical phenograms are cladograms, in which the correlation coordinate is a crude measure of the periods of time during which the various taxa existed. (Løvtrup, 1977a, p. 54)

It should also be pointed out that Løvtrup (1977a) does not use the conventional Hennigian terminology (e.g., plesiomorphy, apomorphy) for classes of taxonomic/phylogenetic characters and their distributions. Rather, Løvtrup (1977a, p. 34) introduces the following terminology: "In a basic classification [a three-taxon problem], the taxonomic characters distinguishing T_j, and those superior taxa in which the latter is included, are plesiotypic, those of the non-terminal taxon T_{j+1} are apotypic, and those of the three terminal taxa are teleotypic." Thus, in this context plesiotypy is approximately equivalent to plesiomorphy, apotypy is approximately equivalent to synapomorphy, and teleotypy is approximately equivalent to autapomorphy.

Numerical Cladistic Methods of Inferring Phylogeny

A number of investigators have advocated various numerical and statistically based methods of inferring phylogenetic trees and cladograms (reviewed by Felsenstein, 1982, 1984). As Felsenstein (1984, p. 170) points out, using Hennig's (1966a) method of cladistic analysis, ideally each synapomorphy should define a monophyletic group, all of the monophyletic groups defined by all of the putative synapomorphies should be compatible (congruent), and all of the monophyletic groups (some nested inside of others) define the topology of the phylogeny. However, with real data and utilizing numerous putative synapomorphies, some supposed monophyletic groups will frequently be incompatible (incongruent) with supposed monophyletic groups defined on other bases (usually using other putative synapomorphies). This problem Felsenstein (1984, p. 170) refers to as Hennig's dilemma.

One way to deal with Hennig's dilemma is to apply strict methodological parsimony (see section on parsimony). Methodological parsimony is the rule that the proposition (in this case, the cladogram or phylogeny) that best fits all of the observed data (distribution of characters among the taxa under consideration), and thus requires the smallest number of ad hoc or a posteriori assumptions, is to be preferred (Kluge, 1984)—that is, ad hoc hypotheses of homoplasy should be minimized. Methodological parsimony makes no assumptions about the evolutionary process (see chap. 3). However, this is not the only way to attempt to resolve Hennig's dilemma. One can make certain a priori assumptions about nature—for example, that evolution is economical or parsimonious; it will occur with a minimum number of steps. This type of proposition Kluge (1984, p. 26) refers to as *evolutionary parsimony*. Often a phylogeny reconstructed on the basis of methodological parsimony will differ from one reconstructed on the basis of some form of evolutionary parsimony.

It is in order to attempt to solve Hennig's dilemma that the various numerical cladistic methods, based on different assumptions about the probability of various kinds of evolutionary events and changes, have been developed. Here (following Felsenstein, 1984) some of the more popular techniques are briefly reviewed. For more information on the specific methods, the reader is referred to the cited papers and to Sneath and Sokal (1973). Virtually all of these numerical methods are designed primarily for computer use; Baum (1984) has described them from the point of view of the computer user. R. A. Arnold and Duncan (1984) provide information on obtaining the actual computer programs commonly used in the types of analyses reviewed below. The basic assumptions of these methods are discussed in chapter 3.

A. W. F. Edwards and Cavalli-Sforza (1963, 1964; Cavalli-Sforza and A. W. F. Edwards, 1967) developed the concept of minimum evolution and constructed computer programs to construct phylogenies following this criterion for continuous-state character data. Camin and Sokal (1965) used essentially the same principle and an analogous method for discrete-state character data. Based on the concept of evolutionary parsimony or minimum evolution, alternative methodologies of phylogeny (often tree) reconstruction have been developed, in which different types of evolutionary changes are allowed to occur and different character changes are minimized (Felsenstein, 1984). If the raw data consists of characters of two character states each, where the 0 state is primitive or ancestral and the 1 state is derived, advanced, or descendant, the assumptions of some of the most popular methods can be summarized as follows.

The Camin-Sokal method (Camin and Sokal, 1965) allows only changes from 0 to 1 and then minimizes the number of such changes. This is the principle of strict evolutionary parsimony.

The Wagner parsimony method (Eck and Dayhoff, 1966; Farris, 1970; Farris, Kluge, and Eckardt, 1970; Jensen, 1981; Kluge and Farris, 1969) permits evolutionary reversals—that is, changes from both 0 to 1 and changes from 1 to 0 are allowed. The total number of changes, both forward and reverse, is minimized.

The Dollo parsimony method (Farris, 1977a; Le Quesne, 1974) allows only one change from 0 to 1 in each character, but allows as many reversals from 1 to 0 as are needed to accommodate the data set. The total number of reversions (1 to 0) is minimized.

The polymorphism parsimony method (Farris, 1978; Felsenstein, 1979) is based on the assumption that there can be polymorphism for any character (i.e, 01: both states of the character are expressed within the terminal taxa under consideration). It is assumed that for any character the change from 0 to 01 can occur only once, but the polymorphism may be retained (01 to

01) or change to either state (01 to 0 or 01 to 1). The "number of instances of retention of polymorphism needed to explain the data" is minimized (Felsenstein, 1984, p. 172).

All of the above methods of analysis can be run as ordered or unordered variants (Felsenstein, 1984). In the ordered variants the ancestral (primitive, plesiomorphic) states of the various characters are known or hypothesized, whereas in the unordered variants the ancestral states are not known. Of course, ordered and unordered variants of the same method may give different results in particular cases.

Another way to attempt to resolve Hennig's dilemma is through compatibility or clique methods (Duncan, 1980; Duncan, Phillips, and Wagner, 1980; Estabrook, 1979; Estabrook, C. S. Johnson, and McMorris, 1975, 1976; Estabrook and McMorris, 1980; Estabrook, Strauch, and Fiala, 1977; Farris and Kluge, 1979; Gardner and La Duke, 1979; Le Quesne, 1969, 1972; Meacham, 1980, 1984). Character-compatibility analysis is based on the concept that if two different characters define incompatible putative monophyletic groups, we can assume that at least one character is not a good guide to phylogenetic relationships; at least one character involves homoplasy or reversal (Meacham, 1984). All true characters (those of value in reconstructing a phylogeny) must be mutually compatible. A clique is a set of mutually compatible characters. Within a data set of taxonomic characters, a large clique indicates that there is strong mutual agreement among the characters involved, and such a pattern of agreement requires an explanation. As Meacham (1984, p. 153) notes, such a clique may result from the logical dependence of the characters involved or from functional, developmental, or ecological correlations among the characters involved. However, the presence of a large clique of diverse characters that bear no obvious correlations to one another may indicate that the compatibility of the characters is the result of evolutionary (phylogenetic) factors, and the clique of characters is then singled out as especially useful in reconstructing the evolutionary history or phylogeny of the organisms or taxa under consideration.

It is sometimes suggested that compatibility analysis is closely related to Hennig's (1966a) method of phylogeny reconstruction based on methodological parsimony, that compatibility analysis is Hennig's method, or that compatibility analysis is superior to Hennigian methodological parsimony-based techniques (e.g., Duncan and Stuessy, 1984b, p. 91; Estabrook, 1979; Sneath, 1982). However, this seems not to be the case. Essentially, as numerous authors (Wiley 1981, p. 192; see also Farris, 1983; Farris and Kluge, 1979; and further discussion in chap. 3) have pointed out, as homoplasy becomes increasingly frequent in a data set, compatibility analysis explains fewer and fewer of the original observations. The "compatibility method recognizes only perfect correlations among characters, sets of fully congruent characters" (Kluge, 1984, p. 28). In the process, compatibility methods discard some of the data, and as data is discarded there is increasingly less connection between the observations and the phylogenetic hypotheses. Any theory or hypothesis that ignores data cannot be considered scientific (Farris, Kluge, and Mickevich, 1982; Kluge, 1984). In contrast, Hennig (1966a) and Hennigian-based cladistic methods (including the Wagner Tree parsimony methods and algorithm: Farris, 1970; see Kluge, 1984, and Wiley, 1981, p. 180) attempt to explain, and find the best fit for, all available data.

As mentioned above, certain numerical and computer-implemented methods may yield methodologically parsimonious Hennigian cladograms (Brooks, 1984; Wiley, 1981). Of particular note are the algorithm for producing Wagner Trees of Kluge and Farris (1969), the Wagner 78 computer program of Farris (see Brooks, 1984, p. 129, and Wiley, 1981, p. 179, for a discussion of this program), and the algorithms of Dayhoff (1969) and Fitch (1971), which are usually applied to amino acid–sequence data. It appears, however, that all of these methodologies could bear more analysis as to whether they will always yield the same results as traditional Hennigian parsimony analyses. The Wagner 78 program, as described by Brooks (1984), utilizes a Manhattan Distance to first construct a network of taxa that are closest neighbors. Essentially the Manhattan Distance is the number of character-state differences between any two taxa divided by the number of characters compared. The network constructed on the basis of the Manhattan Distances is rooted and turned into a cladogram. The cladogram is then optimized in order to arrive at a parsimonious interpretation of the characters (see Brooks, 1984); the process of optimization yields the final cladogram. An important question is whether cladograms arrived at using this and similar methods will always be identical to "true" Hennigian cladograms. Yet even if this question is not answered definitively, such algorithms and computer-implemented techniques are undoubtedly a boon for the researcher dealing with large data sets that are barely manageable by hand.

Another approach to inferring phylogenies is an explicitly statistical-inference approach (well discussed by Felsenstein, 1984). One important type of statistical-inference approach is that of maximum likelihood. The maximum likelihood of a particular phylogenetic tree is the probability of the observed data set given the tree (not the probability of the tree given the observed data set: Felsenstein, 1984, p. 177). The maximum-likelihood method chooses the tree that gives the highest probability of obtaining the observed character distributions. The major problem with statistical-inference and maximum-likelihood approaches to phylogeny reconstruction is perhaps that they are based on assumed (but actually unknown) probabilities of certain paths of evolutionary change and on admittedly unrealistic and oversimplified models of evolution (see admission to this effect by Felsenstein, 1984, p. 188); any statistical methods are based on our absolute understanding of evolutionary processes. There are also basic problems with the complicated mathematics that would be involved in any realistic model of evolutionary processes. To quote Felsenstein (1984, pp. 188–189):

The difficulty we face is that we know too little to specify a realistic model of evolutionary change. Even if we could do so, it would not be mathematically tractable. In this sense the advocacy of total realism is a counsel of despair. The best that we can do is to use the most reasonable model we can find which is also tractable. To the statistical errors of measurement and estimation we calculate under such a model, we must then add an extra dose of uncertainty, corresponding to our uncertainty about the biological processes.

Finally, it should be noted that each of the methods discussed in this section will often give different results, and there are as yet no general rules for predicting when various methods will give similar results; this depends on the particular data sets, exact methods, and precise assumptions used (Felsenstein, 1984, p. 175).

Stratocladistics

Fisher (1980; 1982, p. 176) has introduced (but not elaborated on) the notion of what he terms *stratocladistics*. This is to be a method of phylogenetic inference that "involves an integrated analysis of comparative morphologic and stratigraphic data and uses, in part, process assumptions regarding the relative likelihood of preservation and discovery of members of different lineages within a genealogy." Presumably Fisher's stratocladistics would combine both strict cladistic methodology used to arrive at a first approximation, or set of approximations, of the phylogeny of the group under consideration with analysis of the adequacy of the fossil record and analysis of the reliability of stratigraphic sequences of fossils (see chap. 4) to arrive at a final phylogeny and evolutionary tree.

Doyle, Jardine, and Doerenkamp (1982) have discussed and attempted to apply Fisher's concept of stratocladistics. Fisher's method is not necessarily concerned with reconstructing phylogeny per se, but deals with a combination of phylogenetic relationships and ancestral-descendant lineages; in this respect it is closely akin to what is called here "classical paleontological methods of reconstructing the evolution of lineages" (see below). The stratocladistic method weighs ad hoc hypotheses of homoplasy (parallelisms and reversals) equally with observed contradictions in the predicted stratigraphic appearances of characters. Both sets of data, ad hoc hypotheses of homoplasy and observed stratigraphic contradictions, are taken into consideration in stratocladistic parsimony analysis. Parsimony considerations and judgments on the part of the particular investigator concerning the characters used and the importance of the stratigraphic sequence of known forms are taken into consideration in deciding upon a particular tree or phylogeny.

In cladistics, the choice of hypotheses on relationships among taxa is usually based explicitly or implicitly on relative parsimony: under given assumptions on directions of character change, the hypothesis is preferred which requires the lowest number of evolutionary parallelisms or reversals, on the principle of Ockham's Razor [see chap. 3 of this book]. Such hypotheses may take the form of cladograms, where all species or clades are treated as end-points of a branching diagram, like species existing at one time plane, or phylogenetic trees, where ancestors are specified at the nodes of a cladogram (cf. Eldredge and Cracraft, 1980). Fisher's stratocladistic method is unusual in that it operates at the level of phylogenetic trees and considers not only the number of parallelisms and reversals (morphological parsimony debt, M), but also the number of cases where the observed stratigraphic order of appearance of character combinations (essentially, taxa) contradicts the sequence predicted by each tree (stratigraphic parsimony debt, S). (Doyle, Jardine, and Doerenkamp, 1982, p. 53)

In a specific example analyzed by Doyle, Jardine, and Doerenkamp (1982), the authors prefer a phylogenetic tree that invokes one or two cases of parallelism (= morphological parsimony debt) but no stratigraphic parsimony debt over a phylogenetic tree that invokes no ad hoc hypotheses of homoplasy but does entail a stratigraphic parsimony debt (the second tree suggests that a certain taxon or its hypothetical ancestor should have occurred in certain early strata, yet Doyle, Jardine, and Doerenkamp [1982] failed to find the taxon in the appropriate aged strata). Thus, using the stratocladistic method, phylogenetic hypotheses based wholly or in part on stratigraphic data (i.e., the relative occurrences of taxa that thus determine what is primitive or ancestral, where early equals primitive) may override phylogenetic hypotheses based purely on morphological data. Such a procedure may be justified for groups that are abundant and well represented (both vertically and horizontally) in the rock record, but that have few or extremely variable morphological features of low phylogenetic value (i.e., analysis of the intrinsic characters of the organisms leads to contradictory phylogenies). However, in many cases it is not necessarily justified to posit such dependence on the known occurrences of taxa at specific positions in the rock record; many taxa may be absent (truly not present, or undiscovered) from the known record for reasons other than that they did not exist during the temporal interval under consideration (see chap. 4). Like the stratophenetic method (see below) and traditional paleontological methods, including that of Harper (discussed below), the stratocladistic method is highly dependent upon the assumption that the fossil record for the group under consideration is relatively complete.

CLASSICAL EVOLUTIONARY
TAXONOMIC METHODS

For classical evolutionary taxonomists, the groups composing a classification consist of phylogenetically related species (Steineck and Fleisher, 1978, p. 629) and higher phylogenetically related taxa. Phylogeny and taxonomy go hand in hand, although there is not a one-to-one correspondence between the two (Bock, 1977). To quote Mayr (1969, pp. 77–78):

The raw data permit (1) the reconstruction of phylogenies and (2) the establishment of classifications. Yet, neither is "phylogeny based on classification" nor "classification based on phylogeny." Both are based on a study of "natural groups" found in nature, groups having character combinations one would expect in the descendants of a common ancestor. Both sciences are based on a careful evaluation of the established similarities and differences.

For most evolutionary taxonomists the basic evidence used in determining the phylogenetic relationships of organisms "consists of weighted similarity" (Mayr, 1969, p. 83). Such similarity must be carefully interpreted in order to sort out cases of parallelism and convergence, for "only similarities between homologous characters are of taxonomic importance" (Mayr, 1969, p. 84). But taking this into account, all available evidence from diverse sources (e.g., conventional morphological data, biochemical data, ontogenetic data, physiological data, ecological data, behavioral data, biogeographic distributions, fossils and stratigraphy) and any appropriate methodologies, including those of the other schools, should be brought to bear on the question of phylogeny reconstruction (Mayr, 1969, 1974; Simpson, 1961, 1975; Steineck and Fleisher, 1978). The methodologies of the evolutionary taxonomists are extremely eclectic and thus difficult to characterize. As mentioned in the introduction, most evolutionary taxonomists work within the conceptual context of the synthetic theory; resulting phylogenies may at times have a subjective or intuitive component based on the researcher's preconceived notions of how organisms should evolve.

Szalay's Procedures for Establishing Phylogenetic Relationships

Straddling the classical evolutionary systematic and traditional paleontological methods of reconstructing phylogeny (see below) on the one hand and the cladistic program on the other are Szalay's (1977b, pp. 324–325; see also Szalay 1977a, 1977c, 1981a, 1981b) "procedures for the establishment of historical relationships of taxa." Like the strict cladists, Szalay advocates the use of only shared and derived similarities for the determination of genealogies. Szalay also adopts Popperian (see Popper, 1968) views on the corroboration and falsification of scientific hypotheses. However, unlike strict cladists, Szalay advocates using a system of character weighting, such as that of Hecht and J. L. Edwards (1976, 1977; see section on character weighting) and places a heavy emphasis on functional and adaptive data and criteria. Szalay rejects stratophenetics (Gingerich, 1979) in its extreme form, but does accept the temporal and geographic positions of taxa or organismal remains as characters. Szalay's general approach to the extrinsic data of the fossil record is summarized in the following quote.

When decisive morphological criteria are lacking, the temporal and geographical information of fossil taxa may become significant in deciding whether a condition is more *likely* to be ancestral or derived. In other words, if morphological criteria fail, then we may develop hypotheses of polarity from the fossil evidence. Whenever morphological (developmental and functional) criteria are reasonably clear as to the polarity of a character cline, however, the temporal position of the taxa bearing the characters, even though the primitive characters may appear later, should be overlooked. (Szalay, 1977c, pp. 6–7; italics in the original)

Unlike strict cladists, and like many classical evolutionary systematists and traditional paleontologists, Szalay accepts hypotheses of ancestor-descendant relationships as desirable and scientifically valid and testable.

Those groups which are distributed on a chronocline that corresponds exactly to temporally increasing derivedness of the known character clines are the instances in which we may postulate ancestor-descendant relationships. To deny this is to deny [the] validity of the phyletic methodology based on the determination of genealogies by synapomorphies. To state that the correspondence between chronocline and morphocline polarity criteria is not falsifiable or that it cannot be corroborated by the appearance of new taxa on hitherto unknown time levels or by the discovery of new characters is simply incorrect, by the criteria of the very logic of [the] Popperian philosophy of objective or empirical systems. (Szalay, 1977c, p. 7)

Szalay (1977b, pp. 324–325) has outlined the five basic operations or procedures that he advocates in performing the initial character analysis and arriving at a phylogenetic scheme for a group of taxa. Szalay's operations can be summarized as follows.

Based on numerous assumptions and hypotheses, which are often left unexpressed, the raw data used in the reconstruction of phylogenies is gathered. Putative similarities and differences in characters and features among a certain set of taxa or organisms are observed and recorded.

If the observations made in the first step (above) are repeatable (they can be consistently recognized by the original investigator, as well as by others), they are considered to form a valid empirical data base (Szalay, 1977b, p. 324).

Similarities observed between different taxa can be the result of homology (either shared and primitive [symplesiomorphy] or shared and advanced [synapomorphy]) or convergence. Hypotheses should be posited as to the basis of any observed similarity. If it is suggested that nonidentical similarities are based on homology, then the similarities can perhaps be arranged into a character cline (morphocline) and hypotheses can be posited as to which end of the character cline is relatively primitive and which end is relatively advanced (i.e., morphoclines are polarized). The above hypotheses are arrived at by using, and judiciously weighting, all available data—for example, biological (morphological, ontogenetic, ecological, behavioral, physiological, functional, adaptational), stratigraphical (temporal), and geographical (distributional) data. All working hypotheses at this stage must be rigorously evaluated before being accepted as the basis of analysis in the following steps. "This pivotal phase of analysis requires the use of the biologically most sophisticated methods, techniques, and interpretive schemes" (Szalay, 1977b, p. 324).

Hypothesized polarities of morphoclines should be tested by comparing the polarities of independent character clines to one another.

Finally, on the basis of as many polarized character clines as possible, phylogenetic hypotheses are constructed utilizing the concepts of ancestor-descendant relationships and sister-group relationships (depending on the nature and completeness of the data at hand). If possible, the phylogenetic hypotheses (trees or cladograms) should, on the basis of the dated fossil record, be placed in an absolute time framework. In order to choose between competing, otherwise equally valid phylogenetic hypotheses, the criterion of parsimony is utilized.

Bock's Method of Phylogenetic Analysis

Bock, an avowed classical evolutionary systematist, has briefly outlined his notions of phylogeny reconstruction in several important papers (Bock, 1974, 1977, 1979, 1981). For Bock, as with virtually all classical evolutionary taxonomists, a formal evolutionary classification and a phylogeny (usually in the form of a phylogenetic diagram showing both sister-group relationships and ancestor-descendant relationships) are not equivalent. An evolutionary classification is based on all aspects of evolutionary theory and should in some way reflect the entire evolutionary history (both branching points—that is, phylogeny—and phyletic evolution, along with amount of divergence) of the organisms classified (Bock, 1977). A complete classical evolutionary classificatory analysis should include both an evolutionary classification and a phylogeny; accordingly, a classical evolutionary analysis includes the formation and testing of hypotheses concerning the existence of groups of organisms (taxa), their arrangement in a formal classification, and their phylogeny (Bock, 1977, p. 872). Following his distinction between classification and phylogeny, and thus his distinction between classificatory hypotheses and phylogenetic hypotheses, Bock (1977, p. 877) distinguishes between phyla (singular *phylon,* essentially a clade in cladistic terms, not to be confused with the Linnaean categorical rank *phylum,* plural *phyla*) and taxa (singular *taxon*). The units of a classification are taxa, and within a Linnaean classification the taxa form an internested hierarchy. The units of a phylogeny are phyla, defined by Bock

(1977, p. 877) as "closed descendant groups consisting of a species, at the base of the group, and all of its descendants" (i.e., a monophyletic group *sensu* Hennig, 1966a, or a holophyletic group *sensu* Ashlock, 1971). For Bock and most classical evolutionary taxonomists, phyla and taxa are not equivalent, whereas for most cladists acceptable taxa must be phyla.

Bock's (1977, 1981) basic premise is that classifications and phylogenies are historical-narrative explanations whose formulation must be based on all aspects of evolutionary theory (for Bock, essentially the synthetic approach to evolution). Bock (1981, p. 7) considers the theory of evolution and the mechanisms and processes of evolutionary change to be a nomological-deductive explanation (cf. Nagel, 1961, and Hempel, 1965) that is used to test historical-narrative explanations of evolution (phylogeny and classification). Apparently, according to Bock, it is known that evolution can proceed only in certain ways (or at least usually proceeds in certain ways), and any particular organism (such as might be hypothesized in an evolutionary history) must be an integrated functional organism that can actually survive in the hypothesized environment (Bock, 1981, p. 12; cf. Crowson, 1982). Thus Bock places a heavy emphasis on functional-adaptive analysis in phylogeny reconstruction; for Bock (1981) a phylogenetic sequence must consist of a series of functional organisms and not just a transformation series of characters (cf. Gutmann, 1972, 1975, 1977; Valentine, 1975).

According to Bock, evolutionary history cannot be recognized or reconstructed without reference to mechanisms of evolutionary change, and one cannot use a historical-narrative explanation (e.g., a phylogeny) to test a nomological-deductive explanation (e.g., a theory of evolution). Such an attempt would be clearly circular if the phylogeny were based on the theory of evolution in the first place. According to Bock (1981, p. 9), the theory of organic evolution (both the concept that such evolution has occurred and the theory of particular mechanisms and processes of evolution) is tested against empirical observations (e.g., among extant organisms in natural or laboratory populations) that are independent of phylogeny or evolutionary history. However, Bock seems to be missing the point of some phylogenetic analyses, those of many cladists in particular. A cladogram, for instance, need not represent phylogeny or evolutionary history a priori. Initially a cladogram may be nothing more than a summary of characters among a certain group of organisms. A phylogenetic or evolutionary explanation can then be attributed to the pattern of the cladogram. Many investigators find it useful to attempt to reconstruct patterns among organisms (i.e., phylogenies) without utilizing certain aspects of evolutionary theory, in order to test evolutionary theory (see chap. 7). Furthermore, despite Bock's assertions about how we derive our theories of mechanisms and processes of evolution, many concepts and assumptions of how evolution works are based on preconceived notions of natural (evolutionary) groups of organisms and the phylogenetic relationships of organisms.

Bock (1977, p. 872) points out that when beginning a phylogenetic analysis of some group of organisms, the basic information potentially available (but usually incomplete) to the investigator is a knowledge of the intrinsic attributes of the organisms (morphological characters, physiological characters, behavior, and so on [form of the organisms in the broad sense]); knowledge of the extrinsic attributes of the organisms (biogeography [space], biostratigraphy [time], ecological relationships, and so on); and a knowledge of evolutionary theory (as noted above, essential for Bock's method of analysis). For Bock (1977, p. 879) phylogenetic analysis means "character analysis used to test phylogenetic hypotheses about groups." Since there is presumably a single phylogeny of all organisms, Bock (1977) believes that there is a single valid method of phylogenetic analysis. Initial phylogenetic hypotheses may be arrived at inductively on the basis of the overall similarity of the organisms under analysis, or phylogenetic hypotheses may be adopted from the previous literature on the group under study; it is immaterial how the initial hypotheses are achieved. These phylogenetic hypotheses can then be tested using character analysis.

Testable statements in Bock's methodology (1977, p. 858) are scientific statements, including general theories, laws, mechanisms, hypotheses, and narrative explanations that attempt to explain phenomena and from which predictions can be deduced. The predictions can then be compared against empirical observations in order to test the statement. Predictions may not always be directly comparable to observations; a long chain of hypotheses may have to be followed, but ultimately the chain should be grounded in empirical observations. Empirical, objective observations are things that can be at least theoretically observed (sensed) by any able-bodied person with the correct abilities and training. Observations that are limited to certain persons, perhaps with certain personal or mystical insights, are not empirical, objective, and valid as tests for scientific statements and hypotheses. Any observations may themselves be theory laden; however, the theory upon which the observations are based must be as independent as possible of the hypothesis or statement to be tested in order to avoid circularity.

The first step in the phylogenetic character analysis is to recognize putative homologues (these may be true or false homologies). Bock (1977, p. 881) notes that the concept of homology is basic to all approaches to classification and phylogeny reconstruction. Bock (1977, p. 881) defines homology as follows: "Features (or conditions of a feature) in two or more organisms are homologous if they stem phylogenetically from the same feature (or the same condition of the feature) in the immediate common ancestor of these organisms." He notes further (1977, p. 881):

Homology is a relative concept, hence it is always necessary to state the nature of the relationship when talking about particular homologous features. This statement is the conditional phrase and it describes the presumed nature of the common ancestor from which the homologues stemmed phylogenetically.

Thus, to use Bock's (1977) own example, it is not complete to merely state that the wing of a bird and the wing of a bat are homologous; one should state that they are homologous as the forelimb of tetrapods (a more inclusive group). From this it follows that there is a hierarchy of homologies as determined by their conditional phrases (the conditional phrases referring to more or less inclusive groupings).

In Bock's (1977, 1981) system, the only valid test for homology is shared similarity; however, it is not a very good test of homology in that it has a poor or low resolving power (Bock, 1981, p. 16). In many particular cases, incorrect results may be arrived at (nonhomologous features may be sufficiently similar such that they are incorrectly considered homologous). Evolutionary theory and functional-adaptive analyses are important in Bock's system for estimating the degree of confidence one can have in any individual set of homologues.

Confidence in the correctness of the test of individual homologues is judged by considering factors such as the complexity of the feature, the relationship between form-functional properties of the feature and the selection forces governing its evolution (to ascertain the probability of independent origins, convergence, multiple evolutionary pathways, correlation between morphological and adaptive convergence), etc. (Bock, 1981, p. 16; see also the discussion of character weighting in chap. 2)

In a similar vein, Fisher (1981, p. 54) suggests that we may have a strong case for the homology of structures when "there is a very strong similarity even between aspects of morphology that are under relatively weak biomechanical constraint." In contrast, functional and adaptive analysis may render the hypothesis of homoplasy more plausible if it is also suggested on other grounds; in other words, functional arguments may indicate which of two sets of apparent synapomorphies is more likely to represent convergence (Fisher, 1981). Fisher (1981, p. 60) has made the important point that such arguments can validly be made only if we fully know the functional significances of all of the characters being compared. One must guard against readily accepting the possibility of homoplasy in a particular character just because it is believed that we understand the function of the character, or in contrast, accepting a character as a potential homology when the functional significance is unknown or obscure (cf. Eldredge, 1979a). One can use functional arguments only if one knows the functions of all of the characters being compared. The ultimate test of homology, however, appears to be Patterson's (1982a; see also Cracraft, 1981a, 1981b, and Wilson, 1965) "congruence test," discussed in chapter 3.

Once initial putative homologues are arrived at, the next step is to arrange the homologues into a transformation series, or morphocline (Bock, 1977, p. 885). A transformation series may be analogous to the conditional phrases of the homologues as arranged in a hierarchy and is based on knowledge or judgment of how characters could or should change during evolution. According to Bock (1977, p. 886), transformation series are scientific hypotheses that are subject to testing. Such tests concern the evolution and functional and adaptive significance of the features in the transformation series and must ultimately be grounded in objective empirical observations.

Once a transformation series is established, it must be polarized—that is, it must be determined which end is plesiomorphic and which is apomorphic. Bock (1977, pp. 886–887; 1981, pp. 14–17) accepts only two methods of establishing morphocline polarities as valid. The first is use of information about the stratigraphic distribution of fossils; earlier fossils should possess more primitive features. Bock is careful to note that fossil sequences are not used to formulate transformation series, only to test the polarity of such series. Furthermore, even if this is a valid test, it may often be a very poor test with low resolving power (it may lead to incorrect conclusions). The second test advocated by Bock is the use of knowledge and understanding of how features could evolve (based on adaptive, functional, and evolutionary analyses and ultimately grounded in empirical observations) to test various hypotheses of morphocline polarity. The crux of this method depends on estimating the probability that evolution will cause change in one direction as opposed to change in the opposite direction. Functional morphology is crucial in placing constraints on the way evolution could have occurred (Dullemeijer, 1974, 1980). Unfortunately, as Cracraft (1981b, p. 29) has pointed out, Bock's second test of morphocline polarity lacks an explicit, precise, and logically based methodology; it is not a rigorous, repeatable, and refutable methodology. Rather, it relies primarily on the subjective and authoritarian judgments (really nothing more than a system of beliefs) of particular "experts" concerning particular character transformation series in particular groups of organisms. Such judgments are based on preconceived notions of how evolution works. In contrast to many evolutionary systematists, a majority of strict cladists probably agree that

any prior understanding of the functional relationships or adaptive advantages of features is completely unnecessary (or rather irrelevant; cf. Farris, 1969) for the reconstruction of a phylogeny. . . . To make the reconstruction of a phylogeny a priori dependent upon a knowledge of functional anatomy, physiology, ethology, or other sorts of "adaptiogenesis," is turning the case upside down. Anyway it is unrealistic or even impossible for most organisms, which cannot in any way mean that attempts to reconstruct their phylogeny should not be made. (Bonde, 1977, p. 751)

Bonde (1977, p. 777) has even suggested that the use of functional and adaptive statements in phylogeny reconstruction belongs to a type of science distinct from that of natural science, perhaps a type of hermeneutics, the science of interpretation where the investigator tries to understand an organism from the point of view of the organism (understanding from the "inside"). Of course, functional interpretations are extremely interesting to add to a strictly cladistic-based phylogenetic reconstruction.

Bock (1977; 1981, pp. 14–15) rejects a number of commonly used tests for character hypotheses (specifically for determining morphocline polarities). He rejects any parsimony-based tests because he believes that the concept of parsimony, when applied to character hypotheses, is synonymous with assuming that it takes a minimum number of evolutionary steps to reach a particular character, and this may not be true. Similarly, Dullemeijer (1980, p. 171) states that "unless parsimony is equivalent to efficiency, constraint or survival chance, such a concept is very unlikely" and (p. 176) "evolutionary parsimony is inacceptable [*sic*] in the neo-Darwinian theory of evolution." Elaborating on his concept of parsimony, Bock (1977, p. 859) argues that parsimony is not used to test the validity or internal consistency of theories, hypotheses, or statements, but to choose between still valid theories (i.e., theories that are in agreement with the observations, that do explain the data at hand). A theory could fully explain the observations and be internally consistent but nonparsimonious (noneconomical). Bock (1977, p. 868) suggests that criteria other than parsimony may be of greater value in deciding which of several competing hypotheses or theories to utilize. Such criteria include *boldness,* the concept that the theory explains new, hitherto unexplained phenomena; *content,* the number of things actually explained by the theory; and *testability,* the number of valid tests that can be proposed for the theory. Of course, these criteria are not mutually exclusive.

Bock (1981, p. 14) rejects internal consistency as an empirical test of a hypothesis. Although internal consistency is desirable and even necessary, it may consistently yield an incorrect result. Bock (1981, p. 15) rejects ontogenetic data as bearing on the problem of morphocline polarity, stating categorically that:

No N-D E [nomological-deductive explanation] has been demonstrated in the relationship between these two patterns [ontogeny and phylogeny] of change which is needed before ontogenetic changes can serve as a test for postulated phylogenetic changes.

Bock (1981, p. 15) rejects any tests that are based on the distribution of character states among various taxonomic groups, including commonality and out-group comparison (see section on determination of morphocline polarity), as circular. Such hypotheses are based on prior hypotheses of group membership, and all of the hypotheses involved are historical-narrative explanations. First, according to Bock, a more inclusive historical-narrative explanation cannot serve as a test for an included historical-narrative explanation. Second, observational (concerning the distribution of character states among a group or groups of organisms) and theoretical statements to be tested (hypotheses as to the groups that character states delineate) are intermingled: "theory used in the observational statements stems directly from the theory being described in the theoretical statements" (Bock, 1981, p. 15). Bock does not take into consideration that ultimately it is assumed that all life is monophyletic and thus at some point characters common to all life (completely general, but not necessarily "generalized" in an adaptive sense) must be primitive (plesiomorphic), whereas restricted characters (special, but not necessarily "specialized" in an adaptive sense) must be relatively apomorphic at some level.

Once polarities of transformation series are established, one is concerned with the apomorphous characters only. Following Hennig (1965, 1966a), Bock (1977, p. 888) uses autapomorphies to distinguish monophyletic (holophyletic) phyla and synapomorphies to unite various phyla as more-inclusive phyla (the synapomorphies serve as the autapomorphies of the more-inclusive phyla); finally, convergent features—nonhomologous apomorphies (i.e., homoplasies)—must be distinguished, especially if there are character conflicts in the final phylogeny.

Bock (1977) suggests that ideally all putative homologous apomorphies should be retested, particularly to sort out nonhomologous apomorphies. This final retesting should depend on studies of functional and adaptational analysis, evolutionary theory, ecology, ontogeny, and so forth, all ultimately grounded in empirical observations. In the final analysis the proposed phylogeny must make sense evolutionarily, adaptively, and functionally.

Hanson's Method of Phylogenetic Analysis

Hanson (1977) attempted to formalize a method of classical evolutionary phylogeny reconstruction. In the process he synthesized the classical evolutionary school's assumptions concerning evolutionary processes into a procedure, based on his concept of semes, to calculate what he refers to as the *phyletic distance* between any two species. Calculations of phyletic distance are used, in turn, to help put constraints on the construction of evolutionary trees (dendrograms, in Hanson's terminology).

It must be understood that Hanson's (1977) concepts of phylogeny and the aim of phylogenetic analysis are very different from the view taken in this book. Hanson states: "The aims of phylogeny are twofold: to reconstruct the course of evolutionary events, and to analyze those events so as to understand better the courses of evolutionary change" (1977, p. 94). "The treatment of the problem of evolutionary potential" is "the second major goal of phylogenetic research" (Hanson, 1977, p. 110); and "in that phylogeny is an evolutionary study, its aims and procedures must be derived from and consistent with the known processes of evolutionary change through natural selection" (Hanson, 1977, p. 68). From reading Hanson's work, it is evident that very often he deemphasizes the phylogenetic pattern of species diversification (the taxic approach) and instead is primarily interested in the transformational approach to phylogeny reconstruction. In other words, Hanson is often interested in speculation about the transformation of an ancestral species into a descendant species by means of the causal mechanism of natural selection; he is not particularly interested in reconstructing branching sequences.

Hanson begins his discussion by formulating three procedural guidelines to be followed in attempting any phylogenetic analysis or set of speculations:

1. "Species define the basis for interorganismic comparisons." (Hanson, 1977, p. 68)
2. "Phylogenetic comparisons must examine the exploitive, homeostatic, and reproductive functions [of species] and/or their anatomical correlates whenever possible." (Hanson, 1977, p. 69)
3. "Every surviving innovative step must be selectively advantageous." (Hanson, 1977, p. 70)

In order to actually analyze phylogenetic relationships (in Hanson's sense), Hanson (1977) utilizes his concept of *semes*. He considers his semes to be the basic units of phylogenetic information: each seme corresponds to a structural or functional part of an organism (e.g., Hanson considers the skeletal system to be one of the fundamental 20 or so animal semes). He states (1977, p. 92) that "not all characters are semes, though all semes are characters." Presumably not all characters are semes because not all characters are "sufficiently complex to permit . . . 'meaningful comparison' " (Hanson, 1977, p. 89). He compares semes using the classical criteria for recognizing homology as formulated specifically by Remane (1955, 1956, 1971: these are reviewed in chap. 3). Expanding upon Hennig's (1966a) terminology, Hanson recognizes various types of semes. "A *plesioseme* is an evolutionarily conservative trait and an *aposeme* is one showing innovation relative to a preexisting, plesiosemic condition" (Hanson, 1977, p. 92, italics in the original). "If there appears a brand new innovation, this is called *neosemic*" (Hanson, 1977, p. 93, italics in the original). There are also four different named categories of aposemes: (1) hypersemic: "when the change in character is simply enlarging on something already present"; (2) hyposemic: "when the change is reductive"; (3) polysemic: an innovation that consists of a structure or function being "repeated, either metamerically or as in some type of colony formation"; (4) episemic: "when an old part or process is transformed into something new" (Hanson, 1977, p. 93). Hanson (1977, p. 98) also uses the term *plesiomorph* to refer to the species "which contains the largest number of plesiosemes in the group of species being studied," and the term *eotype* to "refer to [perhaps hypothetical] forms ancestral to the plesiomorph" or the group represented by the plesiomorph.

Hanson firmly believes that it is relatively conservative traits that give the best idea of phylogeny. Without explanation, he states (Hanson, 1977, pp. 94–95):

Using the idea of semic characteristics, it is clear that if there are many plesiosemic traits and few aposemic ones held in common between two species, the phyletic distance between them is small. If, on the other hand, there are few plesiosemic characters and several aposemic characters in common, then the phyletic distance is greater than in the first case. Finally, an obvious third condition, if there are few plesiosemic and aposemic characters in common and many neosemic characters, the two species are only distantly related. . . . Therefore, as a first approximation, to establish phyletic distance through selected similarities, the relative number of plesiosemic entities as compared to aposemic ones is important.

The reader should note that Hanson's method of using primitive traits to determine phylogeny is the opposite of Hennig's (1966a) method of utilizing derived traits to determine phylogeny.

Hanson (1977, p. 101) proposes the following equation for quantitatively calculating the phyletic distance (R) between any pair of species: $R = ((-p + (2a)^2 + (3n)^2)/t) + 1$, "where p is number of cases of plesiosemy between two species, a the instances of aposemy, n the cases of neosemy, and t the total pairs of semes studied."

To what does "number of cases of plesiosemy" refer? If two species, A and B, are being compared and seme #1 is identical in both species, this is considered a case of plesiosemy—that is, neither A nor B is aposemic for seme #1 relative to the other species. Thus "number of cases of plesiosemy" is really just the number of semes compared between the two species that are *identical* in the two species. What does identical in this context mean? Hanson (1977, p. 97) answers the question as follows:

How similar must semes be to be plesiosemic and when are differences sufficiently great to be designated aposemic? This is frankly a matter of judgment. Perhaps only between members of the same species will similarity reach the level of identity, but even here some intraspecific variation must be expected. By and large, in what follows [i.e., the rest of Hanson's book], we have found that if 80% of the pairs of elements (points or subunits) in the two sets being compared correspond to each other, then the sets stand in a plesiosemic relationship.

In the above equation, a, the instances of aposemy, is the number of instances where corresponding semes are found in both species being compared but the semes are not similar enough to be considered plesiosemic; therefore one must be derived relative to the other, and the relationship is aposemic. As far as the calculation of phyletic distance is concerned, it does not matter whether the derived condition is present sometimes in species A and other times in species B. Neosemy refers to a seme being present in one species, but not in the other species (and it is not present in the second species because it was never present in the ancestry of the second species; if it were absent through reduction and loss, then it would be a case of aposemy). Hanson (1977) suggests that between 10 and 20 semes should be compared when calculating phyletic distances between species.

Hanson's (1977) procedure is to calculate phyletic distances between all, or selected, pairs of species under consideration. Next, phylogenetic diagrams are constructed on the basis of the calculated phylogenetic distances. Circles representing the species under consideration are placed within the confines of a two-dimensional diagram, and lines or arrows labeled with phylogenetic distances are drawn to connect them (Fig. 1-10a). The spacing between the individual species is proportional to the phyletic distances between them; species pairs with low phyletic distances separating them are close to each other in the phylogenetic diagrams and resulting evolutionary tree. (Comparison of a species to itself would yield a phyletic distance of zero.) Next the phyletic distances and diagram can be converted into an evolutionary tree (a *den-*

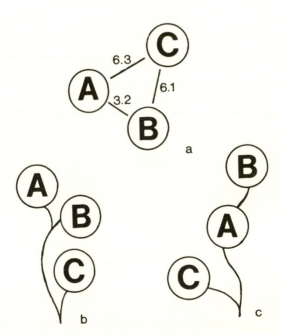

Figure 1-10. Hypothetical example of a Hanson phylogenetic diagram *(a)* for taxa A, B, and C, and two possible evolutionary trees *(b, c)* derived from it (compare to Hanson, 1977, p. 102, fig. 3-8).

drogram: Fig. 1-10*b*, 1-10*c*). Hanson (1977, p. 102, italics in the original) warns:

There are two essential guidelines that must be observed in translating phyletic distances into a dendrogram. One is that *all pairs of comparisons must be made* and the second is that *the position of phyletic branchings must be checked against the original data.*

From any set of phyletic distances, numerous evolutionary trees may conceivably be constructed—the position of phyletic branchings in particular is not specified. (In actuality, if anything is specified by the phyletic distances, it seems to be nothing more than grades of similarity.) In order to determine the positions of the phyletic branchings, Hanson (1977, p. 103) argues that we must go back to the original data and use our best interpretive judgments of the evolution of the organisms under consideration. Of course, this is a relatively subjective matter based on evolutionary scenarios.

Hanson's method of phylogenetic analysis has been outlined in some detail because it is a recent, rather explicit formulation of many of the assumptions of the classical evolutionary school. With its dependence on the equation for calculating phyletic distance (given above), it also shares some aspects with the phenetic school. Under Hanson's (1977) system, phylogenetic hypotheses are dependent on assumptions concerning the mechanisms and processes of evolution; phyletic distance between pairs of taxa (as measured by shared similarity) is given great weight, conservative characters are considered extremely valuable in reconstructing phylogeny, and precise hypotheses of phylogenetic relationships are unattainable using his methods (except as a result of interpretive judgments on the part of the investigator; there is no precise concept of two taxa being more closely related to each other than either is to a third taxon, and there is no real way to recognize monophyletic groups).

Harper's Method of Phylogenetic Inference

Harper (1976, p. 180) has attempted, on the basis of a classical evolutionary paradigm, "to explicitly formulate some of the procedures commonly followed, and at least a few of the ordering principles commonly employed in constructing phylogenetic hypotheses based on fossils." Harper (1976; see also Harper, 1979, 1980, 1981; Harper and Platnick, 1978) bases his phylogenetic hypotheses on morphologic, stratigraphic, and geographic data; he assumes that fossil taxa (of any rank) can be arranged as related according to both ancestry and descendancy and according to common ancestry (i.e., he recognizes and searches for potential actual ancestor-descendant relationships). Harper (1976, p. 180) considers phylogenetic analysis, as he practices it, to be an art rather than a science since the ordering principles upon which it is based "are statements that certain patterns are common, or probable, rather than general statements held to be invariably true."

Harper's (1976, p. 184) recommended procedure

is to construct a proposed evolutionary tree piecemeal as follows:
  (a) select one or more taxa as an arbitrary starting point;
  (b) consider additional taxa one at a time, and proceed step-by-step to apply ordering principles to morphologic, stratigraphic and geographic data to (1) systematically group taxa into subgroups, putting together taxa which are thought to be more closely related to one another than to the others, and (2) develop explicit hypotheses as to how these taxa are related to one another (either as ancestor-descendant or by close common ancestry);
  (c) stop when all of the taxa have been incorporated into a proposed evolutionary tree.

Harper (1976) proposes a number of *ordering principles* as criteria for establishing the probable evolutionary affinities and relationships of taxa.

1. Principle of shared derived character states (Harper, 1976, p. 185). Harper suggests that of a group of taxa, if some subgroup shares the possession of a complex, derived character state, then such a subgroup of taxa probably shares a closer common ancestry among themselves than with other taxa that do not possess the derived character state. This, of course, is the basic principle upon which the Hennigian cladistic and Wagner Groundplan-divergence methods are based. Harper, however, appears to be relatively quick to drop the use of this principle if it significantly conflicts with any other of his principles (described below). He notes that the probability of common ancestry is increased when larger numbers of putative shared and derived character states are shared among taxa.

Harper (1976) briefly discusses several methods of distinguishing between primitive versus derived character states. These are: the character state first appearing in the fossil record is probably primitive (ancestral); if one of several character states in a given group occurs in related groups, it is probably primitive; if a character state is widely distributed among early members of related groups, it is probably primitive; if inferred lineages of related groups exhibit similar morphological trends, then such trends might be used to determine which character states are ancestral for the particular group under study.

2. Principle of inexact convergence (Harper, 1976, p. 186). Harper notes that highly complex character states are less likely to have arisen independently, and thus taxa that share identical complex character states very likely inherited them from a common ancestor. When organisms undergo convergence, such convergence is not exact in its details. Harper (1976) considers Dollo's Law, the principle of irreversibility, to be a special case of this basic principle. Once a complex character state or set of characters is lost, it will not reevolve again in all of its details.

3. Principle of morphologic similarity (Harper, 1976, p. 187). Harper suggests that two or more taxa are probably "more closely related" to each other than they are to other taxa if they have a large number of complex character states in common. He (1976, p. 187) explicitly states that, according to his terminology, two taxa may be considered more closely related to each other even if they do not share a closer common ancestry with each other than either does with any other taxon under consideration. Thus, given that taxa B and C share a common ancestor that is not shared with A, and all three taxa share a common ancestor relative to other taxa under consideration, if taxon C is widely divergent from taxa A and B, and the latter two share many ancestral traits in common, Harper (1976, p. 187) states that "A and B are more closely related to one another than either is to C in the sense that they are both less divergent from the common ancestral group than is C."

4. Principle of minimal morphologic gaps (Harper, 1976, p. 188). Harper (1976, p. 188) suggests that "given a number of alternative phylogenetic hypotheses, other things being equal, the one postulating the least amount of morphologic change between taxa is the most probable." Harper states that this hypothesis can be justified by Occam's razor, but such a suggestion is not necessarily parsimonious in minimizing ad hoc hypotheses of homoplasy (see chap. 3). Minimizing postulated morphologic change among a series of known taxa may necessitate the invocation of many independent origins and reversals of character states.

5. Principle of minimal geographic gaps (Harper, 1976, p. 189). Given alternative hypotheses of the relationships of taxa, if all other considerations are equal (the alternative hypotheses are equally plausible on the basis of other criteria), the hypothesis that involves the minimum number and extent of geographic gaps between closely related taxa is to be preferred as more probable. Harper notes that in assessing the significance of geographic distributions of fossil taxa, the paleogeography existing at the time that the taxa were extant must be taken into consideration; it may be very different from that of the present.

6. Principle of minimal stratigraphic gaps (Harper, 1976, p. 190). Of two or more alternative hypotheses of relationships of taxa, all of which are equally plausible on the basis of other criteria, the one that involves fewer or shorter stratigraphic gaps between putative ancestral and descendant taxa is to be preferred as more probable.

7. Harper (1976, p. 190) cites the commonly held principles that trends are not reversed and that meristic structures may undergo reduction, but rarely increase (see Crowson, 1970, pp. 102–103). However, as Harper (1976, p. 190) notes, if trends are "assumed as an ordering principle, then the real possibility presents itself that any inferred trends are, in fact, imposed on the data."

8. Principle of veracity (Harper, 1976, p. 190). Paraphrasing Hennig (1966a, pp. 130–133), Harper (1976, p. 190) summarizes his approach to phylogenetic inference in stating:

A phylogenetic hypothesis may be considered a probable approximation to reality only if it is consistent with (serves to explain) at least a large segment of the available data—morphologic, geographic, stratigraphic—interpreted within the framework of the ordering principles cited.

Harper (1976, p. 191) cites two common paleontological tests of phylogenetic hypotheses. A phylogenetic hypothesis may be evaluated (tested) in the light of new fossil occurrences. A phylogenetic hypothesis may be evaluated (tested) in the light of new character-state information. Such new information, using Harper's ordering principles, may either corroborate, weaken, or modify a previous phylogenetic hypothesis.

Stratophenetics

The stratophenetic approach to phylogeny reconstruction is an explicit methodology developed by Philip D. Gingerich (1976a, 1976b, 1979, 1980, 1985; Gingerich and Schoeninger, 1977; and references cited therein) using the traditional assumptions of classical paleontology and the traditional evolutionary approach of George Gaylord Simpson and Ernst Mayr. In summary, the stratophenetic approach

combines detailed stratigraphic information with phenetic clustering to give an empirical record of phylogeny. The approach requires a relatively dense and continuous fossil record. Where this record is available, a stratophenetic approach gives the most direct and complete reading of phylogeny possible. (Gingerich, 1979, pp. 49–50)

In other words, stratophenetic methodology is used to attempt to read evolution directly from the rock record.

According to Gingerich (1979, p. 50), the stratophenetic method consists of three basic operations: data organization, phenetic linking, and testing.

Data organization itself involves two different procedures. First, for each fossil fauna, fossil locality, or other fossil sample (preferably from as limited a stratigraphic interval as possible), the actual fossil specimens of all species present "are clustered phenetically to determine the total number of species present at that level and which specimens represent the particular species or groups of species of interest" (Gingerich, 1979, p. 50), those of interest being those for which one is trying to reconstruct the phylogeny. The second aspect of data organization is to arrange all of the fossil samples "in the proper order based on demonstrated superpositional relationships in the field" (Gingerich, 1979, p. 50). Unfortunately, even in Gingerich's own work this ideal is not always upheld. Thus, in Gingerich's (1976b) monograph on the plesiadapids (early Tertiary primates) he was forced to use biostratigraphy rather than observed stratigraphic superposition to temporally order some plesiadapid localities. Further, in certain cases Gingerich (1976b) used the plesiadapids themselves as the basis of his biostratigraphic correlations. The practice of using the particular organisms whose evolution is under study to determine biostratigraphic correlations leads to obvious circular reasoning (McKenna, Engelmann, and Barghoorn, 1977). More recently Gingerich (1979, pp. 51, 54) has rejoined that

if the ordering of samples biostratigraphically is based on taxa other than those under study [and thus the assumption is that these other taxa are evolving independently of those under study], the methodology does not involve circular reasoning. However, a purely lithostratigraphic ordering of samples is to be preferred whenever possible, since this obviously involves fewer assumptions about age correlations or evolutionary independence.

The next stage in the stratophenetic method is *phenetic linking*. Once the phenetic clusters are distinguished and the samples are temporally ordered, "then a species in a chosen level can be linked to other species in adjacent levels based on overall similarity" (Gingerich, 1979, p. 54). Thus species in the lower levels are connected to the most similar species in slightly higher levels in a strict ancestor-descendant fashion. As McKenna, Engelmann, and Barghoorn (1977) have noted, Gingerich's method yields only ancestor-descendant relationships, and it must be assumed that strict ancestor-descendant relationships are actually preserved in the fossil record and in the particular spots where the investigator has sampled that record. Further, it is never clear where morphology leaves off and stratigraphic sequence picks up. Using stratophenetics one could theoretically link any older taxon with any younger taxon in a supposed ancestor-descendant relationship. What apparently constrains the investigator are vague, morphologically based notions of similarity that imply close phylogenetic relationship. In actual practice Gingerich does not link species solely on the basis of their proximity in time and morphospace, but also on the basis of their geographical proximity. Thus Gingerich (1976b) presented separate phylogenies for the North American plesiadapids and the European plesiadapids, the implication being that the two groups had separate evolutionary histories. Yet Gingerich (1976b) recognized plesiadapids common to both Europe and North America at the generic level, considered Europe and North America to be part of the same land mass during at least part of the temporal duration of the Plesiadapidae, and biostratigraphically correlated sedimentary deposits between the two continents on the basis of the plesiadapids. Obviously, either the plesiadapids formed two clades, one restricted to North America and the other restricted to Europe, or they did not. If they did form two clades, then their phylogenies should be distinct, there should be no genera shared between the two continents, and one could not hope to correlate between the two continents using the plesiadapids. If they did not form two such geographically restricted clades, then there can be only one phylogeny for both the North American and European plesiadapids.

According to Gingerich (1979), the final aspect of the stratophenetic method of phylogeny reconstruction is *hypothesis testing*. Here Gingerich refers to the discovery of new fossils, in stratigraphic context, of the particular group under study after a reconstructed phylogeny has been proposed based on stratophenetic methodology. The investigator decides whether the new specimens fit into the stratophenetic picture, or if the story should be modified to take the new specimens into account. In this manner the phylogeny constructed using stratophenetics is said to be tested.

The results of stratophenetic analyses are commonly depicted in the form of evolutionary trees.

Classical Paleontological Methods of Reconstructing the Evolution of Lineages

Harper's and Gingerich's methods described above are two explicitly formulated methods of doing what many paleontologists have traditionally done in reconstructing the phylogeny or evolution of the group of organisms with which they are dealing. That is, much traditional work in paleontology has been concerned with the reconstruction of the actual evolution (as opposed to the mere phylogenetic relationships—degrees of relatedness between organisms) of a particular lineage of organisms, often involving relatively few forms whose putative evolution is studied through a relatively limited temporal duration. In this context, a lineage is generally considered a "temporal continuum formed by a species [or series of species standing in an ancestral-descendant relationship] . . . reproducing itself, generation by generation through time" (Bock, 1979; quoted with approval by Kellogg, 1980, p. 197). The core of such a program is the use of stratigraphy, the temporal dimension, in a search for actual or potential ancestors and descendants and in attempts to reconstruct lineages but not necessarily phylogenetic relationships. "Fossils, stratigraphically arranged, provide historical documentation of the course of morphological evolution of the hard parts and, indirectly, soft parts in a lineage of organisms; this is a paleontological phylogeny. Even the best paleontological phylogenies are probably only approximations to the actual biologic-genetic story" (Savage, 1977, p. 430). (Of course, a correctly reconstructed lineage of actual ancestors and descendants does shed light on the phylogenetic relationships of the forms involved in the lineage, but a reconstructed phylogeny of the same forms may say nothing about the ancestral-descendant lineage relationships of the organisms.) It must again be stressed that the subject of this book is not the attempt to reconstruct actual lineages of ancestors and descendants over short periods of time, but the reconstruction of the detailed and explicit phylogenetic relationships of organisms (i.e., the relative degrees of relatedness between organisms). As such, phylogeny *sensu stricto* (the pattern of degrees of relatedness—the subject of this book) and evolution (phylogeny plus ancestors and descendants, processes, rates, mechanisms, tempos, modes, and so on of evolution) are distinct.

Such traditional paleontological studies of the evolution of lineages may be extremely rigorous and detailed, involving numerous specimens with good stratigraphic (temporal) and geographic (lateral) control, or they may consist of little more than scenarios based on only a trivial number of specimens. As Hallam (1982, p. 355) points out, at the minimum the following requirements should be met in order to obtain adequate results:

1. The fossils should be abundant and easy to collect and should have a high preservation potential so that they will give a fair reflection of the original communities. It is desirable to undertake biometric analysis on large assemblages that correspond to the living populations, to allow adequate assessment of the range of variation.
2. There should be good biostratigraphic control, allowing correlation over large areas and a minimizing of the potential complication of hiatuses in the stratal sequence.
3. There should be good geographic control so that, in alliance with biostratigraphy, one can determine the role and extent of species migrations.

For some abundant, easily preserved groups of macroinvertebrates and microfossils, Hallam's requirements may be admirably met. It may be possible to eliminate all other potential factors and thus attribute morphological changes in a series of organisms up a stratigraphic column(s) to evolution; one may posit actual ancestor-descendant relationships and in some cases even attempt to document speciation patterns, evolutionary tempos, and migrations of the organisms studied. For example, Hallam (1968)

undertook a comprehensive biometric study of *Gryphaea* [a group of Jurassic oysters] samples from the British Lias and concluded that, whereas geographic variation at given stratigraphic horizons was negligible, there were significant changes up the stratigraphic succession which were independent of facies change and hence attributable to evolution. (Hallam, 1982, p. 355)

In later studies (see Hallam, 1982) he expanded his sample of *Gryphaea* to include many more stratigraphic intervals and samples from around the world. He (1982) concluded that he could document numerous evolutionary trends, including some that were paedomorphic, and migrations among various species of *Gryphaea*. Other recent examples using this traditional paleontological method for reconstructing the evolution of lineages include Benson (1977, 1982b, 1983), Boucot (1975, 1983), Cisne et al. (1982), Hallam (1975, 1978), J. G. Johnson (1982), M. E. Johnson (1979), M. E. Johnson and Colville (1982), Kauffman (1977), Kellogg (1975, 1976, 1982, 1983), Lazarus, Scherer, and Prothero (1985), Lohmann and Malmgren (1983), Malmgren,

Berggren, and Lohmann (1983, 1984), Malmgren and Kennett (1981, 1982), M. L. McKinney (1984), McNamara (1980, 1982), Raup and Crick (1981), Rightmire (1981), and Wiggins (1982).

Unfortunately, the traditional paleontological method is easily subject to abuse. In many cases one may be tempted to apply it uncritically to inadequate data, such as to a series of similar fossils in a single isolated stratigraphic section or deep-sea core (here the problem might be lack of geographic, lateral, control). There is also no denying that for life as a whole, the rock-and-fossil record is extremely incomplete; the traditional paleontological method of reconstructing the evolution of well-represented lineages is not a general technique that could be universally applied to the reconstruction of the phylogeny within and among most groups of organisms.

PHENETICS

The primary concern of the phenetic, or numerical, taxonomic school is not to reconstruct the phylogeny of organisms, but to construct natural classifications. For pheneticists such as Sneath and Sokal (1973, p. 24) "the central idea underlying 'natural' groupings [of which is composed a natural classification by successively naturally grouping natural groupings] is the great usefulness of a method that can group entities in such a way that members of a group possess many attributes in common." "A 'natural' taxonomy is a general arrangement intended for general use by all scientists" (Sneath and Sokal, 1973, p. 25), which should maximize the number of propositions that can be made about its constituent elements (the things classified) and thus is said to have a maximum of information content and a maximum of predictive value. This concept of "natural" follows Gilmour's (1940, 1961) concept of a natural classification: that a classification should simultaneously serve as many separate purposes as possible, and it will do so if it describes (by the hierarchical grouping of taxa) the distribution of as many features (characters) as possible among the organisms classified (Farris, 1977b). Any particular taxon should be characterized by a large suite of well-correlated characteristics that are restricted to the particular taxon. If a particular classification is based on only a limited set of criteria, then such a classification is considered an arbitrary or special-purpose classification. Classifications of organisms based purely on phenetic relationships are considered by pheneticists as examples of such special classifications rather than as natural classifications; thus they are not the goal of phenetic taxonomy.

In concert with their deemphasis of phylogeny and evolution, Sneath and Sokal (1973) distinguish between two major kinds of taxonomic relationships. These authors define

phenetic relationship as similarity (resemblance) based on a set of phenotypic characteristics of the objects or organisms under study. While phenetic relationship may be an indicator of cladistic relationship it is not necessarily congruent with the latter. The magnitude of phenetic relationships between pairs of any set of objects depends on the kinds of characters and similarity coefficients employed. (Sneath and Sokal, 1973, p. 29; italics in the original)

The same authors (p. 29, italics in the original) state that "cladistic relationship can be defined as—and represented by—*a branching (and occasionally anastomosing) network of ancestor-descendant relationships.* These treelike networks expressing cladistic relationships are called *cladograms.*" The distinction that Sneath and Sokal (1973) are making between phenetic and cladistic taxonomic relationships is fairly clear, but note that Sneath and Sokal's concept of cladistic and cladogram is different from that accepted by most Hennigian cladists and from the manner in which these concepts are used in this book.

The pure pheneticists, as contrasted with the numerical cladists who use numerical methods to construct variations of cladograms or phylograms, are primarily interested in determining the phenetic relationships between organisms and creating stable phenetic groupings that form the basis of taxonomy.

A basic attitude of numerical taxonomists is the strict separation of phylogenetic speculation from taxonomic procedure. Taxonomic relationships are evaluated purely on the basis of the resemblances existing *now* in the material at hand. These phenetic relationships do not take into account the origin of the resemblance found nor the rate at which resemblances may have increased or decreased in the past. (Sneath and Sokal, 1973, p. 9, italics in the original)

However, "phylogenetic inferences can be made from the taxonomic structures [as discovered using numerical taxonomic procedures] of a group and from character correlations, given certain assumptions about evolutionary pathways and mechanisms" (Sneath and Sokal, 1973, p. 5), and "phylogenies are deduced necessarily from phenetic relationships" (Sneath and Sokal, 1973, p. 313). By phenetic relationships Sneath and Sokal (1973) presumably mean phenetic relationships as they have defined them (cited above), which include patristic similarity (= homology in a phylogenetic or cladistic sense; similarity due to common ancestry, including plesiomorphy [primitive] and apomorphy [derived]), and homoplasy (parallelism, homoiology, and convergence. Pheneticists commonly tabulate and utilize both positive (presence or derived state) and negative (absence or primitive state) occurrences of characters in constructing their diagrams of phenetic relationships of taxa. In contrast, cladists commonly use only positive occurrences of characters (Nelson and Platnick, 1981). The actual calculated numerical (quantitative) coefficients of overall similarity, and levels of similarity, between any particular organisms or taxa under consideration by a particular phenetic worker are a function of the particular numerical methods and algorithms used and of the particular data sets utilized. From what has been said thus far, it is evident that many workers who subscribe to nonphenetic or nonnumerical taxonomic and systematic philosophies would not agree that phylogenies are necessarily deduced from such phenetic relationships. Observable morphological, phenetic data may be the prime data used in reconstructing phylogeny, but it may be considered important to identify cases of homoplasy and symplesiomorphy in order to correctly deduce phylogenetic relationships.

In their studies of organisms, pheneticists commonly utilize a heterogeneous variety of techniques and methods, such as techniques of multivariate statistical analysis, graphical analysis, multivariate morphometrics (Reyment, Blackith, and Campbell, 1984), and so on to arrive at and describe what is sometimes termed taxonomic structure (Sneath and Sokal, 1973). As Farris (1977*b*, p. 825) notes, what exactly is meant by the term *taxonomic structure* is unclear, and the techniques cited above produce neither phylogenies nor hierarchically arranged groupings of organisms and higher taxa (= classifications). Actual phenetic classifications are almost always produced by a hierarchical clustering strategy performed on a matrix of coefficients of overall similarities between the organisms or taxa classified (Rohlf, 1970; Sneath and Sokal, 1973). Farris (1977*b*, p. 825) refers to these techniques as *phenetic similarity clustering* and to the resulting product as a *phenetic classification* by definition.

Phenetic classifications (as defined in the last paragraph) are produced in two major steps (Farris, 1977*b*).

1. A data matrix is constructed listing the character states (in coded numerical form) for all the characters among all of the organisms or taxa under consideration. This descriptive information is converted into a measure of overall similarity between every pair of taxa under consideration. A matrix is constructed where both the rows and columns are terminal taxa (the initial taxa or organisms studied) and the entries are the coefficients of overall similarity between any pair of taxa. The coefficients of similarity are intended to measure the average degree to which the characters of one taxon resemble the characters of another taxon. Typical formulas for calculating coefficients of similarity are given by Sneath and Sokal (1973) and Farris (1977*b*). Among pheneticists there is no agreement as to which one particular formula or method is most preferable for calculating the coefficients of similarity (Farris, 1977*b*). Most pheneticists advocate some form of a coefficient of overall similarity, but numerical (phenetic) techniques can also be applied using a coefficient of special similarity (Farris, 1977*b*; Farris, Kluge, and Eckardt, 1970). If the special similarities used in such a coefficient are set as synapomorphies, then clustering by such special similarities may yield phylogenetically based classifications.

2. Next, a hierarchical clustering strategy is applied to the matrix of coefficients of similarity (Farris, 1977*b*). Usually this procedure is as follows. A pair of taxa are identified that, on the basis of the coefficients of similarity, are more similar to each other than to any of the other taxa under consideration. These two taxa are united into a higher taxon, and the rows and columns for the original two taxa in the initial matrix of coefficients of similarity are replaced by a single column and row of new coefficients of similarity for the new higher taxon. After this procedure, the similarity matrix thus has one less row and column than previously, since two taxa have been grouped. This procedure is carried out repetitively (until the matrix consists of a single row and column), each time uniting the terminal taxa of the current similarity matrix into higher taxa, thus forming a hierarchical classification. There are several commonly used formulas for computing the coefficients of similarity for the newly formed taxon that is constructed from the union of the two most similar taxa. These various formulas and computational methods are reviewed by Sneath and Sokal (1973) and Farris (1977*b*).

How do the numerical taxonomists or pheneticists actually deduce phylogenies from their phenetic similarities? Sneath and Sokal (1973, p. 50) answer this question in the following manner.

Let us return to the question of whether phylogenetic deductions can be made from phenetic similarities. If the resemblances are based on living organisms alone, we can only speculate on the phylogeny; to check our speculations we must have fossil evidence. Yet there are some conclusions that are more probable than others. We believe these are as follows. (1) Phenetic clusters based on living organisms are more likely than not to be monophyletic sensu Hennig. Thus phenetically adjacent taxa represent phyletic "twigs," which usually originate from the same branch; in other words, overall convergence is unlikely. (2) In the absence of direct evidence our best estimate of the attributes of a common ancestor of a cluster must be derived from the cluster itself. In short, if we have no fossil evidence, the existing pattern is our best guide to the past history—though this may often be wrong. An argument similar to the first argument in favor of equal weighting of characters applies here: if we have no evidence that evolutionary rates differed, it is difficult to proceed further without assuming these to have been constant and equal in all the phyletic lines studied. If the reader thinks of a cross section through the top of a shrub with the vertical dimension representing time and the horizontal dimension representing phenetic dissimilarity, he will have a ready, though somewhat inadequate, simile for the situation.

Such phenetic clustering to form groups that "are more likely than not" monophyletic can hope to succeed only for organisms from one time plane. If organisms sampled through time—with the possibility of ancestors and descendants being introduced—are subjected to simple phenetic clustering, then the resulting phenetic groupings may well represent grades of similarity. To deal with the problem of organisms from successive strata Sneath and Sokal (1973) would plot phenetic relationships against time (reminiscent of Gingerich's [1979] stratophenetics). Sneath and Sokal do not elaborate on this problem further, but state that (1973, p. 51) "a diagram of the phenetic relationships (expressed in one or two dimensions) versus time will usually make clear the degree to which we can safely reconstruct the phyletic tree, and where we must indicate by dotted and queried lines our uncertainty as to the course of the descent." Such a diagram would appear to be one form of a traditional phylogenetic or evolutionary tree.

Some paleontologists have interpreted phenetically based dendrograms directly as phylogenetic trees (e.g., Brower, 1973; Pope, 1976).

MOLECULAR PHYLOGENY

While not directly pertinent to the subject matter of this book (although in a few cases such methods may be applied to well-preserved fossil data), phylogenetic studies based on biochemical or molecular data of organisms should be briefly mentioned here for the sake of completeness.

In the last two or three decades a number of techniques have been developed for comparing biochemical constituents of various organisms (for reviews of the subject see Dene et al. 1982; Fitch, 1984; Goodman, 1982; Goodman, Weiss, and Czelusniak, 1982; Hanson, 1977; Novacek, 1982a; Simpson, 1975; Wiley, 1981; and abstracts in Fuller, Nietfeld, and Harris, 1985). These approaches include systematic serology or immunology, in which resemblances in the proteins of different molecules are quantified. Electrophoresis and chromotography have also been widely used to identify and compare molecules (usually proteins) between organisms. There has been recent work developing methods to elucidate amino acid sequences in proteins and in the DNA of different organisms; some have hoped ultimately to be able to compare the actual DNA of various organisms and use such information to deduce phylogeny. Here should also be mentioned DNA hybridization techniques and chromosomal studies of various organisms.

Such studies generally apply some kind of phenetic analysis to determine phyletic distances between the molecules (or other data) under consideration (see, for example, Farris, 1981; Fitch, 1977a, 1977b, 1977c, 1982; Fitch and Margoliash, 1967). The resulting phenetic analysis is often expressed in terms of a dendrogram, which is then interpreted as a phylogram. However, the raw data of such biochemical studies could, in theory, be subjected to any technique of analysis (phenetic, cladistic, classical evolutionary).

Writing in 1972 on the topic of phylogeny reconstruction in paleontology, and also mentioning molecular studies, Ghiselin (1972, p. 145) stated:

In the next few years, we may be confident, the accumulation of sequence maps for proteins will resolve what remain of our difficulties with the relationships between phyla. The anatomical work of the last hundred years will have served its function, in giving a first approximation and a guide; its basic findings in all probability will not prove grossly in error. Anatomists, knowing what the phylogenetic relationships actually are, will then gladly shift to a new series of problems, such as explaining convergence rather than detecting it. Paleontology will retain its value in providing dated materials for systematics, but its phylogenetic interests too will become less descriptive and more explanatory.

Thus far, Ghiselin's prediction has not held true. Molecular phylogeny has not provided all of the answers concerning the relationships of organisms.

CHAPTER 2

The Basic Units of Phylogenetic Analysis

There has been some discussion in the literature of what are the basic or fundamental units of phylogenetic (or more generally, biologic or paleonto-logic) analysis. There are actually two different points of concern here. First, what are the most basic units with which we are trying to reconstruct phylogeny (or if the main concern is not phylogeny reconstruction but general taxonomy, what are the most basic units to be used in the classifications)? And second, what are the basic units, or fundamental bits of information, used in recon-structing phylogeny?

With regard to the first point, Sneath and Sokal (1973, p. 69) suggest that in the majority of cases the fundamental unit to be classified is the individual organism. Hennig (1966a, p. 6) regards the basic element of biological sys-tematics not to be even a single individual, but an individual organism at a particular point of time (during a particular period of its life). Hennig (1966a) calls this fundamental entity a semaphoront. Using the example of an insect that undergoes metamorphosis, he makes the important point that the different stages in the life cycle (e.g., larva, pupa, imago) of the individual will be very different morphologically, physiologically, ecologically, and so on (thus yielding different complexes of information that can be used in systematic studies), and often only as the result of complex studies is it finally realized that various semaphoronts are actually stages in the life cycle of a single type of organism. Applying Hennig's terminology, any single fossil specimen is of necessity representative of a semaphoront, not a complete individual. And since during the fossilization process only certain morphological elements of the semaphoront are preserved (usually hard tissues), the fossil specimen is not even completely representative of the morphology of the semaphoront. The same is true of typical preserved museum specimens of extant organisms.

Hennig (1966a) used the term *holomorphy* (= total form) to refer to all of the physiological, morphological, and ethological (behavioral) properties or characters of a semaphoront. An individual organism, an holomorph, would be composed of the sequence of semaphoronts that make up its entire life cycle. He was careful to point out that an individual organism is a multidimensional totality that must be considered within the multidimensional continuum of intrinsic morphology of the organism (and species), geographic and ecologic space, and time. Morphology of the organism for Hennig (and it is used here in the same way) refers not only to morphology (e.g., anatomy) in the narrow sense, but to all intrinsic aspects of the organism. (Hanson [1977, p. 92] missed this basic point when he stated "the suffix '-morphous' in Hennig's terminology is restrictive in its implication that only morphological characters [in the narrow sense] are under consideration.")

From a practical point of view, Sokal and Sneath (1973, p. 69) have defined their operational taxonomic units (OTUs) as "the lowest ranking taxa employed in a given study." The OTUs may be semaphoronts, individuals, species, genera, or some higher taxonomic category in any particular study. They have pointed out that it would be both impractical and impossible to begin every study anew using truly fundamental biologic units. In the course of certain systematic analyses, some investigators refer to units related to OTUs—hypothetical taxonomic units (HTUs) for postulated ancestors (Farris, 1970; Sneath and Sokal, 1973) and evolutionary units (EUs) for taxa at infraspecific levels (Baum, 1984).

In order to analyze the relationships between the basic elements, units, or OTUs, researchers utilize what are variously termed characters, characteristics, attributes, features, traits, character states, properties, and so forth. In general, characters are often regarded as peculiarities that distinguish a semaphoront, or identical peculiarities that distinguish a group of semaphoronts, from other semaphoronts (Hennig, 1966a)—that is, characters distinguish certain organisms from each other and serve to unite other organisms with one another (Mayr, 1969; Sneath and Sokal, 1973). It is important always to remember that characters are not empirically observable, but that a character is a theory of homology (Platnick, 1979). As Bonde (1977, p. 751) has noted,

features are purely conceptual; they are arbitrarily chosen and delimited; they may overlap or be exclusive of each other; they may be interdependent or independent; they may be simple and indivisible or complicated and divisible with ease or difficulty into simpler (sub-) statements; and very often they are generalizations (e.g. mean values, ranges).

The systematist's characters or features are mere symbols that are hoped to correspond to some real phenomena in nature, to some real knowledge. In this sense, taxonomy cannot be reduced to simple empirical observations. In a slight variation on the theme that characters distinguish organisms from one another, a character can be regarded as some feature that is common to several organisms but varies from one organism to another; the variations seen in the particular character are the various states of the character (Sneath and Sokal, 1973). The *states* of Sneath and Sokal (1973) correspond to the *characters* of Hennig (1966a) and Mayr (1969). Sneath and Sokal (1973, p. 74) define their unit character as "a taxonomic character of two or more states, which within the study at hand cannot be subdivided logically, except for subdivision brought about by the method of coding." In this book the terms *character* and *character state* refer to a feature held in common by two or more organisms and the variations on that feature. As Eldredge and Cracraft (1980, p. 30) point out, these two terms refer to similarities (identity) at two hierarchical levels. The character is similarity (identity) at one level higher (more inclusive) than that of the character state. Any character can conceivably be considered a character state at a different level, and likewise any derived character state (but not a primitive character state) can be considered a character at a different level. For example, having a well-developed head could be considered a character (i.e., common feature) of certain vertebrates and the presence or absence of jaws as part of the head apparatus considered two character states (variations on the common feature). On another level, possession of jaws (derived) would be the character, and having one or another form of teeth the character state.

Given two or more character states for a particular character, these character states may be arranged into a nonpolarized transformation series, or morphocline. A morphocline is a hypothesis as to the probable pathway, but not direction, of change from one character state to another (Eldredge and Cracraft, 1980, p. 54; Masalin, 1952; Schaeffer, Hecht, and Eldredge, 1972). Usually character states are arranged into a postulated morphocline because they grade into one another or form an apparent sequence or continuum. Thus, if a certain character is "number of spines," and various states include "1 = one spine," "2 = two spines," "3 = three spines," and "4 = four spines," we would probably be tempted to order the character states into the morphocline "1 to 2 to 3 to 4." Here it must be emphasized that the morphocline does not imply a particular direction of evolution. In the above example *to* can mean evolution in either direction—that is, any of the character states (1, 2, 3, or 4) may be the primitive character state from which the others are derived. Determining which character states are primitive and which are derived is the problem of determination of morphocline polarity (see chap. 3). In the above example, if the morphocline is correct (the morphocline itself is a hypothesis of homologies) and 4 is determined to be primitive, then in an evolutionary sense we might say that 4 gave rise to 3, which gave rise to 2, which gave rise to the most derived form, 1. If, on the other hand, it was concluded that 2 was primitive, then we could suggest that 2 gave rise to 1 and also independently to 3; subsequently 3 gave rise to 4.

Whenever a character is composed of only two character states, there are only two possibilities: The morphocline is 1 to 2 (i.e., 1 transforms to 2, or 2 transforms to 1), or the two character states do not form a morphocline (both may be derived relative to a primitive character state for the character, and neither known character state is derived from the other). In any phylogenetic analysis, the more single synapomorphies (homologies) can be dealt with at the correct hierarchical level, the less hypothesized (and possibly incorrect or misleading) morphoclines have to be introduced (see discussions on homology and morphocline polarity in chap. 3). Of course, putative characters, character states, or synapomorphies at a particular level that are not homologous at that level are not really characters at that level.

Platnick (1979, p. 542) has been careful to make the point that the concept of a character state, if not used with care, may be potentially misleading. Any character is a hypothesis of homology (synapomorphy). Given a certain character consisting of three character states composing the transformation series A transforms to B transforms to C (i.e., character state A gave rise to character state B, which gave rise to character state C), character states A, B, and C are not alternative and distinct character states, but rather character state B is an addition to A and C is an addition to B. Or, in other words, A is a more general, more inclusive, more primitive (plesiomorphic) character state relative to B and C (and likewise B relative to C). An organism that bears character state B or C also bears character state A. It would perhaps be more correct to state that the alternative character states for this particular character are: A, AB, and ABC. However, in common parlance, when we refer to organisms that bear character state A, we refer only to those with just A and exclude those with AB and ABC. Of course, organisms distinguished solely by A (lacking B and C) form a primitive, plesiomorphic, potentially paraphyletic group that merely lacks certain characters (namely B and C in this case). Thus character state A, if primitive, as used in common parlance is not a positive occurrence of a character.

It must also be pointed out that generally (and this is the way the terms are used here if not explicitly stated otherwise) the terms *character* and *character state* refer to intrinsic morphological features of biological organisms such as size, shape, color, and behavior (what characters are preserved in fossils is usually a function of the fossilization process; in some cases the behavior of extinct organisms may be preserved in the form of trace fossils). However, some workers do consider extrinsic features, such as the geographic and temporal positioning of a recent or fossil organism (i.e., where it was found and how long ago it lived) to be characters of the organism on a par with intrinsic characters. Hennig (1965), for instance, suggests that in some cases nonphysical characters (nonholomorphological characters) such as a total life history that is relatively derived ("apooec") or the geographic distribution of a group may be used to define a monophyletic group of organisms. However, he points out that such intrinsic data may be particularly prone to homoplasy, and hypotheses based on such data should be confirmed by the study of morphological characters. For most forms of phylogenetic analyses the characters used must be intrinsic, inheritable traits of the organism (not ecophenotypic variation, for instance). All separate characters must be logically and biologically independent. Thus, one would not use "number of eyes" as a character and "twice the number of eyes" as a second character.

Cartmill (1982) has perceptively noted that in some cases, especially those involving complex anatomical features, the anatomical descriptors and terminology that investigators utilize to describe characters and character states may not initially be neutral, but may be phylogenetically biased a priori. The use of such biased terminology inevitably leads to a certain conclusion in regard to phylogenetic relationships. Part of the problem is due to defects in terminological systems; using a certain terminology may result in certain characters being unrecognized while others may be scored multiply and redundantly.

Characters that represent critical synapomorphies (or, in Gingerich's "stratophenetic approach," simply phenetic similarities) in one system are not recognized in the other or appear only as interaction effects between two or more differently-analyzed characters; features counted as single resemblances in one system become multiple resemblances in the other; and different partitioning of morphological continua into discrete character states results in different parceling-out of intermediate morphologies. When formalized algorithms (e.g., Estabrook, 1968; Farris, 1970) are used to assess the relative parsimonies of various phylogenetic hypotheses by counting the number of evolutionary changes that each hypothesis implies, the two systems of terminology will yield different assessments. (Cartmill, 1982, p. 282)

Cartmill (1982, p. 282) suggests "it is possible that eliminating these defects would eliminate the differences between the two systems of anatomical descriptors, or at least cause both to yield the same assessment of the most parsimonious phylogeny." However, according to Cartmill, "Unfortunately, this is not necessarily the case. It is possible to devise two systems of descriptors which have the same factual content, but yield different phylogenetic reconstructions for certain collections of taxa" (Cartmill, 1982, p. 282). Cartmill then proceeds to present two different, nonredundant systems of description of the ear regions of primates, applying these systems to the description of five genera (Tables 2-1 and 2-2). Using each system alone, he develops a cladogram for the five genera and finds that the two resulting cladograms are incompatible (Figs. 2-1 and 2-2). On this basis, Cartmill (1982, p. 283) concludes that "two factually equivalent, non-redundant systems of anatomical terms may have different phylogenetic implications." Cartmill (1982, p. 284, italics in the original) further concludes that

Table 2-1. Data Matrix Using Cartmill's System 1 Set of Descriptors for the Ear Region of Five Genera of Primates

| | Character | | | |
	1^a	2^b	3^c	4^d
Taxon				
Lemur (= L)	a	B	c	g
Allocebus (= A)	A	B'	c	G
Tarsius (= T)	A	b	C	g
Cebus (= C)	a	b	C	g
Necrolemur (= N)	A	B'	C	G

Source: After Cartmill, 1982, pp. 279–287.
[a]Character 1—ectotympanic shape: a = ring-shaped; A = tubular.
[b]Character 2—anular bridge: b = absent; B = composed of soft tissues; B' = osseous.
[c]Character 3—ectotympanic position: c = wholly intrabullar (aphaneric); C = not wholly intrabullar (phaneric).
[d]Character 4—transverse septa in bullar floor ("struts"): g = absent; G = present.

Table 2-2. Data Matrix Using Cartmill's System 2 Set of Descriptors for the Ear Region of Five Genera of Primates

| | Character | | | |
	1^a	2^b	3^c	4^d
Taxon				
Lemur (= L)	R	P'	q	g
Allocebus (= A)	R	P	Q	G
Tarsius (= T)	r	p	Q'	g
Cebus (= C)	r	p	q	g
Necrolemur (= N)	R	p	Q'	G

Source: After Cartmill, 1982, pp. 279–287.
[a]Character 1—petrosal contacts: p = petrosal contacts ectotympanic but not (soft) meatal tube; P = petrosal contacts ectotympanic and (soft) meatal tube; P' = petrosal contacts (soft) meatal tube but not ectotympanic.
[b]Character 2—ectotympanic contacts: q = ectotympanic surrounds eardrum only; Q = ectotympanic surrounds drum and recessus meatus; Q' = ectotympanic surrounds drum, recessus, and distal meatus.
[c]Character 3—subtympanic extension of tympanic cavity: r = absent; R = present.
[d]Character 4—transverse septa in bullar floor: g = absent; G = present.

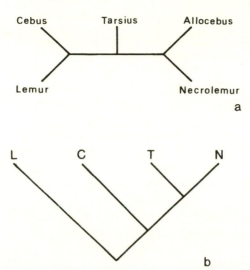

Figure 2-1. Cartmill's unrooted network (a) and cladogram (b) for five primate taxa based on the descriptors of his System 1 (see Table 2-1); see text for further explanation. After Cartmill, 1982, pp. 279–287.

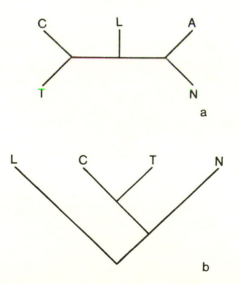

Figure 2-2. Cartmill's unrooted network (a) and cladogram (b) for five primate taxa based on the descriptors of his System 2 (see Table 2-2); see text for further explanation. After Cartmill, 1982, pp. 279–287.

any sufficiently complex system of morphological descriptors is apt to incorporate phylogenetic bias, either deliberately or accidentally, by defining a non-unique morphological space . . . [and] . . . parsimony assessments based on morphological character-state analysis cannot reliably yield neutral judgments of conflicting phylogenetic hypotheses.

This statement is extremely strong and undercuts much of the basis of phylogenetic analysis; therefore, it is important to consider Cartmill's (1982) example in detail.

The first point to be considered is that the method of phylogenetic analysis utilized by Cartmill (1982) is not cladistic analysis (*sensu stricto,* based on strict parsimony: the minimizing of ad hoc hypotheses of homoplasy), but that of unrooted Wagner networks (unrooted trees). Cartmill decided which network to utilize for each system on the basis of "minimum number of transitions from one character state to another" (Cartmill, 1982, p. 283; = minimum lengths, = parsimony in Cartmill's terminology, but not as accepted here; see chap. 3 on parsimony). He converted the unrooted networks into cladograms by assuming that *Lemur* is morphologically primitive and by omitting *Allocebus* from consideration because it was grouped with *Necrolemur* in both networks. Cartmill's (1982) networks and cladograms are shown in Figures 2-1 and 2-2.

Reanalysis of Cartmill's (1982) matrices of character states for each system utilizing cladistic analysis (and assuming *Lemur* is primitive) yields somewhat different cladograms. When only four characters are included in each system, in neither case are the phylogenetic relationships convincingly (without char-

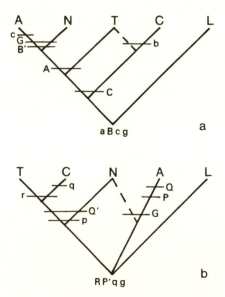

Figure 2-3. Hennigian cladograms derived by the present author from Cartmill's (1982, pp. 279–287) Systems 1 *(a)* and 2 *(b);* compare to Figures 2-1 and 2-2.

acter conflict) resolved (Fig. 2-3). The two systems do not yield the same cladogram, and system II in particular does not produce the same cladogram that Cartmill derived. In both cladograms *Allocebus* (which Cartmill dismisses as always united with *Necrolemur*) is a problematic taxon. In system I *Allocebus* seems to be secondarily primitive for one character. In system II the characters that unite *Necrolemur* with *Tarsius* and *Cebus* are not present in *Allocebus*. It is not the case, as Cartmill seems to imply, that the two systems of terminology give straightforward, incompatible cladograms.

In discussing his two systems of terminology, Cartmill (1982, p. 282) regards each as "adequate to yield an unambiguous picture" of the ear regions concerned. However, it would appear that neither system is a complete description of the ear region in and of itself. Combining the two systems and eliminating redundancy (i.e., character 4 is the same in both systems) will yield considerably more information on which to base a phylogenetic hypothesis. Using the characters of both systems, and again assuming that *Lemur* is primitive for all character states, one can arrive at a single phylogenetic hypothesis that accounts for the character state distributions parsimoniously (Fig. 2-4). Only three apparent instances of homoplasy are present in this cladogram: (1) *Allocebus* is secondarily primitive for one character state (c); (2) *Tarsius* and *Cebus* have independently acquired derived character state b; and (3) *Tarsius* and *Cebus* have independently acquired derived character state r. However, b and r may be redundant; b is the absence of an anular bridge, while r is the absence of the subtympanic extension of the tympanic cavity. Both b and r are essentially the same character (Cartmill, 1982). Therefore there is only one actual instance of homoplasy, and this is the parallel loss of a structure—a character often considered of low weight a priori (see section on character weighting)—and one reversal (c). In sum, it would seem that given Cartmill's (1982) own example, cladistic analysis does yield an unambiguous hypothesis of phylogenetic relationships. As discussed in chapter 3, it can be contended

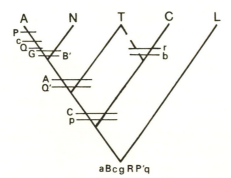

Figure 2-4. Hennigian cladogram for five primate taxa using all information from both of Cartmill's (1982, pp. 279–287) systems of descriptors (see Tables 2-1 and 2-2).

that parsimony analysis per se is neutral with respect to the outcome of a phylogenetic analysis. The data, however, are not neutral; the distribution of a particular character either falsifies (but does not necessarily prove false) or corroborates a particular phylogenetic hypothesis.

Hanson (1977), in relative isolation from most other workers, has suggested that the species is the unit of phylogenetic analysis and has also introduced the term seme for what he regards as "a unit of phylogenetic information" (Hanson, 1977, p. 88). At first sight it would appear that Hanson's seme is equivalent to a character or character state, but on close inspection it seems to be something very different. Elaborating on his concept of a seme, Hanson (1977, p. 89) states: "A seme is an information-containing entity in an interbreeding population of organisms, but it will most commonly be used in reference to a structural or functional part of an organism, starting at the molecular level"; semes must be "functional or structural features which are sufficiently complex and variable to allow meaningful comparisons." In a list of major animal semes Hanson (1977, p. 96) includes such items as breeding patterns, demic structure, mutation rates, muscular system, skeletal system, circulatory system, digestive system, DNA, RNA, and biochemical pathways and products. Hanson's (1977) concept of a seme appears to be less inclusive than Hennig's (1966a) concept of a semaphoront (however, remember that Hennig's semaphoront is the basic unit on which analysis is to be performed, not the basic unit [bit of information] to be used in reconstructing phylogenies) but much more inclusive than the usual conception of a character or feature. Hanson developed an elaborate set of prefixes (e.g., plesioseme, aposeme, neoseme) to be attached to the base word seme and a corresponding system of phylogenetic analysis to go along with the terminology. Hanson's system for calculating the phyletic distance between any two species on the basis of comparing 10 to 20 semes is discussed in chapter 2.

A number of years ago Shaw (1969) put forth an extreme view of the fundamental units of paleontological analysis and the aims and goals of paleontology in general. He suggested that there is nothing more to paleontology than morphological features of biological origin (organisms), time, and geography. According to Shaw it is counterproductive to attempt to construct phylogenies for species (which Shaw regards as purely artificial mental constructs) or even organisms. Thus (Shaw, 1969, p. 1095) the researcher "must literally record the stratigraphy of morphology, feature by feature." Shaw (1969, p. 1096, italics in the original) proposed his "analytic paleontology—the detailed study of the exact form of each morphologic unit, its precise stratigraphic [and geographic] distribution, and the observed combinations with other morphological units"—as a replacement for most of traditional paleontology. Any synthesis of the basic morphologic and distributional data, especially if it meant recognizing species or lineages or actual organisms, spelled disaster in Shaw's eyes. "I suggest to you, therefore, that the attempt to study organisms synthetically, which is implicit in the species approach, is wrong because it hinders the understanding of the organisms" (Shaw, 1969, p. 1096). Virtually no one has ever pursued, or even taken seriously, Shaw's (1969) views.

As characters and semaphoronts are recognized and associated as holo-morphs, the usual procedure is to then group the holomorphs into entities termed species. Species are probably the most common OTUs upon which higher-level phylogenetic analyses are performed. Because of the importance, or at least the putative importance, of species in the evolutionary process and to phylogenetic analysis, various concepts of species and mechanisms of spe-ciation (or more generally, lineage splitting—the basis of phylogeny) are dis-cussed in the next section.

SPECIES CONCEPTS

The species is the lowest taxonomic category that is routinely used in sys-tematic work (Mayr, 1969, p. 27); further, it is the one biological taxonomic category that many workers would consider to be "real"—that is, to have an objective, nonarbitrary basis in nature (but for a contrary view, see Sokal and Crovello, 1970). As has been often pointed out, among particular cases of extant organisms the modern biologist's or naturalist's recognition of distinct biological entities labeled species often corresponds remarkably well with the species recognized by pre-Darwinian naturalists, native peoples, and even nonhuman higher organisms (e.g. Mayr, 1963, p. 17; however, for an alter-native view, see Berlin, Breedlow, and Raven, 1966). In most discussions of the subject, a distinction is made between three natural groups of taxonomic categories: infraspecific categories (usually populations and demes), the species category (often including subspecies as species-level taxa), and supraspecific categories (cf. Mayr, 1969, p. 23). This section considers the species category; the other two groups are dealt with in the sections on dynamics of fossil pop-ulations and monophyly, polyphyly, and paraphyly.

In the original sense, and in nonbiological terminology, the term *species* is used to designate a class of similar objects to which a name is given (Mayr, 1963, 1976c). Thus, for example, mineralogists speak of species of minerals (Mason and Berry, 1978), or gold might be considered a species of metal. Likewise, biological species are considered to be groups of similar organisms (individuals) that share a number of traits in common. Central to the concepts of science and scientific classification of many nineteenth-century pre-Dar-winian (or non-Darwinian) philosophers, such as John Stuart Mill (1879), was the notion or idea of *natural kinds* (Hull, 1981). The members of a natural kind will share a number of traits in common (essential characteristics) that distinguish the natural kind under consideration more or less sharply from other natural kinds. For cases where the boundaries between some natural kinds were relatively indistinct, philosophers such as Whewell (1847) suggested that a typical member of the class or natural kind, the *type* (an individual entity), should serve as a central focal point and standard within the natural kind. In such a context many philosophers have considered biological organisms to be paradigm examples of particulars or individuals and biological species to be paradigm examples of natural kinds, universals, or classes (Hull, 1978, 1981). It is this traditional concept, tracing its origins back to Plato and Aristotle, that Mayr (1963, 1969, 1976a, 1976b, 1976c) has labeled the *typological* or *essentialist [biological]* species concept. According to Mayr (1976c), when a typological species concept is utilized, the actual closely related species are often recognized and defined by a certain arbitrary degree of difference (usually morphological) between them. It appears to have been only recently that philosophers, and perhaps biologists too, have realized that if evolution occurs, then species cannot be natural kinds, at least in the traditional sense. Given the concept of evolution, species should grade into one another rather than being sharply distinguished.

There have been, and continue to be, many different species concepts that pertain to biological organisms (for brief reviews of the various species concepts in a historical context, see Mayr 1963, 1976c, 1982; Simpson, 1961). The two founders of modern systematic and evolutionary biology, Carolus Linnaeus (1707–1778) and Charles Robert Darwin (1809–1882) respectively, had very different views about the nature of biological species. Linnaeus (e.g., 1758) believed that species are constant or fixed, real, and objectively delimitable in the natural world (thus Linnaeus has been considered a typologist: A. J. Cain, 1958; Mayr, 1969). Linnaeus denied the beliefs, widespread in his time, that higher organisms could arise by spontaneous generation, that occasionally an individual of one animal species might literally give birth to a different species of animal, or that the seed of one species of plant might produce another, different species of plant. In general, the ancients and medieval scholastics had viewed species as mutable, changing, and not sharply delimited (Simpson, 1961). Linnaeus further denied nominalist speculation that only individuals really exist and all other categories, including species, are merely manmade abstractions (Mayr, 1976c). There are indications from Linnaeus's writings, however, that in some cases he did believe that certain closely related species descended from a common origin (Greene, 1912; Ramsbottom, 1938).

C. Darwin, on the other hand, was extremely interested in and impressed by the variation of species under domestication and in nature, by the often fuzzy boundaries between species in nature, and by the seemingly subjective delimitation of closely similar species (C. Darwin, 1859). In the end C. Darwin appears to have denied the biological reality of species as an objective unit in nature and considered species to be artificial constructs, or units of convenience (see Wiley, 1981, p. 22; more recent authors who have agreed with C. Darwin's view include Ehrlich and Holm, 1963; Gilmour, 1940). Thus

C. Darwin (1859; 1871*a*, p. 69) subscribed to "the view that species are only strongly-marked and permanent varieties." Accordingly, species would seem on the one hand to have a real basis in nature (the members of a species are united by propinquity of descent); but his views on the actual delimitation of named species by naturalists would seem to verge on nominalist views (cf. Wiley, 1981, p. 41). If the origin, or at least existence, of variation could be explained, and if this variation could give rise to more or less differentiated and distinguishable varieties, populations, forms, or subspecies, then, in C. Darwin's view, this would explain the origin of species (permanent varieties). Therefore, as Mayr (1976c) has noted, C. Darwin did little to address the problem of speciation (as commonly construed today) in *The Origin of Species.*

The species concept most widely current today, a product of the new systematics (see discussion in the introduction), is what Mayr (e.g., 1976c), its foremost proponent, refers to as the *biological species concept* (= biospecies). Other commonly recognized species concepts are what can be termed the morphological (morphospecies or phenetic species), cladistic, evolutionary, arbitrary (e.g., chronospecies, paleospecies, successional species), and asexual (agamospecies) species concepts (see Mayr, 1957; Simpson, 1943; Slobodchikoff, 1976; Sylvester-Bradley, 1956). The classic definition of a biological species reads as follows (or reads as some variation of this): "groups of actually (or potentially) interbreeding natural populations which are reproductively isolated from other such groups" (e.g., Mayr, 1940; 1969, p. 412; 1976c). One important point of this definition is that species are not distinguished by degree of difference, but by distinctness or discontinuity. Two species may appear extremely similar, but if they do not interbreed then they are distinct. All other modern species concepts must address this "biological species concept," and most are actually in more or less agreement with this concept. Mayr (1969, p. 26) points out that "this species concept is called biological not because it deals with biological taxa, but because the definition is biological. It utilizes criteria that are meaningless as far as the inanimate world is concerned." Simpson (1961, p. 150) has noted that Mayr's biological species concept "is strictly a genetical concept of species among contemporaneous, biparental organisms."

As his equivalent of Mayr's biological species, Simpson (1961, p. 153; see also Simpson, 1951) proposed the concept of the *evolutionary species* defined as follows: "An evolutionary species is a lineage (an ancestral-descendant sequence of populations) evolving separately from others and with its own unitary evolutionary role and tendencies." Simpson's evolutionary species concept roughly corresponds to Mayr's (1976c) notion of a "multidimensional, or polytypic, species concept," which is an extension of a nondimensional (i.e., applicable only to the relationship of two naturally existing populations of organisms at a single locality at a single moment in time) version of Mayr's biological species concept in space and through time. Further, Simpson's conception is not necessarily limited to organisms that propagate by sexual reproduction; it is also applicable to asexual and parthenogenetic organisms. A modern version of a species definition that is consistent with Mayr's and Simpson's definitions and also recognizes species as individuals and units of evolution (see discussion below) reads as follows: "A species is a diagnosable cluster of individuals within which there is a parental pattern of ancestry and descent, beyond which there is not, and which exhibits a pattern of phylogenetic ancestry and descent among units of like kind" (Eldredge and Cracraft, 1980, p. 92).

Paterson (1978, 1982; see also Vrba, 1984a) has noted that the common biological species concept for biparental organisms is defined in terms of reproductive isolation (isolation concept). Paterson suggests that it might be more productive and correct (true to the way the actual organisms involved view the situation) to define species in terms of organisms that "*recognize* each other for the purpose of mating and fertilization (the Recognition Concept)" (Vrba, 1984a, p. 123, italics in the original). It is the divergence of specific-mate recognition systems (SMRS) that are all-important in speciation. It may be possible for the human investigator to determine just what features or traits are important to the organisms in mate recognition. If this is possible, one might theoretically distinguish natural species in the same manner that the organisms themselves do. Moreover, in some instances the SMRS may be composed in part of visual cues that manifest themselves as morphological traits that are preservable in the fossil record; in these instances one may be able to utilize fossil data to identify with confidence biological, interbreeding species. In other cases, however, the SMRS may consist totally of nonmorphological characteristics such as auditory, behavioral, or chemical cues; it may then be impossible to distinguish sibling species in the fossil record.

Genetical or biological species concepts may be accepted in principle by many neobiologists, as least as far as sexually reproducing organisms are concerned, but such concepts can rarely, if ever, be applied operationally. Biological species concepts, particularly as products of the modern synthesis, new systematics, and evolutionary classification, have come under increasing criticism in recent years from workers in a number of camps. As has been pointed out by Sokal and Crovello (1970), for many practically oriented taxonomists (whether working with extant or extinct forms) the recognition of groups of morphologically similar organisms (taxa) that are named and assigned to the species category is independent of any kind of biological (or other theoretical) species concepts. In practice most actual taxa labeled as species are recognized on the basis of shared phenetic (usually morphological) characteristics and thus are actually morphological (or phenetic, or morphospecies) species.

Particularly relevant to the theme of this book is the apparent fact that many practically oriented paleontologists utilize (either implicitly or explicitly) a morphological species concept, one that does not necessarily make any pretense to agreement with a biological species concept a priori. Expressing this view Shaw (1964, pp. 72–73) writes:

The paleontologist . . . has to deal with many types of discrete and objectively recognizable organic remains whose homology with living species is in doubt. . . . In order to deal with these fossils the paleontologist must perforce call them something. In this discussion I shall call them "species" for want of another name. I shall use the word to cover any of that large class of objects of organic origin that are of sufficiently distinctive and consistent morphology so that a competent paleontologist could define them so that another competent paleontologist could recognize them. For purposes of communication he would apply a Linnaean name to them. This definition is intended to be sufficiently flexible to cover any sort of fossil that would normally be named. The only essentially biologic part of this "species concept" is that fossils are necessarily derived from organisms.

In a sense it is only coincidental if such morphological species happen to correspond to any kind of real, or theoretically interesting, units in nature.

Sokal and Crovello (1970) further criticize the biological species concept for a number of reasons. Not only are most species-level taxa described solely on the basis of conventional morphological or phenetic data (and thus the biological species concept is unnecessary at this practical level; further, Sokal and Crovello [1970] have demonstrated that even when a worker explicitly adheres to a strict biological species concept, phenetics plays an essential role in the delimitation of actually recognized species), but then the biological species concept imbues such species "with a false aura of evolutionary distinctness and with unwarranted biosystematic implications" (Sokal and Crovello, 1970; reprinted in Sober, 1984*b*, p. 542). This problem can occur even when the careful worker takes into account the difficulties of correlating phenetic similarities with genetic closeness—for example, when Simpson (1961) and Mayr (1969, p. 25) carefully distinguish the "difference between basing one's species concept on morphology and using morphological evidence as inference for the application of a biological species concept." As most careful biosystematists will readily acknowledge, there are numerous cases in which phenetics or morphology may actually be misleading when it comes to recognizing true biological species (*sensu* Mayr); well-known cases include cryptic species, polytypic species, and ecophenotypes.

Sokal and Crovello (1970) also point out that the nature of interbreeding and hybridization is a continuing problem for Mayr's biological species concept. Again, phenetics can be a very imperfect reflection of interfertility, and in actual practice for extant organisms how many times do systematists actually base their species on breeding experiments? Of course, breeding experiments are totally out of the question for extinct species. And what does the phase *potentially interbreeding* of Mayr's definition mean? Mayr (1969, p. 26) dropped this phrase from his definition of a biological species in the text of his book, but retained it in the glossary (p. 412). If Mayr's biological species concept were strictly adhered to, then whenever two species could successfully breed and undergo any backcrossing in nature, they would have to be considered a single species. However, while hybrids between "good" biological species are rare, they are not unknown (Mayr, 1963; 1970, p. 71), and as Sokal and Crovello (1970; reprinted in Sober, 1984b, p. 559) note, for the advocate of the biological species concept there is apparently an arbitrary (but ill-defined) level of hybridization that is allowable between populations before all of the individuals concerned need be regarded as belonging to the same species. On the other hand, what of cases of populations that are separated geographically and never interbreed naturally, yet are indistinguishable? Do these form a single species because they are phenetically identical, or should they be labeled as distinct species because they cannot interbreed? Or is this a case of potential interbreeding linking them into a single species if they can be brought together artificially and made to interbreed? One will never know if populations that are separated temporally could have interbred if they had been contemporaneous. Finally, Mayr's concept of the biological species, with its emphasis on interbreeding, is inapplicable to uniparental species (asexual or agamospecies); Dobzhansky (1937, 1970) even believed that there cannot be a true species category, only pseudospecies, for such organisms.

From a different point of view, it has recently been argued that whether interbreeding and hybridization occur between different natural populations has absolutely no bearing on whether the populations should be regarded as a single species or as distinct species (Brady, 1983; Rosen, 1979). What is important is the cladistic (phylogenetic) relationships between the populations under study. In his study of extant swordtail fishes (Poeciliidae, Teleostei) from Guatemala, Rosen (1979) found that the species *Xiphophorus alvarezi* and *X. helleri* are able to interbreed where their ranges come into contact, but *X. signum* is reproductively isolated from the other two species. To the worker enamored of the biological species concept, this situation would immediately suggest that *X. alvarezi* and *X. helleri* are more closely related to one another than either is to *X. signum,* and furthermore that perhaps *X. helleri* should actually be synonymized with *X. alverezi*. However, on the basis of a clear pattern of distribution of derived and primitive morphological traits, Rosen (1979) concluded that *X. signum* and *X. helleri* are more closely related to each other than either is to *X. alvarezi*. Discussing this case, Wiley (1981, p. 30) states:

We may conclude that *X. signum* has different and apomorphic features incorporated into its reproduction and development whereas *X. helleri* and *X. alvarezi* are similar and plesiomorphic for these features (at least to the point that development is not impaired). Their ability to interbreed is due to the retention of plesiomorphic features, not to the fact that they are close relatives.

At some level interbreeding—genetic compatibility—must be primitive. Sister taxa *sensu stricto* (twin taxa) must have been able to interbreed (primitive) before they underwent divergence (derived) and became reproductively isolated. Such cases (examples of immediate sister taxa reproductively isolated, but more distantly related taxa that can interbreed) may actually be fairly common, but they have gone unrecognized because in the past even the most ardent splitter or typologist often synonymized taxa or prospective taxa when they were observed to interbreed. But such synonymization may not be justified and may impede the advancement of our understanding. If in doubt, one may actually be justified in splitting, rather than lumping, species-level taxa (given that there are adequate characters on which to diagnose the taxa). Taxa that are truly over-split should come back together as closest relatives when a phylogenetic (cladistic) analysis is performed on the taxa in question. Relative to such considerations, and also citing Rosen's (1979) study, Brady (1983, p. 55) writes:

. . . one may wonder just how much clear morphological distinction has been sunk in synonymy by an application of the reproductive criterion. In this case it is obvious that the criterion would demote the morphological pattern to noise, shifting the focus of the taxonomist to an entirely different aspect. If Rosen had been one of the faithful, he would have missed the extension of the cladistic pattern discovered when, in the latter part of his monograph, he compares the morphological pattern to the pattern of geographic distribution within the two groups. The congruence that shows up here is an empirical confirmation of the original pattern, and certainly indicates some sort of causal significance.

From the point of view of the paleontologist there is another, practical aspect to Rosen's (1979) study. Paleontologists, solely on the basis of the morphology of their specimens, are potentially able to discover meaningful patterns of relationships between the organisms they study. It is neither particularly necessary nor desirable to worry about interbreeding between the populations being analyzed. The fact that paleontologists cannot run breeding experiments on their organisms is not an absolute hindrance to recognizing evolutionarily meaningful species.

Simpson's (1961) evolutionary species concept avoids some of the shortcomings of Mayr's definition of the biological species concept. In particular, the evolutionary species concept is theoretically applicable to both uniparental and biparental organisms (Meglitsch, 1954; Simpson, 1961); the important point for both uniparental and biparental organisms is that their species evolve as a unit and retain a unitary role and that each species seems to occupy its own niche (Mayr, 1969). An asexual species, like a sexual species, maintains "a community of inheritance"; the differential spread of genes is controlled by differential reproduction or natural selection (Simpson, 1961, p. 162). However, Sokal and Crovello (1970; reprinted in Sober, 1984b, p. 560) reject the evolutionary species concept as "so vague as to make any attempt at operational definition foredoomed to failure." Again, actual asexual taxa are based primarily on morphological evidence (Mayr, 1969, pp. 31, 47). In a way, Simpson's evolutionary species concept is a classic expression of the evolutionary school of taxonomy and phylogeny reconstruction: A species is defined in terms of a certain hypothesis or conceptualization of the processes of evolution (based, for Simpson, on the modern synthesis).

Sokal and Crovello (1970; reprinted in Sober, 1984b, p. 561) further argue that the biological species concept is not necessary, or even useful, for evolutionary classification (here by evolutionary they evidently mean what is called in this book phylogeny reconstruction) in that the majority of evolutionary taxonomy "is based not on interbreeding but on phenetics and homologies, whether they are morphological, behavioral, physiological, serological, or DNA homologies." Furthermore, most phylogeny reconstruction deals with terminal taxa that are operationally individuals representative of monophyletic taxa; it makes no difference if they are biological species in Mayr's sense.

In their final criticism of the biological species concept, Sokal and Crovello (1970) claim that the biological species concept, and apparently any species concept, is neither especially valuable, necessary, useful, or heuristic for evolutionary theory. Rather, they believe that "most of the important evolutionary principles . . . could just as easily be applied to localized biological populations, often resulting in deeper insight into evolution"; moreover, the "localized biological population [rather than the species] may be the most useful unit for evolutionary study" (Sokal and Crovello, 1970; reprinted in Sober, 1984b, pp. 564–565). But here Sokal and Crovello have not even considered macroevolutionary theory (evolution above the species level) where species may be individuals, historical entities, or particles (that is, real [natural] entities) that are the stuff of higher-level processes. Species may be the units of evolution. Of course, such species may not conform to Mayr's biological species concept, but they may be more than Sokal and Crovello's localized biological populations. These topics are discussed further below.

Mayr (1969, 1976c) and Simpson (1961) undercut the objectivity and nonarbitrariness of their own conceptions of biological or evolutionary species by virtually asserting that when placed within a temporal framework, their species must be diagnosed or defined either typologically or nominally. Thus Mayr (1969, p. 35) states: "Species are evolving systems, and the vertical delimitation of species in the time dimension should in theory be impossible." Mayr's species concept is saved only because "in most fossil sequences there are convenient breaks between horizons to permit a nonarbitrary delimitation of species" (Mayr, 1969, p. 35). Of course these species will become arbitrary when and if the convenient breaks are filled.

In discussing the same problem, Simpson (1961, p. 165, italics in the original) states:

An evolutionary species is defined as a separate lineage (also sometimes called a *gens* in paleontological usage) of unitary role. If you start at any point in the sequence and follow the line backward through time, there is no place where the definition ceases to apply. You never leave an uninterrupted, separate, unitary lineage and therefore never leave the species with which you started unless some other criterion can be brought in. If the fossil record were complete, you could start with man and run back to a protist still in the species *Homo sapiens*. Such classification is manifestly both useless and somehow wrong in principle. Certainly the lineage must be chopped into segments for purposes of classification, and this must be done arbitrarily . . . because there is no nonarbitrary way to subdivide a continuous line. This is simple enough and fully justified on the taxonomic principles already expounded [those of the new systematics and evolutionary classification]. In practice all that is needed is some criterion as to how large (and in what sense of "large") to make the segments. The following criterion is sensible and is accepted by almost all evolutionary classifiers of fossils. (All paleontologists are of course aware of the fact of evolution, but not all are evolutionary taxonomists.) Successive species should be so defined as to make the morphological difference between them at least as great as sequential differences among contemporaneous species of the same group or closely allied groups.

Thus Simpson has returned to a morphological, phenetic, or typological species concept in which species are distinguished (diagnosed or defined) not on the basis of distinctness but on the basis of degree of difference.

When a supposedly continuous lineage is broken up into arbitrary segmental species, as described by Simpson in the quotation above, these species are often referred to as *successive* or *successional* species, allochronic species, chronospecies, or paleospecies (thus a species taxon based on fossil material is not necessarily a paleospecies). Under this concept, when the entire population of species A evolves into species B, species A is said to have undergone taxonomic or pseudoextinction.

Note that Simpson (1961, p. 153, quoted above) in defining his conception of an evolutionary species states that among its attributes is that it has "its own unitary evolutionary role and tendencies," yet he does not utilize this aspect of the evolutionary species in the quote given directly above. If we consider the evolutionary roles and tendencies of the organisms involved, then man and a protist have very different roles and tendencies. Wiley (1978, 1979a) in particular has followed up on this latter line of reasoning.

Under this concept [a revision of the evolutionary species concept of Simpson], a species is essentially a lineage of ancestor-descendant populations with its own identity and evolutionary fate. Species originate when they split from their ancestral lineage and persist until they become extinct via extinction of the lineage or splitting of the species into two or more daughter species which have evolutionary tendencies that are different from the ancestral species (see Wiley, 1978 for examples of nonextinction at splitting). One logical corollary of this concept is that lineages cannot be subdivided into a series of so-called ancestor-descendant "species"; one cannot define species as lineages and then call a particular species only a part of a lineage. (Wiley, 1979a, p. 216)

Of course Wiley's conception of the evolutionary species fits the cladistic assertion (see, for instance, Bonde, 1975; Hennig, 1966a) that "genealogical descent is composed of continua (lineages/species) and splitting of those continua" (Wiley, 1979a, p. 216). In regard to the infinite regression proposed by Simpson, Wiley (1979a, p. 216, italics in the original) suggests that

> as long as we classify by nature's own history of lineage splitting and not the psychologism we impose on nature, no such infinite regression will occur. This is because (a) higher taxa arise concurrently with their own ancestral stem species (Hennig, 1966a; Wiley, 1977) and (b) classifications document past continua as long as they are based on monophyletic taxa (*sensu* Hennig, 1966a). Although it is true that one can run from man to protist in one taxon, that taxon would be Eucaryota, not *Homo sapiens*. There was a clade *Homo* before there was a *Homo sapiens*, just as there was a Eucaryota before any Recent protist or *Homo sapiens*.

As is evident from the quote given previously, Wiley believes that his concept of the evolutionary species "does not preclude a particular ancestral species from surviving a speciation event" (Wiley, 1981, p. 34; see also Hull, 1979). For example, a small, peripherally isolated population may give rise to a new species without disrupting the homeostatic equilibrium or evolutionary role of the majority population, which then remains as the coexisting parental species. On the other hand, the entire population of a species may divide into two large, subequal subgroups, each of which diverges from the evolutionary role and tendencies of the original population (parental species); the parental species thereby ceases to exist. In contrast, Hennig (1966a, p. 63) would delimit any species "by two successive speciation events" and thus as a methodological rule considers the ancestral species to go extinct at a speciation event. Hennig's species concept can be considered the strict cladistic species concept. Bonde (1977, p. 754) labels Hennig's (1966a) concept of a species a *time-bio-species*.

Another way to think of species is as minimal monophyletic groups, as terminal taxa on *a* cladogram (but not any cladogram), or as the lowest natural categorical rank in the Linnaean hierarchy (cf. Eldredge and Cracraft, 1980, p. 88). Of course, some would argue that all three of these concepts could apply equally well to subspecies or varieties; but note that the concept of phylogenetic relationship and the construction of cladograms probably do not validly apply to nondiscrete entities below the species level, such as subspecies that readily hybridize (Bonde, 1977; see discussion in the introduction). And there is the very real problem of whether the concept of monophyly is even applicable to species. Wiley (1979a, p. 214, italics in the original) argued that species "are *a priori* monophyletic by their very nature," but Wiley (1981) and Eldredge and Cracraft (1980) view the notion of monophyly as inapplicable to species. My own preference is to regard a species-level taxon operationally (I do not see any strict distinction between species and subspecies that do not readily interbreed) as a minimal monophyletic group (that is, defined by one or more autapomorphies; in the case of fossil-based organisms, the apomorphies are usually morphologically recognized character states, but they need not be, especially in the case of extant organisms) or as a minimal plesiomorphous sister taxon of a monophyletic group (that is, there are no smaller groups of individuals within the taxon that can be united on the basis of synapomorphies relative to any other individuals included in the taxon—cf. Wiley's [1979a, p. 214] suggestion that a species does not have to be justified as monophyletic by synapomorphies because, given Wiley's view of species, all real species are, a priori, monophyletic).

Thus, on an operational level, I consider species-level taxa to be strictly defined minimal monophyletic groups whenever possible, the terminal taxa on a cladogram that starts with individual organisms (i.e, the smallest possible groupings, based on synapomorphies, of the individual organisms; some of these may be plesiomorphous relative to the remainder of the cladogram—that is, minimal plesiomorphous sister taxa), and the lowest categorical rank of the Linnaean hierarchy. (Cf. Nelson and Platnick, 1981, p. 12, who state "species are simply the smallest detected samples of self-perpetuating organisms that have unique sets of characters. As such, they include as species the 'subspecies' of those biologists who use that term." Likewise, Cracraft, 1983b, p. 170, defines his "phylogenetic species" as "the smallest diagnosable cluster of individual organisms within which there is a parental pattern of ancestry and descent.") Theoretically, I believe that such operationally defined taxa will most closely resemble Wiley's concept of the evolutionary species. Species are recognized by unique combinations of apomorphy and plesiomorphy (a true ancestor, or stem species, will be distinguished by lacking the apomorphies of its direct descendants), which in turn are indicative of the identities and evolutionary tendencies of the species involved. Even on a theoretical level, I reject Mayr's biological species concept, which is based explicitly on the criterion of interbreeding or not interbreeding. Likewise, I must reject a species definition such as that proposed by Eldredge and Cracraft (1980, p. 92, cited above) for they too rely in part on Mayr's concept of the biological species when they state that beyond the cluster of individuals composing a certain species there is no "parental pattern of ancestry and descent" (i.e., there is no interbreeding between species).

Thus far we have discussed several theoretical and operational definitions and concepts of species. A related topic concerns the nature and ontological status of species. Are they merely classes of objects, as many philosophers have traditionally considered them to be (see above), or are they individuals?

Recently a number of philosophers of biology have come to regard species as historical entities and the units of evolution (Ghiselin, 1966b, 1969, 1974; Griffiths, 1974; Hull, 1974, 1975, 1976a, 1976b, 1978, 1980, 1981; Löther, 1972; Mayr, 1976d). Genes are mutated and organisms are selected (although it can, and has been, argued that selection also takes place above and below the level of the organism: see Ridley, 1984, and articles in Sober, 1984b): the end result of these two processes is not that individual organisms evolve, but that species evolve. In this context the term *species* is applied to the "effective units of evolution" or the "supraorganismic entities which evolve" (Hull, 1978; reprinted in Sober, 1984b, p. 624). As we have seen, species can be viewed as lineages of ancestors and descendants that are internally cohesive—that is, copies (and copies of copies) of an organism through time. The individual organisms that compose a species at one instant in time may differ from the individual organisms that compose the species at another point in time (analogously, an individual may be composed of different cells and look very different at two different stages in its life cycle), yet the species remains the same species.

Further, Eldredge and Gould (1972; Gould, 1982; Gould and Eldredge, 1977) have argued that species generally form well-buffered, epigenetic, homeostatic systems that have an amazing stability even under the pressure of external disturbing influences. This coherence of a species is a historical consequence of the origin of a species and may not necessarily be (or need to be) maintained by gene flow between individuals composing the species. According to the same theory (Eldredge and Gould, 1972; Mayr, 1963), new species usually form by the separation of a small number of individuals (a peripherally isolated population) in a possibly new environment that then reaches a new homeostatic equilibrium. Also contributing to the unity and coherence of a species might be relatively constant selection pressures on the members of the species (Ehrlich and Raven, 1969). From these bases it can be concluded (perhaps must be concluded) that species are "spatiotemporally localized cohesive and continuous entities (historical entities)" rather than "spatiotemporal unrestricted classes" (Hull, 1978; reprinted in Sober, 1984b, p. 624; note that Hull, ibid., p. 627, is careful to exclude "from the notion of class those 'classes' defined by means of a spatiotemporal relation to a spatiotemporally localized individual"). As Wiley (1981, p. 23) states, species can be considered "to be natural units in the biosphere that can be objectively studied and have certain characteristics." Furthermore, according to this view, although any particular species is an individual, the taxon concept species is a class concept and a universal (Wiley, 1978, 1980, 1981). That is to say, the taxonomic concept of species is the class composed of all particular species (individuals). If life has evolved elsewhere in the universe independent of life on earth, there may be supraorganismic entities that we would recognize as species and that would fit (be members of) our taxon concept of species.

As noted in the last paragraph, some workers (e.g., Eldredge and Gould, 1972; Gould and Eldredge, 1977) believe that most species are normally found in a state of homeostatic equilibrium; such equilibrium breaks down only during relatively rare speciation events. In contrast, other investigators (e.g., Bonde, 1977, p. 781; Hennig, 1966a) have stressed a *dynamic species concept* that suggests that species are rarely found in homeostatic equilibrium in nature. The temporal durations of segments of lineages representing good or true species may be relatively short as compared to the amount of time it takes for speciation to occur. Perhaps many species persist for long periods of time in a state of speciation. In instances where this were the case, it might mean that often the allocation of individual specimens to one species or another would be difficult or problematic. However, any species would still be an individual—a spatiotemporally localized historical entity with a beginning (at the speciation process), continuity, and cohesion through time via ancestry and descent, and an end (at the extinction of its last included member or speciation into new species).

I believe that by the same reasoning used to deduce that species are individuals, any monophyletic (holophyletic) taxon must be an individual (cf. Patterson, 1982a, who came upon this conclusion independently). In contrast, Wiley (1981, p. 75) considers species individuals because they are units of evolution with a reproductive cohesion, but he considers higher taxa to be merely individual-like historical groups that are units of history that contain and are derived from individuals, but are classlike in not participating in natural processes. Any taxon originates with the origination of its stem species and goes extinct when the last descendant of the stem species goes extinct. Any monophyletic taxon is, if anything, a spatiotemporally localized entity. But following this last line of reasoning a bit further, it seems to me that perhaps it is only monophyletic taxa that are true individuals; and as some species may not be monophyletic—that is they are true ancestral species—perhaps they should not be considered individuals (arguments for the epigenetic coherence of a species aside). I believe monophyletic groups may be the only real or natural entities in nature; some wholly plesiomorphic named species-level taxa (minimal plesiomorphic sister taxa of monophyletic groups) may be artifacts (although perhaps necessary artifacts) of our nomenclatural system. To have a truly natural system (i.e., one that recognizes only monophyletic taxa), perhaps some organisms (those that are true ancestors—not to imply that such could necessarily be recognized; see chapter 4) should not be assigned to species-level or other low-level taxa. For example, the only monophyletic taxon (relative to other vertebrates, not to other organisms in general) to which the common ancestor of all vertebrates could be assigned would be a taxon containing all vertebrates, commonly termed Vertebrata. If we assign the common ancestor of all vertebrates to a species, genus, family, and so on, then this species, genus, or family would be paraphyletic unless we were to include all other vertebrates within the same group (species, genus, or family, and so on).

If species, or any taxa, are individuals, it also follows that once a given species goes extinct, it can never reappear. Even if organisms that were identical to the organisms of the species were to arise, these new organisms would be a second species because they would have a different spaciotemporal localization and also would not share continuity and internal cohesiveness with the first species.

Wiley (1978, 1981) has listed several corollaries of his evolutionary species concept that, although they may seem self-evident, are worth mentioning as his species concept forms the basis of the species concept that I prefer. First, all organisms belong (or, in the case of extinct organisms, belonged) to a species. Unlike some who would argue otherwise (e.g., Dobzhansky, 1937), there are no organisms that do not belong to a natural species; furthermore, higher (supraspecific) taxa arise concurrently with their stem species, not before (*contra* Løvtrup, 1973, 1977*a*, 1977*b*) or after (*contra* Simpson, 1961) due to subsequent evolution of the species or its descendants.

Second, contemporaneous species will be isolated reproductively from each other to the extent that it is necessary to maintain the separate identities and tendencies of the species. In some cases what were good species, or perhaps only incipient species, may unite and hybridize, and either one species may absorb the other species or the two may unite to form a third (thus evolution may not occur by strictly dichotomous branching).

In some cases it may not even be necessary for good species to be reproductively isolated; they may be sufficiently isolated by other barriers, such as geographic ranges, behavior, or ecology. As noted above, Rosen (1979) has described a case where the interbreeding between two species of swordtail fishes (Teleostei) is apparently a plesiomorphous (retained primitive) character; thus two species that can interbreed are not necessarily sister taxa. In the case of the swordtails, one of the interbreeding species shares synapomorphies with (and thus is more closely related to) another species with which it cannot interbreed.

Third, true species (perhaps as perceived by the organisms themselves) may not be recognizable to the investigator morphologically (but that does not mean that such sibling species, cryptic species, or semispecies are any less real species). On the other hand, a single species may be composed of two or more morphs (or phena) that may be mistaken for distinct species. Thus, an investigator may overestimate or underestimate the number of species that he or she is dealing with.

Finally, Wiley (1981, p. 34) argues that "no presumed, single, evolutionary lineage may be subdivided into a series of ancestral and descendant species." But I would argue that a lineage is not necessarily synonymous with a single species. As long as a lineage maintains its unique identity and evolutionary tendencies, it should be considered a single species, but without undergoing splitting a single lineage (that is, the whole population of the species) may undergo a shift in its evolutionary role, identity, and tendencies (this should be manifested by the development of apomorphies). Thus, the lineage could transform into a different, recognizable species. The important point here is that no continuous lineage that is characterized by a constant evolutionary role, identity, and set of tendencies and is undergoing anagenesis should be arbitrarily divided into successive species as advocated by Gingerich (1976a, 1976b, 1979) and Simpson (1961).

PRACTICAL SPECIES RECOGNITION

Theoretically, as I regard species-level taxa, one should perhaps begin with individual organisms and then recognize as species those smallest units of identical and nearly identical individuals that group together as terminal nodes on a cladogram. Such units or groupings would be distinguished from one another by unique combinations of apomorphy and plesiomorphy. One immediate problem, however, is to identify and separate species-level character states (those that are synapomorphies and symplesiomorphies—*shared* derived and *shared* primitive character states) and traits from those that are merely due to relatively random or unique intraspecific variation (i.e., individual derived [or primitive?] character states that are not *shared*—passed on to later generations of the species; and ecophenotypic variation), perhaps intraindividual variation (e.g., different life stages in otherwise identical individuals) or some form of polymorphism within a single population. Even if I do not accept Mayr's biological species concept and the criterion of interbreeding as dogma, I do believe that natural species will have a continuity of parent-child ancestry and descent within themselves (but not necessarily exclusively with themselves). Thus, I would not regard the males and females of a sexually reproducing population to be two different species, no matter how different they might be. Rather, the males and females must be regarded as two stages or forms of the same species, just as the caterpillar and moth must be regarded as two stages or forms of the same individual. In paleontology (and in practice the same is often true in neontology) many species are based initially on only one part of an organism; this adds further complications as we strive to establish the true biological entities. This problem is especially acute in paleobotany where the leaves, fruits, and roots of a single fossil organism may all bear different Linnaean names (MacGinitie, 1969). Likewise, as stated above, species are unique spatiotemporally restricted individuals: Once a species goes extinct, it can never reappear even if an organism were to appear that was absolutely indistinguishable from the species and could even have interbred with the species.

With these considerations in mind, it is beneficial to examine extant populations of organisms in order to get at least a sense of the difference between what previous taxonomists have considered individual (and other infraspecies-level) variation and species-level traits. This subject has been examined in detail by a number of workers; two major reference books on the subject are

Mayr (1969) and Simpson, Roe, and Lewontin (1961). It is beyond the scope of the present work to pursue this subject in detail; we can only hope to touch on a few of the problems. Ultimately, there are no easy answers to many of the problems of species-level taxonomy. In many cases the best decisions about the limits of species-level taxa are made by the experienced taxonomist who has spent considerable time studying one particular group; unfortunately there is still an element of art to taxonomy at the species level. It must always be kept in mind that any statement that a particular individual belongs to a particular species is only an hypothesis, never a fact (Bonde, 1977). Fortunately, however, the precise delimitation of species-level taxa is not critical to phylogeny reconstruction (in the sense of recognizing sister groups). What is critical is that the initial taxa under study—the terminal nodes on the cladogram—be monophyletic. They will be monophyletic if they are individual organisms or groups of organisms united by synapomorphies.

Most taxonomists will begin by sorting out organisms into groups of identical or nearly identical individuals, discrete diagnosable clusters (Eldredge and Cracraft, 1980, p. 95), or what Mayr (1969, p. 5) refers to as *phenotypically reasonably uniform samples* and has designated phena (singular, phenon: the term was originally introduced by Camp and Gilly, 1943). The grouping of organisms into initial phena need not, however, be based solely on intrinsic data, such as morphological criteria; extrinsic data may also be utilized. Thus the paleontologist, when establishing initial phena, may keep as distinct working phena those individual organisms of similar morphology that originated from different localities or stratigraphic horizons. These phena are the "first cut" at a species-level grouping of the taxa. Further analysis often results in combining some phena into a single species-level taxon (e.g., males and females of a single population may represent distinct phena), but in some instances a single phenon may represent more than one species-level taxon (such as in the case of sibling species that are not readily recognizable as distinct; among ammonites in particular homeomorphs may be easily confused with one another; see Kennedy, 1977; Kennedy and Cobban, 1977).

Since within any given species-level taxon there is usually a pattern of parental ancestry and descent, the general rule has arisen among taxonomists that species-level taxa should generally exhibit a relatively continuous distribution of intrinsic and extrinsic properties, while there should be relative discontinuities between species. Specifically, if there is continuous and finely graded morphological variation among a number of individual organisms, these will generally be considered to belong to a single species barring any other evidence to the contrary. Likewise, a single species should have a fairly continuous range both geographically and temporally (through a stratigraphic column); as the size of disjunctions in the range of a potential species increases, the more likely it will be considered to be two or more distinct species. Of course, when dealing with paleontological materials one must take into account the notorious incompleteness of the fossil record; an apparently discontinuous range may actually have been continuous during the lifetime of the species (see chap. 4).

However, as has been pointed out numerous times (see Eldredge and Cracraft, 1980, for a recent review), continuity of features (whether intrinsic or extrinsic) observed by the taxonomist is no guarantee that one is dealing with a single species. Likewise, discontinuity of features does not necessarily indicate that more than one species is present.

In dealing with fossil organisms, where one is usually restricted to exclusively morphological data, it may be impossible to recognize sibling species that are morphologically indistinguishable. Among recent organisms, sibling species have most often been discovered in the course of medical, genetical, or cytological studies rather than during the course of routine taxonomic analysis (Mayr, 1969). These sibling species are usually "good" species that differ from more easily recognized species only in that they have not undergone a great amount of morphological differentiation. Among animals, many such sibling species are found among certain insects. Mayr (1969, p. 183) has noted that "once discovered, and thoroughly studied, sibling species are usually found to have previously overlooked morphological differences." A somewhat different case of sibling species from that discussed above is the problem of polyploids (Mayr, 1969, p. 185)—forms that are closely related to normal diploid forms but have three or more times the haploid number of chromosomes. Such polyploids may be reproductively isolated and morphologically indistinguishable from each other and normal diploid forms. Such forms are relatively rare among animals, but more common among plants.

When relatively incomplete fossil material is under consideration, it can often be the case that several species will be grouped into one phenon simply because the only parts preserved in the record of the particular organisms are not diagnostic at the species level. Thus in life two species of mammals may have been easily distinguished by their coat patterns and limb proportions, but as fossil remains they may be known only from their teeth, which are indistinguishable. Similarly, isolated spores, pollen, roots, stems, and leaves, none of which are diagnostic in and of themselves at low taxonomic levels, and none of which can be unequivocally associated into true botanical entities (whole organisms), may be known in paleobotany. Knowingly or not, different systems of taxonomic nomenclature may be set up for different parts of the same organisms. Paleobotanists routinely utilize *form-genera* and *organ-genera* that strike me as nothing more than convenient wastebasket categories.

One may periodically encounter cases where a set of organisms is characterized by morphological continuity on a very fine level with no discontinuities, but the extremes of the series are very different. If the entire series of morphs is restricted to a single time and place, it may represent an ontogenetic series, but if the morphological gradation correlates with the geographic and temporal distribution of the morphs (i.e., clines are recognized), then something else may be represented instead. The taxonomist may wish to formally recognize various of the different morphs, but may not wish to label them *species* and be forced into a situation where the species are not distinct, but grade into one another. Such consistently recognized morphological variations may be given subspecies status. How much difference is necessary to recognize distinct subspecies? Mayr (1969, pp. 190) states:

A so-called 75-percent rule is widely adopted. According to this, a population is recognized as a valid subspecies if 75 percent of the individuals differ from "all" (= 97 percent) of the individuals of a previously recognized subspecies. At the point of intersection between the two curves where this is true, about 90 percent of population A will be different from about 90 percent of the individuals of population B (to supply a symmetrical solution).

Two or more phena may pertain to a single species. Such phena may be the different parts, different sexes (sexual dimorphism), ontogenetic stages (individual variation with time), allometric variation within a species, or ecophenotypes of a single species. In the case of extant organisms it may be possible actually to observe the whole organism through its life cycle. Thus males and females will be observed to breed, and the various ontogenetic stages will be demonstrated as pertaining to a single individual. Furthermore, organisms from the same interbreeding population may be experimentally grown under various environmental conditions such that it will be possible to distinguish variation induced by the habitat (ecophenotypic variation) from that with an underlying, intrinsically based, genetic cause. It seems that virtually any external influence will cause some morphological variation in some organism. To cite just a few examples, the forms of mollusk shells are often affected by the substrate and water conditions in which the organisms grew; the various forms of a single species of parasite may be determined by the hosts upon which the individual organisms plant; and various morphological features of foraminifera (including their direction of coiling) can vary with the temperature and salinity of the water in which they mature (for references on ecophenotypic variation see Dodd and Stanton, 1981; Hermelin and Malmgren, 1980; Kinne, 1963, 1964, 1971; Malmgren, 1984; Mayr, 1969; Werdelin and Hermelin, 1983). Medioli and Scott (1983, p. 11) have noted that among a certain group of predominantly uniparental shelled amoebas, the Arcellacea (thecamoebians), there is an enormous amount of phenotypic plasticity such that "considerable phenotypic overlapping may exist (especially in natural assemblages) between different genotypes." The same authors (Medioli and Scott, 1983, p. 5) also point out that

in the past many arcellacean species and varieties have been defined according to the nature of their xenosomes (foreign agglutinated particles). While the shape of idiosomes (test particles secreted by the organism) may be a valid taxonomic characteristic, in most cases the nature of the xenosomes depends on the availability of inorganic particles and not on a genome-based selectivity.

One must also be wary of accidental, teratological, pathologic, traumatic, and post-mortem variations and changes of individual organisms (Mayr, 1969).

The paleontologist must depend to a great extent on inference and analogy with extant organisms in coming to decisions that two or more recognizable phena actually pertain to a single species-level taxon. In some cases it may be possible to find intermediates between juvenile and adult fossil forms such that an ontogenetic series can be pieced together and thus several phena synonymized as one species-level taxon. In some cases two distinct phena may always be found together, perhaps always in the same proportions; based on analogy with closely related modern forms, it may be reasonable to presume that the two phena are the males and females of a single species. In the case of ecophenotypic variation, it may be possible to show that certain phena of what appears to be a set of closely similar phena always occur exclusively in certain paleoenvironments. Furthermore, intermediate paleoenvironments may contain what appear to be intermediate phena. Such observations would lend weight to the argument that the phena pertain to ecophenotypes of a single species.

Whenever one attempts to arrive at species-level decisions, it is often extremely useful to use some form of quantitative or statistical analysis of numerical or metric data. Such methods are well covered in Simpson, Roe, and Lewontin (1960), Sneath and Sokal (1973), and Reyment, Blackith, and Campbell (1984). The reader is referred to these works for further information on this topic. The following are just a few elementary statistics of univariate analysis (dealing with a single character at a time) that are quite useful in all forms of taxonomic work.

Perhaps the simplest and most basic statistic is the arithmetic average or mean, M or $\overline{X}$. The mean of any given sample of values is simply the sum of the observed values (X) divided by the number or observations (N):

$$M = \Sigma X/N.$$

For most biological phenomena or features pertaining to a single species, the frequency curve of values will correspond roughly to a normal distribution (Mayr, 1969); the mean is a measure of the central tendency of the distribution (Simpson, Roe, and Lewontin, 1960) around which most of the values will congregate.

The variance and standard deviation are measures of the variability of the values of the observations—how much they tend to differ from the mean. If the variance and standard deviation are large, there will be a great amount of scatter in the values. In contrast, if these are both zero, all of the observations have the same value. The variance of a sample population is given by the formula:

$$s^2 = \Sigma(X - M)^2/(N - 1).$$

The variance for a complete population, rarely used in biological data, is:

$$s^2 = \Sigma(X - M)^2/N.$$

The standard deviation is the square root, s (also represented by SD), of the variance. For a normally distributed data set, approximately 68.3% of the observed values will fall within one standard deviation of the mean, approximately 95.5% of the observed values will fall within two standard deviations of the mean, and approximately 99.7% of the population will fall within three standard deviations of the mean.

From the last paragraph, it should be obvious that the numerical value of the standard deviation is meaningful only in relation to the value of the corresponding mean. In the taxonomic literature the standard deviation is usually expressed as a percentage of the mean; this statistic, the coefficient of variation or coefficient of variability (commonly represented by CV or V), is given by the formula:

$$CV = (s \times 100)/M.$$

The value of the CV is a pure number (the units of the numerator and denominator cancel out one another), and thus CVs are often and readily compared across taxa. For certain measurements in certain groups of organisms, it is generally accepted that the CVs will usually fall below a certain value if only one species is represented in the sample. An abnormally high CV may indicate that more than one species is present.

Utilizing the concept of the coefficient of variation, A. L. A. Johnson (1981) described a model and potential methodology for determining whether two or more closely similar morphotypes represent ecophenotypes or genetically differentiated entities. Unfortunately, when he applied his methodology to a species of Jurassic scallop, the results were inconclusive (perhaps due to a problem in sampling). Werdelin and Hermelin (1983) applied Johnson's methodology to two variants of the extant, chambered, benthic foraminifer *Reophax dentaliniformis* and interpreted the results as indicating that the two forms are genetically differentiated and may represent distinct demes or subspecies (they are not considered distinct species by Werdelin and Hermelin because intermediate forms are known between the extreme variants).

Werdelin and Hermelin (1983, pp. 304–305) have succinctly summarized Johnson's (1981) assumptions and methodology:

Johnson's method takes as its basis the ideal developmental pathways of developmentally flexible and canalized species. In a developmentally flexible species, he hypothesizes, morphology should during ontogeny tend towards a single adaptive type, whereas this should not be so in developmentally canalized species. Thus, the coefficient of variation (V) should tend to decrease during ontogeny for a developmentally flexible species, but should tend to remain constant for a developmentally canalized species. Therefore, if we find two morphotypes which deviate in mean size (or some other criterion) during ontogeny, and whose coefficients of variation decrease during ontogeny, we may be justified in viewing them as ecophenotypes (ontogenetic deviation directly influenced by environment). If, on the other hand, V in these morphotypes remains constant during ontogeny, we may be justified in viewing their differences as being genetically controlled.

In studying the two variants of *Reophax dentaliniformis*, Werdelin and Hermelin (1983) found that they are similar early in ontogeny but gradually deviate as they add additional chambers. This first result is what would be expected if the two variants were true ecophenotypes. However, when these authors plotted the coefficients of variation against ontogenetic age for each of the two variants, they found that the CVs remain nearly constant throughout ontogeny. This second result suggests that the observed differences between the two variants are genetically controlled.

Finally, Werdelin and Hermelin suggest that Johnson's method does not detect ecophenotypic variants per se, but rather is useful in determining which variants are genetically determined and which variants are ecologically determined. They argue that Johnson's two criteria (population divergence during ontogeny; decrease in CVs within populations during ontogeny) are necessary but not necessarily sufficient for the absolute identification of ecophenotypic variants. If both criteria are met, the variants under consideration may be ecophenotypes, but they need not be. However, if one of Johnson's two criteria is not met, the variants cannot be ecophenotypes.

DYNAMICS OF FOSSIL POPULATIONS

Between the level of the holomorph and the levels of subspecies or species lie the levels of demes and populations. The study of population dynamics has been a concern of much modern ecology (e.g., Pianka, 1974; Ricklefs, 1979). Although what can be done with fossil organisms is extremely limited, many paleoecologists have also been concerned with population dynamics of fossil organisms (see summaries in Dodd and Stanton, 1981; Hallam, 1972; Raup and Stanley, 1978). For the purposes of phylogenetic analysis, population studies of fossil material may help determine when a natural, interbreeding population is preserved in a sedimentary deposit. The catastrophic preservation of a putative single population might yield information on the intra- and interspecific variability of certain morphological (taxonomic) characters. Population studies also have the potential to yield additional, nonmorphological characters that may be useful in phylogenetic and taxonomic analysis and in the construction of evolutionary scenarios (such as seasonal spawning patterns, mortality patterns, or growth rates).

A deme, local breeding population, or more simply, a population is a group of individuals living in close enough contact such that they can interbreed freely. Any single species at a certain instant in time may be composed of one or more demes. Sometimes the term *population* may be used in a broader sense than a deme to refer to several closely situated demes of a single species in a certain geographic area. In paleontology the most we can normally hope for is to sample a deme via a death assemblage that is representative of the original life assemblage. Such a death assemblage may be produced catastrophically (e.g., by a mud slide instantaneously burying a surface covered with living sessile organisms) and be preserved in the rock record as a single bedding plane of undisturbed, unabraded, articulated organisms in life position.

As Hallam (1972, p. 63) points out, "the basic data of population dynamics relate to the age of organisms and their rates of growth, mortality and recruitment." To this can be added the rate of reproduction. In many cases such information can be gleaned from the fossil record.

For many fossil organisms it is possible to determine the individual's age, either absolutely or relatively. Concentric growth rings, often (though not necessarily) annual or seasonal in nature, are present on certain hard parts of organisms such as bivalve shells, fish scales, and genital plates of echinoids and in tree rings (Hallam, 1972). Relative wear of hard parts, such as mammalian teeth, can also be used to determine the age of an organism. Growth-rate curves are determined simply by plotting age on the abscissa against some measure of size (usually overall body size) on the ordinate (Fig. 2-5). The growth curve will be more or less convex up if the organisms are characterized by a high initial growth rate (apparently the common situation in most organisms), or more or less convex down if the organisms are characterized by a low initial growth rate.

There are two types of mortality that are commonly distinguished: census and normal (Hallam, 1972; Kurtén, 1964). The census type is catastrophic mortality where an entire living assemblage is suddenly wiped out and an identical death assemblage is created. Normal mortality is the gradual dying off of individuals as the result of old age, disease, accidents, and/or predation. In an environment conducive to fossilization, the bodies of the dead organisms may slowly accumulate. In most cases it would be expected that death assemblages resulting from normal mortality will be very different in composition from those that are a result of census mortality.

Rates of mortality are usually expressed in terms of the percentage of a population that dies each year, and the data are presented in the form of survivorship curves and life tables. Life tables (Deevey, 1947) list for a certain species or population the number of survivors after certain intervals of time given an initial population (in ecological terminology, a cohort—the individuals of a population that are born at essentially the same time), the mortality rate, and the average life expectancy. Essentially the same information is graphically depicted in survivorship curves. A survivorship curve is a graph of the number of survivors of a particular cohort (plotted on the ordinate) versus the time since the origination of the particular cohort (plotted on the abscissa). Often the ordinate is expressed as the number of individuals out of an initial 1,000, but it may be any other arbitrary number, or it may be the percentage of surviving individuals (Dodd and Stanton, 1981). The abscissa may be in any convenient time unit, including (for useful comparisons between species) percentage of the average or maximum life span of the species under study. The slope of the resulting survivorship curve at any point is proportional to the mortality rate at the corresponding cohort age. The ordinate is commonly plotted on a logarithmic scale such that a straight-line plot results for the case in which the mortality rate is constant at all ages (Fig. 2-6b).

Survivorship curves that are convex up (Fig. 2-6a) indicate that the mortality rate increases with age. Such survivorship curves are seen for a number of higher vertebrates. In contrast, survivorship curves that are concave up (Fig. 2-6c) indicate that mortality is high at the outset of a cohort but decreases with age. Such curves are characteristic of many invertebrates, especially where larval stages experience extremely high mortality rates. Likewise, some plants that disperse many seeds, few of which ever reach maturity, may show the same pattern.

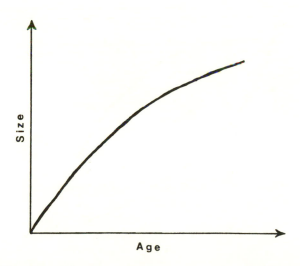

Figure 2-5. Schematic growth-rate curve; age on the abscissa is plotted against some measure of size on the ordinate.

As individuals die, decreasing the number of organisms making up a standing population, the organisms may be replaced. Such replacement may take the form of the birth of new individuals from members within the standing populations or of the recruitment of new members of the species from outside to within the area of the local population. Local population growth or decline can be plotted in terms of total number of individuals in a given area, or density of individuals per unit of area, against time (Fig. 2-7). Various patterns of population change with time can be recognized (Dodd and Stanton, 1981), but two in particular appear to be commonly encountered in nature, although often in combination or in modified form: *J*-shaped patterns and *S*- or sigmoidal-shaped patterns (Odum, 1971). *J*-shaped patterns result when a given population starts small, grows rapidly, and then decreases abruptly; the pattern might be repeated numerous times for a given population. Such a pattern is associated with favorable conditions in an area that is wide open to the species in question. But as the species increases in numbers, the conditions become less favorable for the organisms. The changing condition may be a changing physical factor that is independent of the size of the population (such as a change in the external environmental regime) or it may be brought about by the expanding population itself (such as an overutilization of resources: cf. Schoch and Meredith, 1984). A sigmoidal-shaped curve may result when a population starts small, at first increases gradually, and then accelerates in its growth rate, but as the upper limit of the number of individuals that the particular area can maintain is reached, the growth rate decreases and finally reaches some equilibrium value.

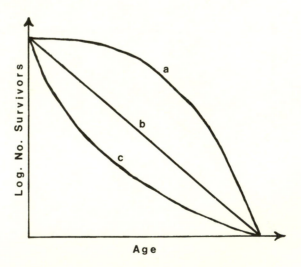

Figure 2-6. Schematic survivorship curves: *(a)* convex up, mortality increases with age; *(b)* straight, constant mortality; *(c)* concave up, mortality decreases with age.

These common types of growth patterns have been mathematically modeled (see Hutchinson, 1978, for the history of development of such models). In the simplest case, a *J*-shaped population growth with no upper limit can be described by the following differential equation of exponential growth (assuming that the growth of the population is proportional to the initial size of the population; Hallam, 1972):

$$dN/dt = rN$$

where N is population size, t is time, and r (constant) is the intrinsic rate of increase of the population (essentially birth and recruitment rate minus death rate). In integrated form this equation is:

$$N_t = N_0 e^{rt}$$

where N_t is the population size at time t, N_0 is the initial population size ($t = 0$), r is as before, and e is the base of the natural logarithm.

The primary drawback of the above equation is that it predicts astronomical population sizes in relatively short periods of time. In other words, it does not take into account any limiting factors that will affect the growth rate of a population; if no other factors limit the growth rate, eventually some form of density-dependent factor will do so (Dodd and Stanton, 1981, p. 343). Consequently, the following general equation was developed that takes into ac-

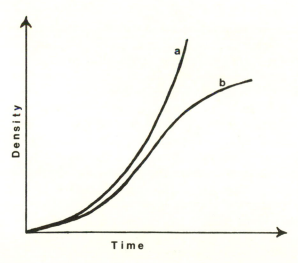

Figure 2-7. Schematic representation of local population growth patterns: *(a)* *J*-shaped pattern; *(b)* *S*-shaped pattern.

count density-dependent factors and approximates the sigmoidal-shaped curve discussed above:

$$dN/dt = rN((K - N)/K)$$

where K is the equilibrium limit or carrying capacity that a certain population size approaches but cannot exceed (the other symbols are as before). As K approaches N, the rate of population growth, dN/dt, approaches zero.

MacArthur and Wilson (1967) have distinguished between environments that are r-selective and environments that are K-selective. R-selection favors organisms that have the ability to rapidly increase their population size under favorable conditions. Such r-selected species have been termed opportunistic species; such species may show a pattern of rapidly increasing in numbers, then crashing, followed by another rapid increase in numbers. In contrast, K-selection favors those species that are able to maintain a large carrying capacity. K-selected species have been termed equilibrium species; such populations are in relative equilibrium with their environment and show relatively little variation in numbers of individuals over time.

It is somewhat problematic as to how survivorship curves, and especially population growth curves, can be applied to fossil populations. According to Hallam (1972, p. 69),

for an assemblage of dead shells accumulated from a living population through normal processes of mortality, modal analysis of growth rings (Craig and Hallam, 1963) will give the necessary age data for the construction of a survivorship curve. Where growth rings are not clearly discernable, a "size-survivorship curve" can be constructed from the size-frequency histogram.

That is, for many fossil assemblages—say, brachiopods from a single bedding plane—all we know is that we have so many brachiopods in one size class, so many brachiopods in another size class, and so on (where the size classes can be in any units, perhaps simply all those specimens between 0 and 0.99 cm in length, between 1 and 1.99 cm in length, and so forth). A histogram of number of individuals in each size class can be generated, and from this a *size-survivorship curve* plotting numbers of survivors (all those of a certain size class or larger) against size class units. A size-survivorship curve can be converted into a standard survivorship curve by knowing (or assuming) the relationship between size and age of organisms (i.e., the growth curve).

Size-frequency histograms (or smooth curves derived therefrom) can also shed light in some cases on population growth and recruitment patterns of fossil organisms. Hallam (1972, p. 71) states that "as a general rule assemblages of dead shells derived directly from living populations tend to give rather smooth unimodal size-frequency distributions, always provided of course that the samples are suitably large." This observation might provide one test of whether one is dealing with a single biological population. The latter information is important for assessing intra- versus interpopulational variability of character states. However, normal, bell-shaped distributions of fossil clasts may be due to current sorting (Dodd and Stanton, 1981), and there may be other (sedimentological) indicators of current sorting, transport, or other biasing factors (see review in Hallam, 1972, pp. 76–78); one must be wary in posing biological interpretations. In contrast, bimodal or multimodal size-frequency curves may indicate seasonality or longer-term variations in recruitment, breeding, or mortality (Hallam, 1972). Dodd and Stanton (1981, p. 361) suggest that "the number of modes in polymodal distributions can give an estimate of the maximum age of individuals in the population."

In sum, although studies of the dynamics of fossil populations do not seem to be on the whole directly applicable to phylogeny reconstruction *sensu stricto,* such studies should not be ignored. Potentially they may yield valuable phylogenetic information. Furthermore, studies of fossil population dynamics assuredly add to evolutionary stories and scenarios that may be built heuristically around a strict phylogenetic analysis.

SPECIATION PATTERNS

Speciation is the mechanism by which lineages split and differentiate, thus producing new clades (initially represented by single species). More rarely, speciation may involve the joining of previously independent lineages to form a new lineage (this seems to occur primarily in some plants). Speciation accounts for the diversity of organisms, and the relative timing—the historical sequence—of different speciation events accounts for the varying degrees of relatedness between organisms: the phylogenetic pattern. Having a basic knowledge of potential and theoretical patterns of speciation can shed some light on what sorts of phylogenetic patterns might reasonably be expected in various cases. Of course, it is important not to force our phylogenetic interpretations to fit some set of a priori assumptions about the patterns and mechanisms of speciation. With these considerations in mind, the general patterns or modes of speciation are reviewed here, especially as they may be influenced by extrinsic factors. We will see in later sections that the sorts of speciation patterns accepted can greatly influence one's thinking in regard to historical biogeography and the formulation of evolutionary scenarios from cladistic hypotheses in particular. How far, and in what direction, to pursue a certain cladistic analysis may even be influenced by the sorts of speciation patterns adopted by an investigator for a particular group of organisms—for example, whether character-state conflicts should be accounted for by ad hoc hypotheses of parallelism or by pursuing the possibility of reticulation. In what follows the general patterns (primarily geographic) of various modes of speciation are reviewed; intrinsic mechanisms (genetic, epigenetic, and other; see, for example, Løvtrup, 1974; Mayr, 1963, 1970) that actually cause speciation to occur are not important for the subject of this book. It is assumed that all continuing species have the basic potential to speciate. What is important to the topic of this book is that given the ability to speciate, different groups in a certain geographic area may respond to extrinsic events with congruent patterns of speciation. Congruent patterns of lineage splitting can be responsible for similar patterns in the biogeographic distribution of organisms in independent clades (see chapter 6). Recent discussions of patterns and mechanisms of speciation can be found in Bush (1975), Endler (1977), V. Grant (1971, 1977), Lazarus (1983), Mayr (1963, 1970), White (1978a, 1978b), and Wiley (1981).

To the traditionally trained paleontologist, the first mode or conception of speciation that comes to mind may be the transformation of one species wholesale into another species, either gradually or in a short burst or series of bursts of evolution. This mode, termed *phyletic evolution* (Simpson, 1961) or *phyletic speciation* (Mayr, 1963), which gives rise to successional species, paleospecies, or allochronic species, has been mentioned elsewhere (see section on species concepts). While not denying phyletic character transformation, certain proponents of the evolutionary species concept have been strongly opposed to the notion of phyletic speciation. Specifically, Wiley (1978, 1979a, 1981) considers it a hindrance to phylogenetic, systematic, and evolutionary research. Wiley points out (as discussed in the section on species) that the recognition of successional species is arbitrary and thus must lead to arbitrary mechanisms and patterns of speciation. Real (i.e., caused by lineage splitting or possibly fusion) speciation and extinction events are confused with taxonomic speciation and extinction events (when one species is transformed into another through time without splitting or fusion). Taxonomic (or phyletic) speciations and extinctions will not necessarily be correlated with geographic or other extrinsic events. When the concept of phyletic speciation is used, speciation and extinction rates in various groups of organisms may reflect nothing more than how finely specialists working on various groups split up their taxa.

It must be pointed out, however, that even with Wiley's evolutionary species concept, speciation could conceivably (even if unlikely and never yet demonstrated) occur via a transformation mode, without any lineage splitting. The crux of the evolutionary species concept is that a species has its own identity, tendencies, and evolutionary role or fate. Conceivably, a single population (species) could have a certain evolutionary tendency, role, fate, and identity and with phyletic character transformation (accepted by most who reject phyletic speciation) could change its tendency, role, and so on. Who is to say when its role is changed enough to warrant identifying a new species? This seems to return us to the problem of arbitrariness. If we adopt the method-ological rule that splitting of lineages has to occur in order for change to be called speciation (and likewise, splitting always causes speciation), then any degree of change can occur within a single species as long as no daughter species splits off. But when this rule is invoked, a series of fossils, for instance, that when analyzed can be ordered in a sequence that shows the progressive accumulation of synapomorphies must be lumped into a single species if none of the less-derived forms have any demonstrable autapomorphies or syna-pomorphies with other forms outside of the considered series (that is, they are all wholly plesiomorphous relative to the more derived forms under con-sideration) no matter how much character transformation is recorded between end members of the series. Once an autapomorphy has been found in one of the intermediate forms in the series, then the putative evolutionary species can be broken up into two species because suddenly a splitting event (spe-ciation event) has been demonstrated. In reality, autapomorphies may have been present in all of the plesiomorphous forms when they were living, but are just not preserved in the fossil specimens. Or it may be that by chance only totally plesiomorphous forms were found, but that does not mean that no splitting occurred between one form and the next. The relevant fossil may simply not have been preserved, found, or identified. Thus in Figure 2-8, A and C may be considered members of a single evolutionary species as long as B is not known. If B is identified and placed in the cladogram at the ap-propriate point, however, then a splitting event has been identified between A and C, and they are separate evolutionary species. Of course, even if B is

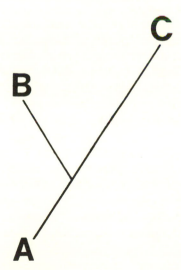

Figure 2-8. Hypothetical evolutionary tree for species A, B, and C; see text for discussion.

never found (or perhaps never existed), that does not mean that A and C are necessarily members of the same evolutionary lineage. The absence of apomorphy does not prove ancestor-descendant relationships (see section on ancestor-descendant relationships). Yet the evolutionary species concept is dependent on the concept of lineages of ancestors and descendants, which seem ultimately (at least in most cases) to be undemonstrable. Again, it is preferable to recognize species empirically by unique combinations of apomorphous and plesiomorphous character states.

Leaving the problem of phyletic speciation, there are several different patterns and outcomes of speciation involving the splitting or fusion of lineages. One way to classify speciation patterns is by whether the net result is an increase (additive speciation) or decrease (reductive speciation) in the number of species (Mayr, 1963; Wiley, 1981). Such a scheme is perhaps useful but relatively arbitrary, and in some instances there may be the same number of species after (either immediately or shortly after) a speciation event as there were before the speciation event.

True reductive speciation is a theoretical possibility that has never been demonstrated to have actually occurred (Wiley, 1981, p. 42). The basic concept behind reductive evolution is that two formerly independent species may come into contact, interbreed (hybridize), and completely fuse into a new species that swamps both old species over their entire ranges. In place of the two formerly independent species will be a single, new species that differs from both of the original species. Hybridization is known to occur between some distinct species, resulting in the formation of a third species (especially common in plants; V. Grant, 1971), but cases where both original species went extinct are undocumented.

Wiley (1981) has categorized all other patterns of speciation as additive. Following our discussion of reductive speciation through hybridization, it is logical to point out that the same mechanism, hybridization, can and does produce additive speciation (not uncommon in plants—perhaps a third of all vascular plants are polyploids: Baum, 1984; V. Grant, 1971, 1977; Stebbins, 1966; reticulate phylogenetic patterns due to hybridization in animals may only be 1 in 1,000 according to Wiley, 1981, p. 37, based on the work of M. J. D. White, 1978a, 1978b). In such cases, certain members of two previously distinct species hybridize to form a third species; the net result is three species (the two original species plus the new species) where formerly there was only two. It is also possible that through the mechanism of hybridization the entire population of one species may hybridize with only part of another species, and the end result would be no net increase or decrease in numbers of species, although one species will have disappeared and been replaced by another species.

Allopatric speciation patterns involve the physical separation or splitting (usually by some kind of geographic barriers) of populations of a once single and continuous ancestral species. In the vicariance, or geographic speciation, model the total population of an ancestral species is split into two (or more) large subpopulations by the rise of some sort of physical barrier within the species range. Once gene flow is significantly decreased or cut off completely between the subpopulations, they will be free to follow different evolutionary trajectories, thus instituting a permanent lineage split (speciation event). If the barrier should be removed and the populations come into contact once again, then they may either interbreed and merge (if true or total speciation had not yet taken place, or perhaps even if they were good species but still reproductively compatible), or they may be reproductively isolated. In some cases, coming back into contact may help reinforce the differences between them. As a rule, vicariance speciation should cause dichotomous lineage splits.

Another major pattern and model of allopatric speciation involves peripheral isolates. In this model it is suggested that a deme or several closely related demes on the boundary of an ancestral species may be easily separated from the parent population (either through an actual geographic disjunction or due to a decrease in gene flow between the central population of the species and the peripheral population) and then be prone to evolve rapidly because of both small population sizes and the fact that the deme (or demes) may be occupying a marginal habitat. Thus, numerous peripheral isolates on the edge of a large species may undergo speciation and become full-fledged species. This could mean that a single ancestral species might give rise to several daughter species. The several daughter species would share only the synapomorphies inherited from the ancestral species, but none among themselves. The phylogenetic relationships of the daughter species will not be resolvable into a set of nested dichotomies—that is, no two daughter species will be more closely related to each other than either is to a third.

Both forms of allopatric speciation described above, geographic and peripheral isolate, grade into one another, and both involve barriers dividing members of an ancestral species into two or more groups (subpopulations). Two basic mechanisms can be proposed to account for the origination of barriers between the members of the initial population. The ancestral species may have originally inhabited a certain area that was subsequently divided by the rise of barriers within it (the vicariance hypothesis). Alternatively, an ancestral species may have originally occupied a more restricted area, and subsequently some members of the ancestral species dispersed (perhaps by chance) across preexisting (often, but not necessarily, geographic) barriers (the dispersal hypothesis). These alternative hypotheses are important considerations in the study of historical biogeography (see chap. 5; Platnick and Nelson, 1978).

Sympatric models of speciation refer to potential mechanisms and patterns where an ancestral species can give rise to a new species without any form of geographic segregation of populations as in the allopatric models. Hybridization, discussed above, can be regarded as a form of sympatric speciation. Apomixis, the process where a biparental species gives rise to an asexual species, can occur in sympatry. The most discussed form of sympatry is ecological sympatric speciation, or sympatric speciation in the restricted sense. The mechanism hypothesized here is the segregation of subpopulations of a species by habitat or microniche, without geographic segregation being necessary (Bush, 1975); such segregation may lead to distinct species. Regarding this mode, Wiley (1981, p. 57) states:

> Sympatric speciation via ecological segregation requires a very specific phylogenetic hypothesis. The postulated species must be each others' closest relatives. If one of the species is actually more closely related to a third species, then the sympatric model is falsified.

Here Wiley is wrong. All that is required is that one species gave rise to—is the ancestor of—the other (or that both arose from a common ancestral species). They need not necessarily be each others' closest relatives. For example, species A may have given rise to B via ecological segregation, and then B gave rise to C via allopatric speciation. In this case, B and C are more closely related to each other than either is to A, yet the sympatric model is not falsified.

There are several named models of speciation that are neither strictly allopatric nor sympatric. Parapatric speciation refers to the situation in which two (or more) relatively distinct populations of an ancestral species differentiate from one another geographically, resulting in the formation of distinct species, without a complete disjunction (barrier) ever occurring between them. Alloparapatric speciation takes place when two populations of an ancestral species are initially segregated geographically and differentiate to a certain degree (but not enough to be considered good species) and subsequently come into contact again (parapatry or limited sympatry); during this time they finish differentiating and become independent species (Wiley, 1981, p. 54).

In certain cases it may be hypothesized that the only thing that maintains a species identity is some form of intrinsic homeostasis or evolutionary stasis, rather than homeostasis via gene flow (Eldredge and Gould, 1972; Gould and Eldredge, 1977; Wiley, 1981). This hypothesis or assumption may be a necessary one in the case of truly asexual species, as there is no gene flow between populations (individual lineages). For some sexual species there may be no gene flow between demes, yet the demes consistently maintain their common species identity over long periods of time. For some reason, hypothesized or unknown, this homeostasis may suddenly (on a geologic time frame) break down in one or more lineages or demes, causing a speciation. Such a speciation event might be considered allopatric, sympatric, or microallopatric. One form of this type of speciation is what has been called stasipatric speciation (White, 1978a, 1978b; Wiley, 1981). Here the specific mechanism of speciation is a chromosome mutation that arises in one deme that has relatively little gene flow with other demes of an ancestral population. If the chromosome mutation is fully viable when homozygous but has a reduced viability when heterozygous it may become fixed in the deme and lead to the formation of a new species. One of the assumptions, which may be correct in some cases, is that "chromosomal rearrangements are sufficient to cause lineage splits, with or without phenotypic differences" (Wiley, 1981, p. 56). This instance is one more where, when dealing exclusively with fossil material, two separate lineages (even if only at the species level) may not be morphologically distinct.

Sylvester-Bradley (1977) has suggested an additional model of speciation, one that is not completely separate from some of those outlined above, termed reticulate speciation. In this model an ancestral species is thought to increase in numbers of individuals and migrate to new areas. Populations in the newly invaded areas will become increasingly variable genetically and morphologically and may begin to adapt to the separate habitats, but gene flow is maintained throughout the species. This first phase Sylvester-Bradley calls the eruptive phase. The various populations then form geographical races and alternate between being isolated and hybridizing, and there is a marked increase in morphological differentiation. This is the polytypic or reticulate phase. Finally, during the divergent phase, the geographic races are isolated for a long enough time so as to be genetically incompatible; full speciation has taken place.

Intimately related to the topic of speciation patterns is the question of the driving force behind the divergence of populations of individuals that initially were members of a single species-level taxon. West-Eberhard (1983, p. 156) asserts:

Ecological divergence plays a key role in all current theories of speciation. Theories differ over whether divergence occurs in geographic isolation, at different positions along a cline (Endler, 1977), or in different subdivisions (habitats) of the same locality (Bush, 1975). But they agree in describing the critical divergence as ecological: "The geographic variation of species is the inevitable consequence of the geographic variation of the environment." (Mayr, 1963, p. 311; see also Endler, 1977, p. 7, and his citations of Huxley, Dobzhansky, and Grant)

In contrast, following C. Darwin (1871*b*), West-Eberhard (1983, p. 155) suggests that

rapid divergence and speciation can occur between populations with or without ecological difference under selection for success in intraspecific social competition—competition in which an individual must win in contests or comparisons with conspecific rivals in order to gain access to some resource, including (under sexual selection) mates.

(Note that social competition is not limited to animals, but may also occur among plants; see West-Eberhard [1983] for examples and references.) Accordingly, for social organisms, vicariance events that split single populations into subpopulations may cause divergence and speciation to occur even in the absence of ecological or habitat differences between the areas occupied by the subpopulations.

CHARACTER WEIGHTING

The evaluation of the relative importance of certain characters in recon-structing phylogenies has been, and continues to be, a topic of lively discussion. Ideally, in a data set where there is no character conflict (all characters define the same or compatible groupings of taxa or organisms), there is no need to weight characters. Character weighting is often seen as a means of eliminating character conflict by eliminating (or at least deemphasizing) the use of certain characters (Humphries and Funk, 1984). However, constructing a phylogeny on the basis of only a subset of the data at hand is nonparsimonious (see chap. 3) and necessarily involves assumptions, which may be ad hoc, relative to the phylogenetic significance of particular characters.

In a simplistic sense, it seems obvious that noninherited (of course, it may not always be possible to determine which characters are inherited and which are not) characters of organisms, and in some cases redundant characters (those that are logically redundant, and also perhaps those that are redundant due to being part of a functional or developmental complex), should be ex-cluded (weighted as zero) from phylogenetic analysis. But the problem is much more complex. On the one hand, certain pheneticists (numerical taxonomists, e.g., Sneath and Sokal, 1973) and some workers of a cladistic bent utilize, or weight, all characters (that is, all acceptable characters—and what is ac-ceptable [intrinsic versus extrinsic, inherited versus noninherited] may vary from investigator to investigator) equally (e.g., Eldredge and Cracraft, 1980, p. 12, state, "Cladists avoid the confusing issue of 'weighting' by recognizing that all nonconvergent characters are relevant to defining monophyletic groups at some level"). On the other hand is the opinion (Mayr, 1969, p. 217) that "experienced taxonomists have always insisted that characters differ in the contribution they make to the soundness of a classification [i.e., hypothesis of phylogenetic relationships]." Certain workers who utilize a weighting scheme advocate a priori weighting—that is, the relative weights of characters should be evaluated before a phylogenetic analysis is attempted (Bock, 1974; Hecht 1976; Hecht and J. L. Edwards, 1976, 1977). The rationalization is that i circular reasoning is to be avoided, then weighting must be accomplished prior to the use of the characters in recognizing taxa and phylogenetic rela-tionships. However, as Patterson (1982a, p. 44) has pointed out, merely "to recognize a feature (a homology) is to recognize a group." There seems to be no way in which one can recognize features, and subsequently weigh them, prior to and separately from recognizing taxa. Furthermore, any concept of a priori weighting must be based on some concept or assumption of how evolution works, and this assumption in turn is probably based on some pre-conceived notion of taxonomic groups and phylogenetic relationships. Kluge and Farris (1969; see also Cracraft, 1981a) criticized commonly used criteria for prior selection of good taxonomic characters (i.e., weighting), such as the functional importance of a character, the biological role and significance of a character, and the logical or probable course of evolution, as conjectural and subjective. Such weighting is untestable, nonrigorous, and left to the under-standing, opinion, and authority of an individual investigator and is thus subject to endless argumentation. In support of their opinion, Kluge and Farris (1969, p. 2) gave a specific example:

An example of the perils of using subjective weighting is the work of Ghiselin (1966a), who produced quite plausible sounding "biological" arguments to the effect that torsion must have preceded coiling in the evolution of gastropods, since otherwise some intermediate form would pass through an adaptively impossible stage. Batten, Rollins and Gould (1967), however, noted that the subclass Cyclomya consists entirely of proto-gastropods with coiling but no torsion.

As Mayr (1969, p. 217) correctly points out: "Neither function nor conspic-uousness nor any other known aspect of a character gives it a priori a greater weight than other characters. Indeed the very same structural difference may have high weight in one taxon and low weight in a related taxon."

The classical, a posteriori approach to weighting has been to work back-wards from phylogenies and classifications that produce natural groups, studying and giving weight to those particular characters that consistently de-limit particular natural groups. Mayr (1969, p. 219) gives a pertinent example of this method:

The number of cervical vertebrae (seven) is a class character in mammals, while it is not even a generic character in birds; for instance, it fluctuates from 23 to 25 in the genus *Cygnus* (swans). The only reason why high weight is given to certain characters is that generations of taxonomists have found these characters reliable in permitting predictions as to association with other characters and as to the assignment of previously unknown species.

In this context Mayr (1969, p. 218) defines weighting as "a method for de-termining the phyletic information content of a character."

In what is really an elaboration on, and a logical outgrowth of, Mayr's thoughts on a posteriori weighting, Patterson (1982a, p. 44) suggests that "we do not need to weight homologies [= characters]: they weight themselves, by associations which are beyond the bounds of chance." Characters or homologies to which we assign a high weight are those that are constant, dependable, reliable—that is, not contradicted by other homologies. According to Patterson (1982a, p. 44), the real criterion is lack of contradiction, which is, in turn, only to say that these characters are repeatedly corroborated as homologies by others which are congruent with them. I [Patterson speaking, but the author of the present volume agrees] suggest that when we weight homologies highly, it is not because of complexity, or functional importance, but because the groups they form are congruent with those formed by many other characters, and are contradicted by few or none.

Riedl's (1978) concept of burden corresponds in a general way to Patterson's concept of the weighting of homologies. A character of high burden is one that is relatively constant at high rank and is subject to relatively little incongruence (e.g., convergence). Patterson (1982a, p. 45) also points out that as homologies (characters) necessarily have higher weights or burdens within more inclusive groupings (at higher taxonomic ranks), homologies confined to the species level (species characters) will be of low weight, and consequently phylogenetic analysis and systematics may be most difficult at the species level. The distributions of putative homologies (characters) that are not inherited (such as ecophenotypic and individual variation) will be incongruent with the distributions of the majority of characters studied and thus exposed as nonhomologies (noncharacters in a phylogenetic sense). In other words, such characters are notoriously inconstant and unreliable in forming taxonomic groupings.

Kluge and Farris (1969) have suggested a method of weighting characters that they believe is truly objective. These authors argue that the variation of a character within an OTU can serve as an index of the relative evolutionary rate of change of a character. Their reasoning is as follows:

The relationship between intra-OTU variability, evolutionary rate, and character weighting is easily expressed. The rate at which a character can change in evolution is necessarily limited by the variability of the character within populations. Selection, no matter how intense, cannot change the average character state of a population in some unit time by a greater amount than the range of values available in the population in that unit time. If a character has high variability within OTUs, then a large difference between two OTUs does not imply lack of close relationship, since the variable character could have changed rapidly. If on the other hand, a character has low variability within OTUs, then a large difference between OTUs is probably indicative of lack of close relationship, since the highly stable character probably could not evolve rapidly (Kluge and Farris, 1969, p. 4).

Therefore Kluge and Farris (1969, p. 16) use the following system of weighting:

Characters are weighted according to the concept of conservatism as estimated through within- and between-OTU variability. Multiplying each character by the reciprocal of the within-population standard deviation transforms characters onto scales in which rate of evolution would be approximately equal over characters. With equal rate of evolution, the range of variation between OTUs should be approximately equal; characters prone to convergence would tend to exhibit less between-OTU variability and they are weighted accordingly. The advantage of weighting characters by their OTU variability is that the weighting is fixed objectively by the data itself.

(Of course, it could be argued that the weighting would be fixed equally objectively by the data itself, by differentially weighting all characters that meet any arbitrary criterion.)

The fallacy of Kluge and Farris's (1969) objective method of weighting characters is the major assumption they make about the source of evolutionary innovations. These authors assume that variability of a character within an OTU (such as a species) is inversely related to the conservatism of the character within the species. Unfortunately, such may not be the case. Out of developmental and epigenetic considerations, Rachootin and Thomson (1981, p. 184; see also Løvtrup, 1974) suggest that observed phenotypic variability is not the source of evolutionary change; rather, observed phenotypic constancy may be underlain by a large reservoir of developmental and genetic variability that can be tapped as a source of significant evolutionary (macro- or mega- evolutionary) change, while phenotypically variable characters may be the "result of a single genetically invariant epigenetic process with a large component of self-organization." Furthermore, Kluge and Farris (1969, p. 4) appear to equate close relationship with phenotypic similarity, especially for characters of low variability. However, no matter what the variability of a particular character is, just because two taxa are more similar to one another than either is to a third taxon does not mean that they have a close relationship phylogenetically. Phylogenetically, three taxa, A, B, and C, may be related as follows: (A (B, C)). Yet in most characters A and B may be more similar to each other, while C is wildly divergent from either. The characters in which A and B are similar may simply be plesiomorphies.

In some cases putative homologies may lead to contradictory groupings of taxa, and there may be an equal number of putative homologies supporting each alternative hypothesis. Such situations might be resolved by weighting the characters involved in the sense of Hecht and J. L. Edwards (1976, 1977). For various reasons, the investigator may prefer one set of characters over another, and thus the nonpreferred set of characters (putative homologies) may be dismissed as evidence—that is, considered nonhomologies (cf. Paterson, 1982, p. 44).

Hecht and J. L. Edwards's (1976, 1977) formal system of character weighting recognizes the following five categories of characters, listed in order of increasing information content (Hecht, 1976, p. 344):

I. Derived (and shared) character states which are the result of loss. Such loss states, by definition, do not have developmental markers and therefore have zero information content.

II. Derived (and shared) character states that are the result of simplification or reduction of complex structures as indicated by comparative or developmental anatomy have higher information content and greater reliability than the preceding.

III. Derived (and shared) character states that are the result of growth and developmental processes dependent on size, age, hormonal, allometric and other physiological relationships, and therefore have greater information content than the preceding.

IV. Derived (and shared) character states that are highly integrated functionally and are coevolving due to directional selection for greater functional efficiency have great information content if properly evaluated.

V. Derived character states that are unique and innovative in determining new dichotomies and lineages. The more complex the character in development and structure, the greater reliability in indicating polarity and new dichotomies.

Note that all of the above criteria for assigning weights to characters are based on assumptions about the evolutionary process or prior knowledge of the specifics of a particular phylogeny (e.g., loss, reduction, unique and innovative). Certain "sorts of characters are deemed axiomatically to be more (or less) useful in phylogenetic analysis than others" (Eldredge and Cracraft, 1980, p. 66). Such lines of argument, including those outlined by Bock (1977; reviewed below) are primarily subjective and rely on authoritarianism (Cracraft, 1981b). Fisher (1981, p. 60) also points out that Hecht and J. L. Edwards's (1976, 1977) weighting categories are not mutually exclusive and require the comparison of very heterogeneous entities.

Hecht (1976, p. 344) notes that the above quoted scheme of Hecht and J. L. Edwards (1976, 1977) is basically qualitative, but that it can be quantified in order to facilitate comparisons and weightings of various hypothesized phylogenies. One might assign a value of 1 to characters in the first category, a value of 2 to those in the second category, up to a highest weight of 5. Such a scheme could be interpreted as indicating that characters in category V carry five times more information than those in category I. Hecht (1976, p. 344), however, argues that the information content of characters in category V compared to those in category I "is probably a thousand-fold or greater because of character complexity, morphogenesis, and uniqueness." Thus he suggests that those characters of category I be given a value of 1, those of II a value of 10, those of III a value of 100, those of IV a value of 1,000, and those of V a value of 10,000. Finer subdivisions within categories are possible. Hecht (1976) apparently believes that such a system could be refined and put into practice such that an investigator could evaluate objectively proposed phylogenies.

In a vein similar to the work of Hecht and J. L. Edwards on the weighting of characters, Bock (1977, p. 890) has suggested that in the case of conflict between two putative synapomorphies, the following weighting criteria can be used to help distinguish between synapomorphy and homoplasy. (1) A complex feature is more likely to represent a homology (synapomorphy) than a simple feature. The probability of a complex feature evolving independently two or more times is less than that of a simple feature evolving two or more times. (2) Newly evolved features or structures are more likely to be homologous than reduced or lost features. (3) Features that serve a wide range of biological roles are more likely to represent homologies than features that serve a single biological role. (4) Features whose evolution is the result of complex patterns of relationships (both structural and functional) are more likely to be homologous than those whose evolution was dependent on few such relationships. (5) Features with identical complex ontogenetic bases are more likely to be homologous than features with simple ontogenies.

Functional Studies in Phylogenetic Analysis:
A Basis for Character Weighting?

As discussed in the last section, certain schools have advocated using all available characters. Most notably, the pheneticists believe that any and all available characters should be used as raw data without any a priori weighting or analysis. Likewise, many cladists believe that characters should be used without any a priori assumptions that certain characters are better than other characters. However, as part of the cladistic analysis, the character distributions are analyzed, and certain characters may be rejected a posteriori as symplesiomorphous at the level of analysis or as homoplasy. Many evolutionary taxonomists, on the other hand, subscribe to the view that potential characters should be analyzed in biological, functional, and adaptational terms before being used in constructing a phylogenetic hypothesis. For example Szalay (1981a, p. 44) states

It is taken as a basic tenet of the phylogenetic method that both the establishment of homologies and the sorting of the relative usefulness of these for taxon phylogeny (weighting) depends entirely on sound biological (developmental, functional and adaptational) appraisal of the characters or preferably character complexes. It is maintained that mere enumeration without weighting of different kinds of characters and without attempting to determine the relative time of shared homologies cannot result in biologically (i.e., historically) meaningful taxon phylogeny.

Szalay (1981a, p. 37) contrasts this view with

the post-Hennigean cladistic approach which either disavows or actually scorns any attempt to incorporate functional analysis into phylogenetic assessment, and rather strives for logical consistency between a list of characters (one being considered as "good" as another) and a cladistic (i.e., sister group) hypothesis.

Taking the opposite viewpoint, in direct criticism of Szalay's work, McKenna (1981, p. 99) has stated

Szalay and Drawhorn [1980] believe that they have transcended "mere computer-like recording of form" by applying knowledge of function to the analysis of character distributions. In short, they mix scenarios about the process with analyses of pattern and attempt to convince the reader that they know what the "gestalt" of various specimens really means. I find such additional baggage unnecessary and potentially authoritarian.

In arguing for the importance of functional analysis in order to choose (weight) characters used in phylogenetic analysis, Szalay (1981a, 1981b) contends that some characters are never used because they are well known to be a result of functional or developmental relations that do not bear on phylogeny (this argument seems to be the equivalent of the traditional argument that it is conservative characters, as opposed to environmentally influenced adaptive characters, that must be used in determining phylogenetic relationships between organisms; see, for instance, Borissiak, 1945). As an example, Szalay (1981a, p. 39) puts forth that

> one would have a difficult time finding the use of large sagittal crests [in mammals] as a character signifying a particular recency of relationships. The reason, quite simply, is that several functionally and developmentally oriented contributions have shown (see review by Biegert, 1963) that the size of the crest depends on the complex relation between body size, the relative size of the temporalis, and the relative size of the brain.

Allometric effects may well determine the morphological/phenotypic expression of certain traits, and several characters may show a strong correlation due to allometric and developmental processes, but this does not necessarily mean that such traits are not useful in phylogenetic analysis. Perhaps several characters that are dependent upon one another developmentally should, for the purposes of phylogenetic analysis, be treated as a single character. But this seems not to be Szalay's point in the above quote. Rather, since a character such as large sagittal crest is relatively prone to environmental influence, then it should be eliminated a priori from further consideration in phylogenetic analysis. But if we should happen to include such a character in our data matrix when performing a cladistic analysis, and if it indeed is not synapomorphous at a certain level, then it will not fit into the hierarchically nested set with other character distributions analyzed, but will show character conflict at a certain level. If this is the case, then the character will be regarded as homoplasy. However, the same characters (even if developed independently numerous times) may still be useful for distinguishing low-level taxa and thus be useful at some level of phylogenetic analysis. Indeed, this seems to be the case for large sagittal crests in mammals. All mammals with large sagittal crests do not share a more recent common ancestry with each other than any do with a mammal with a small sagittal crest (this is demonstrated by the distributions of numerous other characters); however, within a low-level taxon, such as a genus, certain species may be recognized by the development of large sagittal crests, among other features. Biological and functional analysis may help us understand why a presumed character is a homoplasy at a certain level of analysis, but such analysis is not necessary in order to determine that it is. For how many characters of fossil organisms do we actually have even a firm idea as to their biological significance? For every character complex that has been well studied, there are hundreds that are enigmatic. Yet we can still attempt to reconstruct the phylogenies of the organisms bearing the biologically enigmatic characters.

Detailed developmental, adaptational, and functional analysis of certain characters may help to weed out convergences (*sensu* Patterson, 1982a: characters that are superficially similar, but can be demonstrated to be different with detailed analysis) before a phylogenetic analysis is attempted. When available, developmental data in particular is relished by those utilizing cladistic analysis. Functional studies may also contribute new information, new insights, and more possible synapomorphies (homologies). And as Szalay (1981a) suggests, in some cases it may be that it is a total working unit—a functional complex—that forms the legitimate unit of analysis, the legitimate synapomorphy, as opposed to the isolated characters making up the unit, which may easily be convergent in themselves. In this sense functional studies may help determine the proper level of analysis of characters; in any particular instance, these studies may determine the smallest (most finely divided) character and the furthest legitimate subdivision of a character or character complex. Fisher (1981) has made a similar point: Functionally dependent characters should perhaps not represent separate characters in a parsimony analysis. However, in commenting on functionally dependent or functionally integrated characters, Cracraft (1981b, p. 27) notes that there is often no way to define objectively the structural limits of such integration. In an extreme case, it could be argued that all the parts of the organism must be characterized by such integration, and thus perhaps the whole organism should be viewed as a single character. Szalay (1981a) also makes the important point that in phylogenetic analysis one should never dogmatically assume a particular presumed homology or set of homologies a priori and then erect a tenuous phylogenetic scheme on the basis of the presumed homologies while ignoring other data that bears on the problem. Attempting to explore and fully understand one's favorite characters in a biological/functional context can help guard against such pitfalls.

CHAPTER 3

The Conceptual Foundations and Bases of Phylogeny Reconstruction

HOMOLOGY

The concept of homology is fundamental to any attempt to reconstruct phylogeny using morphological characters. At the most basic level, it is only by homology that we recognize that certain organisms belong to the group whose phylogeny we are attempting to reconstruct.

The foundations of the modern concept of homology were laid by Owen (1843) when he made the first clear distinction between what was meant by analogous and homologous anatomical parts or organs in organisms (Patterson, 1982a; Schoch, 1984b). In 1866 Owen (1866, p. xii) wrote

A "homologue" is a part or organ in one organism so answering to that in another as to require the same name. Prior to 1843 the term had been in use, but vaguely or wrongly [e.g., Sainte-Hilaire, 1825]. "Analogue" and "analogy" were more commonly current in anatomical works to signify what is now definitely meant by "homology." But "analogy" strictly signifies the resemblance of two things in their relation to a third; it implies a likeness of ratios. An "analogue" is a part or organ in one animal which has the same function as a part or organ in another animal. A "homologue" is the same part or organ in different animals under every variety of form and function.

Note that two organs can be both analogues and homologues (the same organs with the same function), or they can be only analogues (different organs with the same function) or only homologues (the same organ with different functions). Elaborating on the same subject further, Owen (1866, p. vii, italics in the original) wrote

"Homological Anatomy" seeks in the characters of an organ and part those, chiefly of relative position and connections, that guide to a conclusion manifested by applying the *same name* to such part or organ, so far as the determination of the namesakeism or homology has been carried out in the animal kingdom.

Owen's concept of homology as distinguished in the quote immediately above is what Owen (1866, p. xii) termed *special homology*. Owen (1866) also distinguished *general homology* and *serial homology*. General homology referred to the correspondence of a particular structure to that of a general type of a higher taxonomic group (an archetype). "When the 'basilar process of the occipital bone' is determined to be the 'centrum' of the last cranial vertebra, its *general homology* is enunciated" (Owen, 1866, p. xii, italics in the original). Serial homology referred to the repetition of corresponding structures in a single organism, such as the vertebrae or ribs of a vertebrate, or the leaves of a tree. Owen (1866, p. xiii) termed serially repeated parts *homotypes*. The concept of serial homology is now usually referred to as homonomy (Patterson, 1982a; Simpson, 1961). Concerning homonomies Reidl (1978, p. 38) states: "These are structural similarities or identicalities between the building blocks of one and the same individual. The differences are thought of as divergences from identical basic forms, several of which occur in the same organism."

Eldredge (1979b) and Patterson (1982a) have pointed out that there are two basic approaches to evolutionary theory in general and to homology in particular: the taxic approach and the transformational approach. The taxic approach is concerned with the diversity of organisms and their groupings. Owen's concept of special homology and virtually all modern concepts of homology (discussed below) are framed within a taxic approach. Unless otherwise specified, the discussions presented in this book assume a taxic approach. The transformational approach is concerned with change in a series, lineage, or otherwise; it does not imply groupings and it need not be evolutionary. Owen's concept of general homologies and archetypes is an example of the transformational approach. In contrast, the cladistic concept of synapomorphies (= homologies) and primitive morphotypes for monophyletic groups is an example of the taxic approach. As Patterson (1982a, p. 35) states: "An archetype is an idealization with which features of organisms may be homologized by abstract transformations which entail no hypotheses of hierarchic grouping. A morphotype is a list of the homologies (synapomorphies) of a group."

Owen's conception and definition of special homology is what has been called the classical approach to the problem (Patterson, 1982a). This approach is independent of any theory involving ancestry and descent (i.e., evolution); homologous structures are the same structures as operationally recognized by applying identical names to them, but *same* is left otherwise undefined.

With the general acceptance of evolution, the concept of homology took on a new meaning. Two structures in different organisms were considered to be the same structure, and thus deserving of the same name, if they could be traced to the same structure in a common ancestor of the two organisms. This concept has been termed the homogenetic, Darwinian, or evolutionary explanation of homologies, or more simply homogeny as opposed to the classical, nonevolutionary homology of Owen (Hanson, 1977). Lankester (1870; see Kaplan, 1984) advocated the use of homology in the traditional (pre-Darwinian) sense for structural correspondence regardless of any phylogenetic implications. He used the terms *homogeny* and *homoplasy* for similarities due to inheritance from a common ancestor and similarities not due to inheritance from a common ancestor (i.e., convergence or parallelism), respectively. More recently Hunter (1964) has proposed the term *paralogy* to describe structural similarities in the absence of any phylogenetic or evolutionary conclusions concerning their origins. In this context Mayr (1969, p. 85, italics in the original) gives the following definition: "*Homologous* features (or states of the features) in two or more organisms are those that can be traced back to the same feature (or state) in the common ancestor of these organisms." Mayr (1969, p. 85, italics in the original) would also restrict the use of analogy: "*Analogous* features (or states of the features) in two or more organisms are those that are similar but cannot be traced back to the same feature (or state) in the common ancestor of these organisms." Thus two structures cannot be both homologous and analogous, and homologous structures need not be similar as long as they can be traced back to the same structure in a common ancestor. Finally, the same name may not necessarily be applied to homologous structures in different organisms. As an example, Mayr (1969) considers the ear ossicles of mammals and certain bones of the jaw articulation of nonmammal tetrapods to be homologous but not similar, and the elements are given different names in mammals and nonmammal tetrapods. Simpson (1961, pp. 78–79) defines homology and analogy in terms similar to those of Mayr (1969): "Homology is resemblance due to inheritance from a common ancestry," and "analogy is functional similarity not related to community of ancestry." Likewise, Riedl (1978, p. 33) considers analogies to be "structural similarities for which we have to suppose that they arose convergently" and homologies to be "structural similarities which force us to suppose that any differences are explicable by divergence from an identical origin." However, note that both Simpson (1961) and Riedl (1978), in contrast to Mayr (1969), maintain that homologies retain some kind of resemblance or similarity (if they did not, we could have little hope of recognizing homologies; of course, such resemblance may not be immediately obvious and may consist of a transformation series of intermediates that link radically different end-member morphologies).

Besides the classical and evolutionary approaches, Patterson (1982a) has distinguished several other ways of defining homology. Pheneticists have generally attempted to avoid what they regard as circularity if homology is defined in terms of common ancestry and common ancestry is recognized by shared homology. Instead, a strictly operational definition of homology is attempted, returning to the spirit of Owen's classical definition. Sneath and Sokal (1973, p. 77, italics in the original) state:

Homology may be loosely described as compositional and structural correspondence. By *compositional correspondence* we mean a qualitative resemblance in terms of biological or chemical constituents; by *structural correspondence* we refer to similarity in terms of (spatial or temporal) arrangement of parts, or in structure of biochemical pathways or in sequential arrangement of substances or organized structures.

As Sneath and Sokal (1973) point out, such an operational definition of homology can lead to a quantitative (as opposed to the traditional categorical) concept of homology. Thus, the homology of two features need not be a yes-or-no proposition, but degrees of homology may be recognized among structures. A purely utilitarian approach to homology is closely related to the classical and phenetic/operational approach to homology. Homologous structures are simply structures that "can be compared usefully to each other" (Blackwelder, 1967, p. 141; quoted in Patterson, 1982a, p. 29).

Recently, Roth (1984, following Ghiselin, 1976) has advocated a definition of homology that would include what she terms *iterative* homology (= serial homology, including sexual homology [the concept that the male and female genitalia of a single species may be homologous], antimeric homology [symmetry within an organism]), and *phylogenetic* homology (= special homology, or genealogical homology). According to Roth's (1984, p. 27) formulation, "homology is based on the sharing of pathways of development which are controlled by genealogically related genes." Iterative homologues are the result of the same underlying developmental process; as these developmental processes are more similar, the strength or degree of homology increases. Thus with the pheneticists, Roth does not view homology as a simple yes-or-no proposition. As an example, Roth (1984, p. 17, italics in the original) proposes the following as a hierarchy of increasing degree of iterative homology:

VERTEBRAE
  Cervical Vertebrae
    Cervical Vertebrae Exclusive of Axis and Atlas
      —vertebra c5—.

Likewise, she would apparently distinguish different strengths or degrees of homology among phylogenetic homologues. Roth (1984, p. 27, italics in the original) also argues that

because different aspects of structures are controlled by distinct developmental programs (as one infers from the uncoupling of different features during development in different organisms), it is sometimes necessary to speak of homologies of different *attributes* of specific structures, rather than to homologize the structures per se.

Roth's conception of homology raises the problem of whether identical genetic and developmental pathways that have arisen independently can be distinguished, and whether they should be considered homologous. Roth (1984, p. 23, italics in the original) states

Obviously, similar genetic variants which act by different mechanisms cannot be considered homologous. But *identical* changes which arise in different individuals within a freely interbreeding population might as well be. . . . Because genealogy plays a key role in my definition of homology, I should argue that characters must not only be identical at the level of the gene itself, but also identical by descent (*sensu* Falconer, 1981:57) to be considered homologous. But in doing this I would risk making the definition non-operational: one could never, with certainty, distinguish valid homologues in nature. By the strict definition, unless the origin of a character can be traced to a mutation within a single progenitor individual, which ultimately gave rise to *all* the individuals currently bearing the trait, the character cannot be homologized between individuals within a species, or for that matter, between species. This is clearly absurd.

Thus, according to Roth (1984, p. 27), "for the definition of homology to be at all operational, one must in practice be willing to tolerate some ambiguity between parallelism and homology. The level at which this ambiguity arises depends upon the detail or resolution of one's knowledge of genetic and genealogical relationships."

As is evident from these quotes, Roth (1984) must assume some knowledge of genealogy (phylogeny) a priori in order to distinguish parallelism (homoplasy) from homology, thereby eliminating the use of homology as a means of reconstructing phylogenetic relationships. (She does not propose any means of arriving at knowledge of phylogeny.) She ignores the congruence test of putative homologies (Patterson, 1982a; discussed below) as a way of distinguishing homology from parallelism without assuming a particular phylogeny a priori.

Similar to Roth's (1984) concept of parallel mutations arising within a single species, the concept of *collateral evolution* was proposed by Shaw (1969). As Shaw (1969, p. 1096) noted, the usual conception is that the interfertile organisms that compose a species could not have arisen separately; Shaw dubbed this assumption monophyletic evolution, or Adam-and-Eve evolution. In contrast, he suggested that among a series of widely distributed organisms that share a common character, under similar or identical selection pressures the character may independently change to a new, and everywhere identical, mutant. The organisms bearing the independently derived mutants may be interfertile. In Shaw's (1969, p. 1097, italics in the original) words,

I suggest that it is far more likely that organisms bearing the new mutant would appear and survive *everywhere that the ancestral character and the altered selection pressure co-existed* than that the new mutant will appear only once. Collateral Evolution would give rise to identical transitional series and to identical descendant characters [independently in different areas].

Shaw (1969) did not propose any way to test his hypothesis of collateral evolution, nor any method of distinguishing collateral from monophyletic evolution.

From a cladistic point of view, homology may be synonymized with synapomorphy, at least at a certain hierarchical level. Since synapomorphies distinguish monophyletic groups, then cladistically homology can be defined as "the relation characterizing monophyletic groups" (Patterson, 1982a, p. 61). This cladistic argument is pursued further below.

There are a few terms related to the concept of homology, in that they refer to similarities seen between organisms that may be confused with homology, that must be introduced. These terms are defined, in an evolutionary sense, by Simpson (1961, pp. 78–79):

1. "Homoplasy is resemblance not due to inheritance from a common ancestry. . . . Homoplasy includes parallelism, convergence, analogy, mimicry, and chance similarity."

2. "Parallelism is the development of similar characters separately in two or more lineages of common ancestry and on the basis of, or channeled by, characteristics of that ancestry."

3. "Convergence is the development of similar characters separately in two or more lineages without a common ancestry pertinent to the similarity but involving adaptation to similar ecological status."

4. "Mimicry is similarity adaptive as such and not related to community of descent."

5. "Chance similarity is resemblance in characteristics developed in separate taxa by independent causes and without causal relationship involving the similarity as such."

Also adopting a primarily evolutionary stance, Riedl (1978, pp. 36–39) discusses a few additional terms:

6. Homoiology (= homeology: Humphries and Funk, 1984, p. 341) "covers similarities of structure when they include both analogous and homologous substructures. Homoiologues can also be called analogies on a homologous base" (Riedl, 1978, p. 36).

7. Homodynamy "refers to causes [e.g, physiological causes or development commands] which result in homologous effects" (Riedl, 1978, p. 36). It seems that homodynamy is merely a term for "homologous" development processes that may not, however, yield strictly homologous morphological structures.

8. Isology refers to the chemical similarity or chemical kinship of biochemical compounds such as cytochrome and hemoglobin. Isologues may be homologous or analogous.

9. Riedl (1978, p. 38) regards the bilateral and radial symmetries often seen in organisms to be another form of similarity akin to homonomy.

Classical Criteria Used to Recognize Homologies

Classically, structures have been regarded as homologous if they resembled one another in a number of minute details, the basic concept being that it is unlikely that two organisms would arrive at precisely the same structure (in all of its details, many of which may be nonadaptive or maladaptive) by convergence alone. The closer the minute structural similarities, the more likely that the structures are homologous (Hanson, 1977; Remane, 1971; Riedl, 1978; Simpson, 1961). Bock (1977, p. 882, italics in the original) states that "shared similarity is the *only valid empirical test of homology.*" Bock (1977), however, believes that this test has low resolving power; in many cases nonhomologous features may be similar enough that they may be mistaken as homologous. Topographic or positional data is often used to recognize homologies; structures in different organisms that bear the same mutual relationships to other structures thought to be homologous between the two organisms may be homologous (Hanson, 1977; Remane, 1955, 1956, 1971; Riedl, 1978; Schaeffer, Hecht, and Eldredge, 1972). Similarities in the morphogenesis (ontogeny) of structures may indicate that they are homologous. Finally, transitional forms (perhaps seen in a series of systematically intermediate organisms or in an ontogenetic series) may link two otherwise quite dissimilar structures, indicating that they are homologous (Hanson, 1977; Remane, 1971; Riedl, 1978).

Hanson (1977), Remane (1971), and Riedl (1978) also cite conjunctional criteria for recognizing homology. The occurrence of similar structures in a number of closely related organisms, especially if correlated with the distribution of other similar structures in the closely related organisms, may be indicative of homology. In contrast, if certain similar structures occur commonly or randomly in organisms that are not otherwise indicated as being closely related, then the structures are probably not homologous. These conjunctional criteria for recognizing homology relate directly to the cladistic concept that homologies define natural groups while nonhomologies do not, and homologies are congruent with one another while nonhomologies are not.

A criticism of the classical criteria for recognizing homologies is that homologous structures cannot be identified until comparable systems are established on which to base the analyses. There must be some starting point where broad features are accepted as homologous (that is, worthy of comparison) without evidence of the nature used to establish homology according to the classical criteria given above. Stated another way, there is an inherent circularity in the concept: Structures in related organisms may be worthy of comparison, and thus homologous, but it is the presence of homologous structures that indicate the relationships between organisms in the first place. Hanson (1977, p. 80) suggests that this circularity can be broken by remembering that at the lowest levels the most closely related organisms generally are members of interbreeding populations and species. In such cases it is the occurrence of reproductive compatibility, not homologies, that is the criterion for relatedness between the organisms involved (but see discussion of the criterion of interbreeding in the section on species). At the highest level the circularity is also broken because it is assumed that life on our planet has a monophyletic origin, and all organisms are ultimately related.

Cladistic Criteria of Homology

Patterson (1982a) has proposed three tests of homologous, or possibly homologous, structures: the similarity, congruence, and conjuction tests.

First, homologous structures usually possess a shared similarity, and this similarity may be of any form: gross morphological, ontogenetic, topographic, and so on. This similarity criterion is essentially the same as that used by classical evolutionary taxonomists, described above. Cracraft (1981b, p. 26) has suggested, however, that similarity is not a test of homology per se, but a factor that compels us to postulate homology.

The second test is to consider homologies synapomorphies "and since . . . synapomorphies are the *only* properties of monophyletic groups, tests of a hypothesis of homology must be other hypotheses of homology—other synapomorphies" (Patterson, 1982a, p. 38, italics in the original). Hypothesized homologies that are consistent (congruent) with a certain hypothesized homology help to corroborate that hypothesized homology. That is to say, the hypothesized homologies stand in hierarchical relationship to one another in defining nested sets and subsets that do not contradict or violate one another. The best corroboration of a certain hypothesized homology is perhaps a number of other hypothesized homologies that specify the same, or nearly the same, grouping. A certain hypothesized homology that is incongruent with a hierarchical scheme based on numerous other hypothesized homologies is rejected as a nonhomology. In this connection, Cracraft (1981*b*) suggests that the true test of a putative homology is whether it is compatible with the distribution of other putative homologies (the homologies should form a hierarchically nested pattern of sets and subsets of taxa and ultimately of organisms). Patterson's (1982*a*) congruence test is essentially the same as Wilson's (1965, p. 214) consistency test for phylogenies.

Third, Patterson (1982*a*, p. 38) has proposed what he calls the conjunction test: "If two structures are supposed to be homologous, that hypothesis can be conclusively refuted by finding both structures in one organism." In an evolutionary sense, truly homologous structures were a single structure in a common ancestor and thus define a monophyletic group. All forms of the single structure, as found in the common ancestor, are a singular homologue—and only one form of the homologue should be found in any descendant.

Structures that pass all three of Patterson's tests are considered homologous. Structures may initially be hypothesized to be homologous but fail to pass one or more of these tests. In regard to the three tests, there are seven possibilities for nonhomologous structures, as discussed in the following paragraphs.

According to Patterson, a potential homology that fails to pass the congruence test (but does pass the similarity and conjunction tests) is a parallelism, whereas a potential homology that passes the conjunction test but fails the congruence and similarity tests is a convergence (here the intention is that the structures compared initially seemed similar enough to be potential homologies, but under detailed scrutiny proved to be "not closely similar, or 'not really the same' ": Patterson, 1982*a*, p. 46). Thus, for Patterson at least, the terms *parallelism* and *convergence* have very specific meanings, and parallelism is not one type or subcategory of convergence (cf. Eldredge and Cracraft [1980, pp. 70–74] who consider parallelism to be a form of convergence and would drop use of the term completely). Parallelism is a potential homology that is rejected solely because it fails the congruence test. Under convergence Patterson (1982*a*, p. 45) includes Simpson's (1961) analogy, mimicry, and chance similarity.

A pair of structures that are potentially homologous and pass the congruence test but fail the similarity and conjunction tests may actually be two homologies that distinguish the same or similar groupings. If they pass both the congruence and similarity tests, but fail the conjunction test, then Patterson considers them to be a case of homonomy.

Patterson (1982*a*, p. 48) defines the *complement relation* as the "presence of a homology versus absence." An example of it would be the comparison of something to nothing (the something in one organism compared to nothing in another organism). If the organisms are viewed at only one time (at one stage in their life cycle: semaphoront), the complement relation occurs when the potential homology passes the conjunction test and the congruence test but fails the similarity test. As Patterson (1982a, p. 48) points out, for organisms with ontogeny "the conjunction test is not passed if the whole life history is considered, since most homologies present later in life are absent in the zygote."

The last two possibilities are cases of initially potential homology that (1) pass only the similarity test or (2) fail all three tests. Both are left unnamed. Patterson could find no good cases of relations that fit either possibility. The first might be considered a form of nonhomologous homonomy. For the second, Patterson (1982a, p. 48) suggests, "If one mistook an endoparasite for part of its host, comparison between part of the parasite and part of another organism would fail all three tests."

Patterson makes the point that only those relationships that pass the congruence test (homology, homonomy, complement relation, and two homologies) are useful to the systematist in reconstructing phylogenetic relationships. The other relations will be a hindrance (see the section on parallelism and convergence). Finally, it should be noted that all of these tests and accompanying discussion apply only to potential homology in the taxic sense. Transformational homologies are immune to all three tests: They do not specify groupings, and therefore the congruence test does not apply; they need not necessarily be particularly similar to one another; and two structures that are homologized in a transformational context may occur in a single organism (for example, according to Owen the basilar processes of the occipital bone and the centra of the cranial vertebrae are homologous in a general sense, yet both occur together in a single organism).

DETERMINATION OF MORPHOCLINE POLARITY

One of the major problems for the phylogeneticist is to distinguish primitive (plesiomorphic) from derived (apomorphic) character states at a given level of relationship given a morphocline of two or more character states for a character. Commonly used methodologies are the commonality principle, outgroup comparison (argumentation scheme), morphogenetic (ontogenetic) sequences, and biogeographic and biostratigraphic data (older is generally more primitive: the paleontological method). For general recent discussions on various aspects of the problem of morphocline polarity, see E. N. Arnold (1981), Crisci and Stuessy (1980), De Jong (1980), Eldredge and Novacek (1985), Stevens (1980), and Stuessy and Crisci (1984).

Before discussing the various methodologies utilized to determine morphocline polarities, it is important to be explicit as to what it means to distinguish a character or character state as plesiomorphic (primitive) or apomorphic (derived). Our criteria for distinguishing morphocline polarities do not demonstrate that one character state, the plesiomorphic character state, is primitive in real evolutionary history (although we may infer such to be the case); rather, tests of relative plesiomorphy and apomorphy merely show that some character states (those that are plesiomorphic) are more general than their corresponding derived character states (Nelson, 1978b; Platnick, 1979, p. 544).

Many workers consider the commonality principle to be the most widely used and accepted criterion for determining character state polarity (Eldredge, 1979a; Watrous and Wheeler, 1981, and references cited therein). To put it simply, the commonality principle is the notion or a priori assertion that common equals primitive. For any particular group, such as the members of a higher-level taxon that is accepted as monophyletic, the commonality principle states that "a primitive state is more likely to be widespread within a group than is any one advanced state" (Kluge and Farris, 1969, p. 5), or that "the relatively more primitive states are likely to be distributed more generally throughout the group under study" (Estabrook, 1972, p. 439). As Wheeler and Watrous (1981) have pointed out, and many authors in the past have failed to realize, the commonality principle is strictly a principle of in-group comparison. For example, in contrast to Watrous and Wheeler (1981), Crisci and Stuessy (1980) distinguished (incorrectly) two categories of the commonality principle: one of in-group comparison (commonality) and one of out-group comparison (to be discussed below).

Most workers have agreed that commonality does not necessarily equal primitive (Estabrook, 1977), yet the principle is still generally held dear. For many groups of extant organisms, it is known or hypothesized that the most primitive members are relict species that are poorly represented in terms of diversity; relatively more derived members make up the majority of the group or taxon, and thus the relatively more derived character states borne by the more derived members may be more common. Therefore, the commonality principle fails to hold. As a specific example, among the living mammals the Monotremata are generally considered to be primitive in many character states relative to all other living mammals (i.e., marsupials and placentals: see McKenna, 1975; Schoch, 1984b, p. 10), yet the Monotremata are known from only two genera and four species restricted to New Guinea, Tasmania, and Australia (Walker, 1975). This compares to a total of approximately 4,170 extant species of mammals (Honacki, Kinman, and Koeppl, 1982). Obviously, the commonality principle does not work for this specific case of using the extant Mammalia as an in-group; however, as will be discussed below, if the monotremes are considered a sister-group of all other extant mammals, the Theria, and out-group comparison is used, then correct character-state polarities may be assessed. When dealing with fossil organisms, or with a mixture of recent and fossil forms, one may easily come across similar problems. Often, the most primitive members of a particular group may not last long; they may often be replaced by their relatively derived cousins and consequently will be numerically less common in the total fossil record.

More important than anecdotal examples such as the one above, Watrous and Wheeler (1981) have pointed out the theoretical weaknesses of the commonality principle. First, when dealing with the relationships between three taxa, the principle *common = primitive* will make it impossible to resolve relationships between the taxa. Arguing that two of the three taxa are more closely related to each other than either is to the third requires that some synapomorphy be shared between the two taxa and thus for at least that character state *common = derived*. Furthermore, given any three taxa, two of the taxa must be more closely related to each other than either is to the third (unless the extremely unlikely case is encountered that the three taxa under analysis arose simultaneously from an ancestral taxon), and thus there must be some character state that is common to two of the taxa and is also derived relative to the character state of the third taxon. Likewise, if one is dealing with a group composed of only four taxa, it will also be impossible to state that any two of the taxa are more closely related to each other than either is to either of the other two taxa. Using only the commonality principle one cannot start to resolve relationships between taxa until one is dealing with at least five taxa such that there is a clear concept of a character state being common (i.e., occurring in more than half of the taxa, but not in all of the taxa or in only one taxon).

Second, as Watrous and Wheeler (1981) point out, the commonality principle applied to in-groups will tend, on the average, to produce balanced phylogenies. That is, if a number of characters and character states are used, and the taxa being analyzed vary from taxa that are primitive in all of their character states to taxa that are primitive in approximately half of their character states and derived in approximately half, to taxa that are derived in all of their character states, then the resulting phylogeny will tend to divide the taxa into clades composed of approximately equal numbers of taxa. Each of these clades will in turn tend to be divided further along the same lines. This result is simply a function of considering the common character state always to be the primitive character state; thus every clade within the group under consideration must be composed of less than half the taxa within the group under consideration. If a clade were to be composed of half or more of the taxa under consideration, the clade would be united by a character state that was common within the group, and the principle common = primitive would be violated. Moreover, using only the commonality principle for in-group analysis will often tend to yield unresolved multichotomies because, as demonstrated above, it will be impossible to resolve relationships between only three or four taxa or clades.

Third, as should be obvious, adding further characters and character states to an in-group analysis of taxa using the commonality principle will not change or mollify the defects noted above in the commonality principle. And adding taxa to the in-group under analysis may change the polarity of some morphoclines and thus change the phylogeny, but the same types of errors will persist.

There is one hypothetical argument that might save the commonality principle, at least philosophically if not in practice. As discussed by Platnick (1979) and Watrous and Wheeler (1981), a character is composed of a group of character states "that are considered to be modified or alternate forms of the same thing (i.e., states that are homologous)" (Watrous and Wheeler, 1981, p. 4). Thus for a particular character we may think of a transformation series where character state A leads to character state B leads to character state C leads to character state D leads to character state E. Imagine three taxa, *A*, *B*, and *C*, where *A* is characterized by the possession of character state A, *B* is characterized by the possession of character state B, and *C* is characterized by the possession of character state C. Using simple commonality, nothing can be deduced regarding the relationships between these three taxa. But, since A leads to B leads to C, we can think of taxon *C* as possessing not only character state C but also necessarily character states A and B. Likewise *B* possesses A and B, and *A* possesses only A. Thus A is the most common character state among the taxa and is the primitive end of the morphocline, and taxa *B* and *C* are united by the common possession of the derived character state B. In practice, however, it may not be possible to recognize that *C* possesses A, B, and C (although sometimes it may be possible to recognize this, perhaps through ontogenetic analysis) and likewise that *B* possesses A and B. Further, if such could be consistently recognized, we might dispense with the need for the commonality principle in the first place.

As mentioned above, a method that is often associated with, and often confused with, the in-group commonality principle is that of out-group comparison. Watrous and Wheeler (1981, p. 5) have formulated the operational rule for out-group comparison as follows: "For a given character with 2 or more states within a group, the state occurring in related groups is assumed to be the plesiomorphic state. If the character contains 2 states, the alternative state is assumed to be apomorphic, thereby forming a more restricted character." Thus apomorphic character states have relatively restricted distributions, being confined to only some of the in-group taxa and not found in any of the out-group taxa. In contrast, the plesiomorphic state is found both among the out-group taxa and among some of the in-group taxa (in this sense the plesiomorphic state could be said to be more common, and thus the principle of out-group comparison has been confused with the principle of in-group commonality).

Out-group comparison, of course, is dependent upon the initial in-groups and out-groups used (which are particular hypotheses of relationships—the in-group, the relationships of the members of which are being analyzed, must be monophyletic relative to the out-group), and some higher-level phylogeny is assumed for the sake of the analysis. This higher-level hypothesis is itself subject to analysis and testing at another level (Engelmann and Wiley, 1977), and further, as a starting point, it is usually assumed that at least some of the gross relationships of the taxa under consideration are understood. As Eldredge (1979a, p. 171 footnote) states: "We must assume that we have gotten somewhere during the last few hundred years of research in systematics." If the in-group whose members are being analyzed is not monophyletic relative to the chosen out-group, then out-group analysis may lead to erroneous results. Thus, if we were to analyze the relationships of the members of a taxon composed of only monotremes and marsupials, and we used the placental mammals as our out-group, then character states shared in common by marsupials and placentals would be viewed as primitive relative to the derived character states in monotremes. (But unless the monotremes and marsupials are indeed monophyletic relative to placentals, we will be unable to find true synapomorphies uniting the marsupials and monotremes relative to the placentals). If instead we analyze the relationships of the members of the Theria (marsupials and placentals) using the Monotremata as an out-group, then characters shared between monotremes, some marsupials, and/or some placentals would be considered primitive relative to specializations in some clades of marsupials and placentals. Further, the marsupials and placentals should be united by synapomorphies relative to the monotremes.

Of the three taxa, Monotremata, Marsupialia, and Eutheria (placentals), we could decide which two are more closely related to each other relative to the third by initially assuming that all three form a monophyletic clade, Mammalia, within the larger clade Amniota and then use a nonmammalian group of amniotes, such as turtles, in out-group comparison to determine the polarities of various characters and character states within the Mammalia. If marsupials and placentals are monophyletic relative to monotremes, there should be at least one character (call it character X) common to all four groups that has at least two character states (call them A and B). Comparing equivalent semaphoronts, the same character state should be found in turtles and monotremes (call it A) relative to the other character state that is found exclusively in marsupials and placentals (call it B). Thus B would be the derived character state for character X and character state B would be a synapomorphy distinguishing the taxon Theria (marsupials + placentals). The working out of specific phylogenies must be based on formulating, and testing with characterstate distributions, competing hypotheses at both higher and lower levels.

Ontogenetic data is often used in neontology to polarize morphoclines, but its application is limited for fossil organisms. Ontogeny is often considered a direct method (perhaps the only direct method) of determining character-state polarity (Rosen, 1984). The ontogenetic hierarchy reflects the historical, phylogenetic hierarchy; transformations of character states can be directly observed in ontogenetic change. According to Rosen (1984), ontogeny can be used as an arbiter of what is noise (nonhomology or errors) and what is signal (homology) among patterns of characters. Ontogeny may refute polarities and patterns based on comparative anatomical investigations of adults, but not vice versa. Unfortunately, for many (most) extinct groups we lack adequate ontogenetic data. The use of ontogenetic data actually is the one case where the commonality principle is validly applied. By observing the ontogenies of actual organisms we can potentially observe the actual character-state transformation series of a character. Thus, in the embryonic condition a certain organism or group of organisms may exhibit character state A for character X, but in maturing to the adult form A is transformed to character state B. In comparing the character states of character X in a number of organisms that bear character X (here the mere possession of character X is a synapomorphy that defines a higher-level group), some of the organisms will be found to bear both character states A (early in their ontogeny) and B (in the adult forms) while other organisms will be found to bear only character state A (both in early ontogenetic stages and in the adult form). Thus character state A would be common to all of the organisms, but character state B would define a more inclusive group within the larger group. The more widespread character state (A in this case) would be considered primitive relative to the more restricted and apomorphous character state (B).

This concept of the use of ontogenetic data is fundamentally different from Haeckel's famous biogenetic law that ontogeny recapitulates phylogeny (for discussion and historical review, see de Beer, 1948; Eldredge and Cracraft, 1980; Gould, 1977; Humphries and Funk, 1984; Nelson, 1978b; Nelson and Platnick, 1981; Wiley, 1981). Taking Haeckel's formulation literally, the ontogenetic stages of a descendant organism are said to mimic the adult ancestors of the organisms. It is the embryologist von Baer's (and von Baer [1828] worked before Haeckel [1866, 1868]) formulation of the problem that is used today. Von Baer suggested that in the development of the organism general characters appear prior to special characters, and subsequently the special (advanced or derived character states) characters develop from the general characters; during its development an organism progressively departs in form from that of other organisms (i.e., the embryos or young of organisms appear more similar than do the adults). (Strictly speaking, certain general characters need not, and sometimes may not, appear prior to other special characters in ontogeny, but less general characters develop from more general characters: Patterson, 1983b, p. 25.) Thus, it follows that the young stages of an organism are not similar to the adults of its ancestors (or organisms lower on the scale of being) per se, but are similar to the young of the ancestors.

One major drawback to the use of ontogenetic data in attempting to polarize morphoclines is the problem of phylogenetic neoteny (Eldredge and Cracraft, 1980) or paedomorphosis (Gould, 1977). In this special case, through phylogeny, either juvenile characters of ancestors are secondarily retained as adult characters in the descendants (the final ontogenetic stages may be deleted) or characters may revert (or converge) toward the juvenile or ancestral state. Such deletions of final (terminal) ontogenetic stages can pose serious problems, introducing errors that may be exaggerated if only adult characters are studied (Rosen, 1984). For example, given our character X, with two character states, A and B, where A is primitive and ancestral and precedes B developmentally, B may be transformed developmentally and phylogenetically into A' (where A' is a character state that is indistinguishable in isolation [i.e., without knowing its position in the developmental sequence] morphologically from A) or some form that had developed B may completely lose B and the developmental pathway to B. In the first case B may still appear in the developmental sequence of the organism only to be transformed to A', and thus A' will be recognized as a character state distinct from, and derived relative to, both A and B. In the second case, however, there may be no remaining trace of B and no direct evidence that there has been a loss of character state B. Here the loss of a character state (B) is a derived character state in and of itself (and we can label it B'), but organisms with the derived character B' will be indistinguishable from organisms bearing the primitive character state A. Hence the neotenous and derived condition converges on the primitive condition, and character states A and B' may mistakenly be considered homologous. The only way to distinguish such homoplasy may be to analyze a number of other characters and character states for the organisms involved. Such further, independent characters when analyzed should indicate that A and B' are not homologous but independent character states, and that taxa bearing both A and B' (remember, B' is derived from A) belong to a group less inclusive than the group composed of all taxa bearing A.

Conceptually related to the utilization of ontogenetic data for polarizing morphoclines is the use of atavisms for the determination of morphocline polarity. In rare instances where a true atavism is discernable (an example might be extant three-toed horses), such an atavism—literally a throwback to an ancestral condition—will directly indicate the primitive end of a morphocline for the particular character concerned.

The geological and stratigraphical context of fossil remains has traditionally been interpreted as being directly indicative of the polarity of morphoclines. The basic operational rule is that those fossils, and the character states they bear, that are found lower in the stratigraphic record (i.e., are older) are generally more primitive than those fossils and character states only found higher in the stratigraphic record (Gingerich and Schoeninger, 1977; Harper, 1976; Szalay, 1977*a*, 1977*b*, 1977*c*). This operational rule is directly related to the traditional assumptions that the fossil record is the only actual historical documentation of evolution, that a fossil record is necessary to provide the essential time dimension necessary for the reconstruction of evolutionary history, and that ancestral species are identifiable in the fossil record. As is discussed elsewhere in this book, all of these assumptions are open to criticism (see in particular the section on ancestor-descendant relationships).

Nelson and Platnick (1981) assert that there are two distinct, though related, ways in which the fossil record is used in phylogeny reconstruction. First is the notion that phylogeny (evolutionary sequences of ancestors and descendants) can be read directly from the rocks. As Nelson and Platnick (1981, p. 333) point out, this concept is fallacious (or in their words, "superstition and nothing more"). We cannot conclude that fossils belong to the same lineage merely on the basis of their stratigraphic occurrences (if this were not the case, we might conclude that any fossil in a lower stratum may have given rise to any fossil in a higher stratum). Rather, fossils must be somehow ordered or organized on the basis of morphology and higher-level systematic hypotheses before putative ancestor-descendant sequences are hypothesized on the basis of the stratigraphic sequences of fossils. This idea leads to the second use of the paleontological-stratigraphic record: that within a monophyletic group, for various character states of a single character, plesiomorphic character states occur earlier (lower) in the fossil record. That is, actual character-state transformations or character phylogenies are recorded in the fossil record. In this context, the paleontological argument is subservient to the morphology of the organisms and a postulated or assumed higher-level phylogeny. Nelson and Platnick (1981) argue that the paleontological technique is an indirect method (being dependent on higher-level assumptions) of reconstructing morphocline polarities.

Potentially the fossil record should be able to falsify a hypothesis of character phylogeny. For example, if, for a given character, it is hypothesized that character state A is primitive relative to character state B, this hypothesis should be falsified by the fossil record if fossils bearing B are known that are older than any fossils bearing A. However, as Nelson and Platnick (1981) note, any hypothesis of character phylogeny can always be saved from falsification by the stratigraphic sequence of fossils merely by positing the ad hoc (but potentially true) suggestion that the fossil record is not as complete as it was previously believed to be.

From an operational point of view, in many instances ancestor-descendant relationships may not be objectively recognizable, and in many particular cases species bearing derived character states may occur in strata below the earliest known occurrences of the primitive states of the same characters (Eldredge and Cracraft, 1980; Schaeffer, Hecht, and Eldredge, 1972). However, in other instances primitive character states do indeed appear to occur lower and earlier than the corresponding derived character states. Most workers would probably agree that at a very crude level the broad outlines of evolution are recorded in the fossil record; for example, aquatic vertebrates preceded terrestrial vertebrates, and reptiles appear prior to the rise of the first mammals (Cartmill, 1981). The probability that the sequence of appearance of character states in the fossil record will correspond to a true primitive-to-derived transformation series depends to a great extent on the completeness of the stratigraphic and fossil record at a particular level of analysis (see chap. 4). It seems that initially it is perhaps best neither to take stratigraphic/temporal data at face value and axiomatically accept that early equals primitive, nor to ignore it completely. Rather, as with any methodology, one should use such data and methodology with due caution.

Chorological (essentially geographical and ecological) series of organisms or character states are sometimes considered useful in determining morphocline polarities (e.g., Hanson, 1977; Hennig, 1965). It is sometimes assumed or speculated that a certain area is the center of origin of a specific taxonomic group and morphologies that are found in the center of origin will generally be primitive relative to morphologies found in areas progressively farther away from the center of origin (or more extensive departures of descendant organisms from the original, ancestral habitat and ecological niche should correlate with more extensive evolutionary change in the descendant forms). The very notion of centers of origin is under question (Nelson and Platnick, 1981; and see chap. 5). To use this method the basic assumption has to be made that the taxa in the center of origin evolved more slowly (perhaps because their environment remained stable), thus retaining more primitive morphologies, than taxa progressively farther away from the center of origin.

The correlation of character states in hosts and parasites can sometimes be used to successfully determine the polarities of morphoclines for one group of organisms (either hosts or parasites) if the phylogeny is reasonably well reconstructed for the other group (Hennig, 1965, 1966a; for examples, see Brooks, 1979a, 1979b, 1980, 1981a, 1981b; Brooks and Glen, 1982). Similarly, different transformation series for different characters can be compared within a single group of organisms. If the polarity of one morphocline is well established, this may be useful in determining the polarities of other, parallel morphoclines in the same series of organisms.

PARSIMONY

The concept of parsimony was introduced in the Introduction and chapter 1; here it is discussed with special reference to phylogenetic analysis.

Parsimony is one of the principal methodological rules, one of the most basic ideas, in philosophy and science. Put simply, the idea of parsimony is that if there are two or more explanations, hypotheses, or solutions to a given problem, the simplest solution (that involving the smallest number of ad hoc hypotheses, auxiliary conditions, or logical steps) is to be preferred (assuming all other factors being equal). The problem of why the simplest solution is to be preferred, or more generally how we can ultimately justify any criterion of acceptance (e.g., parsimony, falsifiability, informativeness, boldness) used to choose one hypothesis over another, is a problem that is as yet unresolved (see Sober, 1983b; also Dunbar, 1980). However, most science is based on, or incorporates, some general principle(s) that is (are) used in deciding between alternative hypotheses. Often this takes the form of a parsimony/simplicity principle, and there appears to be general agreement among scientists "that science makes sense" (Sober, 1983b, p. 39).

The principle of parsimony (also referred to as the principle of economy or simplicity) is usually attributed to the Catholic philosopher William of Ockham (c. 1285–1349) and thus is called Occam's razor (Copleston, 1972; Hill and Crane, 1982); however, it has its roots at least as far back as the writings of Aristotle in the fourth century B.C. (Brown, 1950; Cohen, 1960; Dunbar, 1980). Original statements of William's, after translation from the Latin, read something like this: "Never is multiplicity to be postulated without necessity" (Kneale and Kneale, 1962, p. 234), and "It is vain to do with more what can be done with fewer" (B. Russell, 1946, p. 243: the two quotes above are given in Hill and Crane, 1982, p. 349; Gaffney, 1979a, and Kluge, 1984, give further quotes from William's writings). Kohlberger (1978, p. 375, footnote) pointed out that originally William's concept of parsimony included divine revelation, whereas now it is used to refer exclusively to hypotheses of maximum corroboration/minimum contradiction without consideration for the gods; Kohlberger has proposed substituting the word "economy" for the latter meaning. However, the term *parsimony* is well entrenched in the literature, and there seems to be little confusion of the modern concept with William's original idea that included divine revelation.

As presently used by many phylogeneticists and systematists, and as used here, parsimony is usually used in a very restricted and specific sense: that ad hoc hypotheses of homoplasy be minimized. In the context of a cladistic analysis, grouping terminal taxa strictly on the basis of synapomorphies is the essence of parsimony (Humphries and Funk, 1984). A certain trait or character in a phylogeny (genealogy) should not be required to originate independently more than once (Farris, 1983; reprinted in Sober, 1984b, p. 687). (Here the word *independently* is important. If numerous terminal taxa show a single feature, and the putative phylogeny distributes these taxa into two distantly related monophyletic groups, then the feature in question need only have originated twice independently, once in each monophyletic group [Farris, 1983]. One need not hypothesize that the feature in question arose independently in every terminal taxon.) Hennig's (1966a, p. 121) " 'auxiliary principle' that the presence of apomorphous characters in different species [i.e., corresponding apomorphies, putative synapomorphies] is always reason for suspecting kinship (i.e., that the species belong to a monophyletic group), and that their origin by convergence should not be assumed a priori" is another statement of the parsimony principle as specifically applied to phylogenetics.

The concept of parsimony outlined in the last paragraph is called methodological parsimony by Kluge (1984). Methodological parsimony is merely a rule that advocates accepting the theory, hypothesis, or proposition that best accords with all relevant data and observations and requires the fewest number of ad hoc or a posteriori assumptions. Unfortunately, systematists sometimes use the term *parsimony* to refer to what Kluge (1984) calls evolutionary parsimony. Evolutionary parsimony involves assumptions about how evolution works—for example, that evolution is economical and will involve a minimum number of steps. Evolutionary parsimony is not the parsimony that is referred to in this book.

Cladists have generally accepted parsimony as a basic aspect of scientific methodology and have generally believed that the acceptance of parsimony "requires no assumptions about the contingent properties of the evolutionary process" (Sober, 1983a, p. 336). This is to say, we do not need to specify any particular models of evolutionary processes, mechanisms, or modes to utilize the concept of parsimony. In contrast, others have believed that the idea of parsimony implies assumptions about the evolutionary process. For example, if it is held that parsimony minimizes the number of homoplasies in any particular case, then the assumption is that homoplasy is relatively rare in nature (Sober, 1983a). In general, noncladistic methods of phylogenetic reconstruction are not parsimonious (Farris, 1983, and see discussion below).

Sober (1983a) has pointed out that what phylogeneticists and systematists observe is "patterns of sameness and difference in the characteristics of taxa." On the basis of this information and using certain principles, of which parsimony may be one, they may attempt to reconstruct phylogenetic relationships between the taxa. Samenesses and differences do not mean anything in and of themselves until some principle of inference is applied to them. Identical (as far as the investigator can discern) derived characters that are shared between taxa (shared derived characters = synapomorphies) and identical primitive characters that are shared between taxa (shared primitive characters = symplesiomorphies) may result from any number of causes—independent evolution (parallelism or convergence = homoplasy), common ancestry, or the whims of the gods. It is by the adoption of the principle of parsimony that synapomorphies and symplesiomorphies come to imply homology and are believed to be the result of genealogical ties and that synapomorphies count as evidence of phylogenetic relationship. Characters that were initially thought to be synapomorphies, but that contradict the distribution of the majority of synapomorphies for a given group of taxa, are relegated to the status of homoplasies, and the preferred hypothesis of relationships is that which invokes the fewest number of homoplasies (independent evolutionary events of any character state) to account for the known distribution of the character states among the taxa under analysis.

Farris (1983; also described in Sober, 1983a) has provided a simple but powerful argument against the criticism that if parsimony asserts that homoplasies should be minimized, this must imply that parsimony assumes homoplasies are rare. Given three taxa, A, B, and C and that there are ten putative synapomorphies that unite A and B, but only one putative synapomorphy uniting B and C, parsimony favors the grouping ((A, B) C). It is assumed for simplicity of discussion that the characters used are of equal weight and the data is relatively complete (an honest search for putative synapomorphies has been made, and no known putative synapomorphies have been omitted). It is also assumed that reticulate evolution has not been involved in the formation of the taxa under consideration. Farris (1983) notes that if the one putative synapomorphy of B and C is correct (a true synapomorphy), then the grouping (A (B, C)) must be phylogenetically correct and the ten putative synapomorphies uniting A and B must be homoplasies. On the other hand, if only one of the ten putative synapomorphies uniting A and B is correct, then the grouping ((A, B) C) must be phylogenetically correct and the putative synapomorphy uniting B and C must be a homoplasy. If none of the putative synapomorphies are correct, this does not imply that either grouping is phylogenetically false. In other words, the only requirement of the phylogenetic grouping ((A, B) C) is that the putative synapomorphy uniting B and C is a homoplasy, and the requirement of the grouping (A (B, C)) is that the ten putative synapomorphies uniting A and B are homoplasies. For either phylogeny to be correct, the only ad hoc hypotheses are those concerning the homoplasies; the putative synapomorphy (or synapomorphies) in either case may be true or false. As ((A, B) C) requires only one homoplasy to have occurred, but (A (B, C)) requires ten homoplasies to have occurred, no matter how abundant homoplasies are, the hypotheses requiring fewer homoplasies cannot be less probable than the hypotheses requiring more homoplasies (Sober, 1983a). By this demonstration, parsimony does not assume rarity of homoplasy.

Felsenstein (1978; reviewed in Sober, 1983a) has developed a specific model of evolutionary processes and relationships in which parsimony will not converge on the true phylogenetic grouping of taxa as more characters are used. Felsenstein assumes a phylogenetic tree as shown in Figure 3-1 for the taxa A, B, and C and a situation where the plesiomorphic character states are labeled 0 and the apomorphous character states are labeled 1, the ancestral taxon at the root of the tree is plesiomorphous for all character states considered, characters can evolve only from 0 to 1 (character reversals are impossible), characters evolve independently of each other, and the probability of the transformation (evolution) from character state 0 to 1 for all characters is P on certain of the branches of the tree and Q for other branches of the tree (as indicated on Fig. 3-1) where P>>Q. From these basic assumptions, a pattern of 101 (for any character, taxon A bears character state 1, taxon B bears character state 0, and taxon C bears character state 1) is more probable than the pattern 110. As one examines more and more characters, the frequency of characters with character state distributions of 101 will be greater than 110. But parsimony would interpret a predominance of 101 distributions to indicate the phylogenetic grouping ((A, C) B), which is not correct in this case.

Felsenstein (1978) has demonstrated that whether parsimony works depends on how evolution operates. He has identified a case where parsimony will not yield the true phylogeny unless homoplasy is extremely rare or distinguishable as such (that is, we could distinguish the 1's in 101 patterns as convergence rather than synapomorphy). However, Felsenstein readily admits that his model of evolution is unrealistic. As Farris (1983, reprinted in Sober, 1984b, pp. 681–682; see also Farris, 1973) notes in discussing Felsenstein's analysis, the fact that the usefulness of parsimony appears to depend on the relative rarity of homoplasy in particular cases as postulated by Felsenstein is a direct consequence of Felsenstein's models. Felsenstein's models include extreme assumptions about biological evolution that no one, including Felsenstein, actually maintains are realistic. If Felsenstein's assumptions do not apply to real cases of evolution and phylogeny, it cannot be claimed that his criticisms of parsimony methods are valid criticisms of parsimony as applied

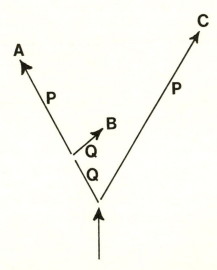

Figure 3-1. Hypothetical evolutionary tree for three terminal species (A, B, and C); see text for further discussion. After Felsenstein, 1978, pp. 49–62 and fig. 1.

to the reconstruction of the phylogeny of real biological organisms. At best, Felsenstein has succeeded in demonstrating that if homoplasy were extremely abundant, and if the conditions and assumptions of his models held true in nature, then one should prefer other, nonparsimonious schemes of phylogeny reconstruction. As Farris (1983) has shown, the extreme abundance of homoplasy by itself does not warrant or justify the use of nonparsimonious methods. Therefore, it is solely the supposition that Felsenstein's admittedly unreal assumptions and model of evolution might apply to real cases in nature that is presented as the potential reason for favoring nonparsimonious methods. The irony of this entire argument for nonparsimonious methods is that parsimonious methods were themselves originally criticized as making unrealistic assumptions about nature. However, as it turns out, it is nonparsimonious methods that require unrealistic assumptions to be made (Farris, 1983).

Felsenstein's (1978) argument seems reducible to this: Methodological parsimony is invalid because one can never be certain that it will lead to the truth, that it will not lead to errors. However, one can never be certain of obtaining error-free truth in science; one can never be certain that one's conclusions are not wrong. Methods in science cannot always be evaluated scientifically. If we knew the answer (the truth), we could determine how well our methodology approached the truth, but in real nature the answer, or truth, is not known outside of our methods used in trying to approach it (Platnick, 1979). Moreover, it is always possible to concoct scenarios wherein scientific conclusions attained utilizing a particular methodology will be incorrect, as Felsenstein (1978) has done. Such scenarios may not be ultimately disprovable, but there may also be no reason for accepting such scenarios, and there may be good reasons for not accepting such scenarios. The abstract possibility that a scientific conclusion might prove incorrect is an entirely empty argument that would not be disputed by any scientist. The attitude that all conclusions that cannot be established with absolute certainty should be rejected leads to a situation where no scientific conclusions can ever be arrived at (Farris, 1983).

Cladists use the term *falsify* or *falsifier* in reference to character distributions that contradict a particular cladogram and, adopting the principle of parsimony, prefer the hypothesis of relationships that is least falsified—that is, has the fewest character-distribution conflicts or requires the fewest ad hoc hypotheses of homoplasy. The cladistic method has been argued against by some workers (e.g., Cartmill, 1981) in the following simplistic manner: Given a large set of data of putative synapomorphies, it is often the case that every possible cladogram is falsified by at least one character-state distribution (every cladogram requires at least one ad hoc hypothesis of homoplasy), and since every cladogram is falsified at least once, then every cladogram has been demonstrated false. But obviously, if evolution occurred, one phylogeny must be true;

therefore, the problem is in the method, not in the data. Both Farris (1983) and Sober (1983a) have addressed this point. Apparently, workers who argue as above have misunderstood the basic concept of falsification. Falsifiers are statements or observations that test a hypothesis, but if a falsifier falsifies a hypothesis, that does not prove that the hypothesis is false. Rather, it implies that either the theory or the observation is false. Proof of falsity, like any other proof, is not literally achieved empirically. For phylogenetics in particular, if every phylogeny is falsified by the distribution of at least one putative synapomorphy, then at least one putative synapomorphy must be a homoplasy.

Sober (1983a, pp. 353–354) has attempted "to show that, in the absence of specific evolutionary information, parsimony indicates which cladistic groupings are most strongly supported by the observed character distributions." This he has done by setting up a scenario where he compares the phylogenetic groupings ((A, B) C) and ((A, C) B) to each other for all possible tree topologies with which they are consistent. Sober (1983a) found that relative to the observation of the character-state distribution 110, the grouping ((A, B) C) has a higher likelihood of corresponding to a phylogenetic arrangement of the taxa than does the grouping ((A, C) B) given the following assumptions: (1) The probability of a character state transforming from 0 to 1 or vice versa is between 0 and 1 (this must be true if chance plays any role at all in evolution—that is, evolution is not predetermined); (2) the probability of transition events in different parts of the evolutionary history (evolutionary tree) are statistically independent from one another (Sober notes that this may be a somewhat idealized case, but probably it holds as long as the analyzed taxa are well separated spatially or temporally and do not interact); and (3) at any event (on any branch) the probability of the transition 1 to 1 is greater than the probability of the transition 0 to 1 (this is to say that intuitively "if evolution is pushing in a certain direction, it cannot hurt your chances of ending up there that you are there already": Sober, 1983a, p. 351). In this context "likelihood may be regarded as a measure of the degree to which a hypothesis explains the observations" (Sober, 1983a, p. 344). Sober found that the character-state distributions 001, 111, and 000 are uninformative. Thus, given his three assumptions, which do appear to be fairly reasonable in terms of most any evolutionary models, Sober was able to demonstrate that likelihood and parsimony do coincide, and his analysis is independent of whether homoplasy is common or rare.

Sober (1983a, p. 353) is careful to note—and this should always be kept in mind—that he has not considered or been concerned with the weighting of evidence, specific information concerning a specific group of organisms, or "independently confirmed empirical evidence" that contradicts an otherwise technically parsimonious cladogram. Nor has he considered possible reticulation or hybridization. In any real case when an investigator is considering the phylogeny of a particular group of organisms, there may be a vast amount of empirical data, over and above the raw matrix of character-state distributions, on which to draw. One must be extremely careful not to utilize unfounded assumptions, or intuitive feelings, about a particular group of organisms a priori. Sober (1983a, p. 356) also points out that parsimony does not tell us which cladogram is to be accepted absolutely—it merely "records which cladistic groupings are best supported by observations" [the observations we have at hand of putative synapomorphies]. Careful workers will not accept a single parsimonious cladogram as the true phylogeny, especially when there are one or more alternative cladograms that are only slightly less parsimonious.

The cladistic program is based on the two principles: (1) that putative synapomorphies provide the only evidence of phylogenetic relationships (unless we directly observe ancestry and descent) and (2) that hypotheses of phylogeny should be parsimonious—that is, they should minimize hypotheses of homoplasy (which are ad hoc assumptions) and be maximally explanatory (phylogenetic hypotheses should agree with the observation at hand; the only way phylogeny explains similarities between organisms is by accounting for them in terms of inheritance from common ancestors). In contrast to the cladistic program, many other methodologies proposed for reconstructing phylogeny are nonparsimonious (Farris, 1983; see chap. 1).

Some methods of phylogeny reconstruction include the assumption that the evolution of individual characters is irreversible; examples include the techniques of Camin and Sokal (1965) and Farris (1977a). As Farris (1983; reprinted in Sober, 1984b, p. 691) notes, such methods may, in particular cases, result in the same phylogenetic hypotheses as do parsimonious analyses, but they need not. In general, the number of hypotheses of homoplasy will be increased given the restriction of evolutionary irreversibility. Why should one wish to assume that character evolution is irreversible? Farris (1983) suggests that this notion has been arrived at inductively from putative lineages and phylogenies, but these putative phylogenies were arrived at by observing and ordering data (character distributions and organisms) while presupposing irreversibility. This method involves obviously fallacious and circular reasoning. Otherwise, it is not known that evolution is irreversible. In order to avoid such fallacious reasoning, it must be allowed that character distributions may support conclusions of reversal, if such reversal is more parsimonious (explains more of the observations with fewer ad hoc assumptions). Farris also notes that it might be argued that concluding that reversal has taken place itself requires that an ad hoc hypothesis be made—that reversal has taken place. But, to quote Farris (1983; reprinted in Sober, 1984b, p. 693),

If a particular conclusion of reversal could be legitimately criticized as presupposing the possibility of reversal, then any scientific conclusion whatever could be dismissed as requiring the supposition of its own possibility. The argument outlined is seen in that light to be simply another rationalization for discarding evidence.

Phenetic clustering, grouping organisms together on the basis of raw similarity, has often been suggested as a means of reconstructing phylogenetic relationships (see, e.g., Sneath and Sokal, 1973). The basic assumption justifying this method is that rates of evolutionary change (divergence) are nearly constant, or at least constant enough that in general the degree of raw similarity between different organisms will reflect the relative recencies of common ancestry. As in the case of irreversibility, Farris (1983) notes that in particular cases phenetic clustering may produce the same results as cladistic analysis, but it need not. Theoretically, there is no limit to the number of hypothesized homoplasies that phenetic clustering might require. As for the assumption of irreversibility, the assumption of constant rates of evolution seems to be based on putative phylogenies that were constructed on the basis of character distributions while presupposing the constancy of evolution. The argument that parsimony analysis presupposes that rates of evolution can vary is analogous to the argument that parsimony analysis presupposes that evolution is reversible. Such an argument sets up a situation where the hypothesis (assumption) of constancy of evolution is not empirically testable.

Farris (1983) also points out that if phenetic clustering requires otherwise unnecessary hypotheses of homoplasy, then the theoretical relationship between raw similarity and recency of common ancestry must include some mechanism to explain the homoplasies, or itself take the form of an ad hoc covering assumption that discounts evidence en masse.

That mechanism would have to have the property that organisms would come to possess features in common for reasons other than inheritance, and in just such a way as to maintain the correlation between raw similarity and recency of common ancestry. As no known natural process appears to have this property, it would seem that use of a postulated correlation between raw similarity and kinship to defend clustering by raw similarity rests necessarily on an ad hoc covering assumption. (Farris, 1983; reprinted in Sober, 1984, p. 696)

Compatibility, or clique, methods resolve character conflicts by discarding certain incongruent characters and then interpreting the remaining, mutually compatible characters parsimoniously (see, for example, Estabrook, 1979; Estabrook, Strauch, and Fiala, 1977; Gardner and La Duke, 1979; LeQuesne 1972: such methods are reviewed by Farris and Kluge, 1979, and discussed by Farris, 1983, and Wiley, 1981). The resulting phylogeny will, not surprisingly, often be quite unparsimonious for the excluded characters. It seems that the basic assumption of clique analysis is "that excluding a character—concluding that it shows some homoplasy—implies that all points of similarity in that character are homoplasies" (Farris, 1983; reprinted in Sober 1984, p. 698). Such an assumption regards all occurrences of a character that in one case is a putative homoplasy to be independent homoplasies. Such a blanket or covering assumption may not be valid, but rather may be an ad hoc hypothesis for dismissing certain evidence that might lead to different conclusions from those based on the selected clique of data. For example even if it is concluded that the hypocone (a cusp on the upper molars of some mammals) is not homologous in all mammals possessing a hypocone, this does not mean that a hypocone is an independent acquisition (a homoplasy) in every species of mammal bearing a hypocone. Some monophyletic groups of mammals are, in part, characterized by the synapomorphy of possessing hypocones, even if hypocones have been acquired independently in various monophyletic groups of mammals. Equating the hypocones of certain species of mammals as a synapomorphy might be in error (where the two species compared bear hypocones that were independently derived in different ancestral species), but in other cases the occurrence of hypocones in two species may be a synapomorphy, indicative of an ancestry common to both. Clique techniques might exclude the use of hypocones altogether, thus discarding the latter, valid evidence of common ancestry. In contrast, a parsimonious analysis would ideally recognize both cases where the possession of hypocones is a homoplasy and where the possession of hypocones is a synapomorphy.

The Wagner Groundplan-divergence Method is, in its original formulation, based on shared derived characters and therefore is parsimonious. Kluge and Farris (1969) formulated an automatic technique of the Wagner method that is parsimonious and in which numbers of independent origins of features on a branch of a proposed phylogenetic tree were treated as the lengths of the branches (Farris, 1970, 1983; Wiley, 1981). Thus, the most parsimonious tree (with the least number of required origins) would be that with the smallest summed lengths of branches. However, many putative phylogenetic trees can be constructed that are not based on parsimony methods but in which lengths of branches are minimized. As Farris (1983; reprinted in Sober, 1984, pp. 687–688) states:

Once ideas have been reduced to formulae, it is easy to forget where the formulae came from, and to devise new methods with no logical basis simply by modifying formulae directly. . . . The lengths arrived at by such calculations are generally incapable of any interpretation in terms of origins of features, and the evaluation of trees by such lengths consequently has nothing to do with the phylogenetic parsimony criterion. What is worse, the trees produced by these methods frequently differ in their grouping from parsimonious genealogies, and to that extent the use of these procedures amounts to throwing away explanatory power.

In conclusion, parsimony does not make any assumptions about the mechanisms or processes of evolution, and this is its strength. If parsimony makes any assumption, perhaps it is simply that character distributions among organisms may permit the reconstruction of phylogeny (cf. Sober, 1983*b*). Parsimony is only a tool used to find agreement among patterns. Because parsimony is presupposed does not mean that evolution is presupposed to be parsimonious; the concepts of parsimony and historical (evolutionary) truth are perhaps best divorced from one another (Nelson and Platnick, 1981). The only assumption that parsimony makes is that the data at hand, the evidence, should conform to hypotheses in as complete a manner as possible. Phylogenetic hypotheses should be maximally explanatory. If adopting some assumption, such as the irreversibility of evolution, requires the discarding of parsimonious explanations, then the assumption is placed in a position where it is assumed to be true despite the evidence at hand. Parsimony analysis may potentially corroborate any theory, given that the evidence supports the theory. Phylogenetic analysis as an empirical study must rest on the principle of parsimony (Farris, 1983).

PARALLELISM AND RETICULATION
IN CLADOGRAMS

Brady (1983) has demonstrated that the construction of cladograms depends upon two assumptions or expectations: (1) The expectation that homologies (synapomorphies) can be ordered hierarchically into proper subclasses (Hill and Crane, 1982, p. 284) where each subclass or taxon recognized is distinguished by attributes (character states) unique to that taxon at that level. "Once the decision to use cladograms has been made, the question is not *whether* the data set can be represented in hierarchical form but *which* hierarchy best represents it" (Brady, 1983, p. 51, italics in the original). (2) The basic assumption is made that the data set used to construct the cladogram accurately and adequately reflects the taxa it represents. Once these assumptions are made, it is on a purely rational and logical basis that one searches for the cladogram that contains the fewest violations (i.e., fewest character states with contradictory distributions). The cladogram with the fewest violations is the most parsimonious cladogram, and on a logical basis (given the initial assumptions), the preferred cladogram.

Once a preferred cladogram is settled upon, there may be some character-state distributions that contradict the preferred cladogram—that is, such a contradictory character state is found in two or more mutually exclusive groups and thus does not distinguish a proper subclass in the context of the schema of the preferred cladogram overall. To state it another way, what was initially hypothesized to be a single (identical) character state in two or more taxa appears to arise independently in two or more monophyletic groups of a purely dichotomous or multichotomous version of the preferred cladogram (i.e., a cladogram without reticulation).

There are several ways that one can deal with violations stemming from contradictory character-state distributions. Either or both of the basic assumptions necessary for erecting cladograms can be discarded. Discarding assumption one, that the character states can be ordered hierarchically, is in this context equivalent to denying that phylogeny can be reconstructed on the basis of the character-state distributions (data) at hand. Perhaps not enough information is recorded in the morphologies of the particular organisms under consideration, or the data set is inadequate, to reconstruct degrees of relatedness, and the investigation must be terminated pending a better data set. (Or a creationist might argue that phylogeny—evolution—has not occurred, and thus there is no reason why the data set should be orderable hierarchically; of course the data may be orderable hierarchically on some basis that reflects something other than phylogeny.)

Another way to deal with the problem is to reject some of the initial hypotheses as to the polarities of the character-state transformation series (or to reject a whole transformation series as not being a transformation series) within the data set. Those character states that have contradictory distributions can be rejected as synapomorphous character states at the level they were believed to be synapomorphous (this level would be the level including all the groups, and only those groups, under consideration some of whose members possessed the character state in question). A character state that appears to be confined to two or more mutually exclusive groups may actually be a primitive (plesiomorphic) character state at that level that has been retained in certain entities within the two or more mutually exclusive groups. Such a character state may be reevaluated as a synapomorphy at a much higher (perhaps using out-group comparison), more inclusive level that has been lost in many derived groups at lower levels. The character state would distinguish a proper subclass after all, but at a higher level. Of course, the opposite end of the transformation series for the character in question will then have to be interpreted as a synapomorphy at lower levels rather than at higher levels as it previously was, and this may result in new violations. Alternatively, it may be hypothesized that the character state under consideration is a true synapomorphy uniting all of the taxa at the specified level, but that it has been secondarily lost in some clades and thus has what appears to be a disjunct (contradictory) distribution.

What had initially been considered to be an identical character state in a number of organisms may, when found to have a contradictory distribution, be hypothesized to be actually several independent characters and character states that had been incorrectly regarded as a single character state—that is, an ad hoc hypothesis of homoplasy may be invoked to explain the contradictory character state distribution.

Finally, it may be hypothesized that hybridization has occurred, resulting in a reticulated cladogram.

Parallelism and Convergence

The two major forms of homoplasy that are distinguished by many authors are parallelism and convergence. These topics have been discussed in some detail in the section on homology, and they are further elaborated upon here.

Traditionally, and among many evolutionary and other systematists, it has been argued that while convergence is a hindrance to phylogeny reconstruction, parallelism may shed some light on phylogenetic relationships. In this context, parallelisms are usually considered "similarities resulting from joint possession of independently acquired phenotypic characteristics produced by a shared genotype inherited from a common ancestor (similarity through parallel evolution)" (Mayr, 1969, p. 202: cf. Patterson's, 1982a, concept of parallelism). Similarly, some workers of a cladistic bent have recognized forms of parallelism that they believe might be useful in recognizing phylogenetic relationships. Brundin (1976a, p. 140: quoted in Patterson, 1982a, p. 49; see also Brundin, 1976b) recognized unique inside-parallelism as "the presence of a unique parallelism inside a group that is supposed to be monophyletic, meaning that a unique character has been developed independently within each of two subgroups making up the group," and Saether (1979a: quoted in Patterson, 1982a, p. 49; see also Saether, 1979b, 1979c, 1983) has distinguished a special type of parallelism that he calls underlying synapomorphies, which is due to "common inherited genetic factors including parallel mutations." (The other type of parallelism recognized by Saether [1979a] is parallelism due to parallel selection; it is not considered to be useful in reconstructing phylogeny.)

The basic idea is that such forms of parallelism hint at, or indicate in an imperfect manner (in the sense that the parallelism may not be present in all of the groups derived from a certain common ancestor), otherwise unrecognized and unseen synapomorphies of a group (the "underlying synapomorphies" of Saether or the "shared genotype inherited from a common ancestor" of Mayr). If the parallelism is not seen in all of the derived groups, the underlying synapomorphy must be an ad hoc hypothesis for some of the groups that are united. This seems very flimsy evidence on which to base phylogenetic interpretations. In actuality, as Patterson (1982a) has discussed, other, observed characters are used to distinguish the monophyletic groups in the first place, and the parallelisms could potentially be used to reinforce any of many different groupings; therefore, they really add nothing but confusion to attempts at phylogenetic reconstruction. If the parallelism is seen in all of the derived groups under consideration that are being united as a monophyletic group, then the parallelism is not a parallelism at all, but merely a synapomorphy.

Eldredge and Cracraft (1980, pp. 70–74) have approached the problem of parallelism and convergence from a slightly different angle. These authors consider convergence to be any similarity that cannot be used to define a set within the context of a certain cladogram (= convergence or parallelism, or both, *sensu* Patterson, 1982a, depending on how one defines similarity in this context [grossly or closely similar, or indistinguishable in isolation, see section on homology]). Eldredge and Cracraft (1980) argue that classically (but expressed using cladistic terminology) the difference between parallelism and convergence is as follows: In both cases the common ancestor of the forms under consideration is primitive for the character state considered, and in both cases some descendants of the ancestor independently evolve identical (that is, indistinguishable as different character states by the investigator) derived forms of the character state. In the case of parallelism the descendant taxa with the derived character state are strictly sister taxa (that is, they are more closely related to each other than to any other taxa, known or unknown), whereas in the case of convergence the descendant taxa with the derived character state are known not to be sister taxa.

As Eldredge and Cracraft (1980) allude to, the only feature distinguishing parallelism from synapomorphy when treated in the above manner is the ad hoc hypothesis that the last common ancestor of the descendant taxa with the derived character states did not have the derived character state also, and so it must have been derived independently in the descendant lineages. Unless we can know the common ancestor (which may be impossible; see section on ancestor-descendant relationships) we must consider the supposed parallelism a synapomorphy.

Eldredge and Cracraft (1980) also point out that in arguing for parallelism, evolutionary systematists in particular usually stack assumptions that may be epistemologically, and therefore scientifically, impossible to evaluate. Closely related species (sister groups) probably have very similar genotypes, which would mean that they would have similar developmental potentials and therefore would respond to similar environmental selection pressures in similar ways, leading to indistinguishable independently derived traits in close relatives that underwent similar selection regimes. Again, it must be assumed that the derived trait was not present in the latest common ancestor, and it must also be assumed that the genotypes, developmental potentials, environments and selection pressures, and responses (in the form of the observed indistinguishable derived traits) were all closely similar. In other words, ad hoc hypotheses are stacked upon ad hoc hypotheses, and an assumed set of phylogenetic relationships for the organisms under study (including assumptions about ancestors) is used to hypothesize and interpret the scenario of parallel evolution in the descendant lineages. Thus, this concept of parallelism does not contribute to the elucidation of phylogenetic relationships per se.

Reticulation

Nelson (1983a) has discussed hypotheses of hybridization, resulting in reticulated cladograms. The following discussion received its initial impetus from Nelson's (1983a) work.

The most parsimonious cladogram without reticulation is the cladogram that allows the maximum number of character states to be represented by

Table 3-1. Hypothetical Data Matrix Illustrating Reticulate Evolution

	A	B	C	D	E	F	G	H	I	J	K	L	M	N	O
								Character							
Taxon															
1	+	+	+	−	−	−	−	−	−	−	−	−	−	−	+
2	+	+	+	+	+	+	−	−	−	−	−	−	−	−	+
3	+	+	+	+	+	+	+	+	+	−	−	−	+	+	−
4	+	+	+	+	+	+	+	+	+	+	+	+	+	+	−
5	+	+	+	+	+	+	+	+	+	+	+	+	−	−	−

single lines. Given the data in Table 3-1 (where "+" = derived condition and "−" = primitive condition) on the state of characters A through O in taxa 1 through 5, the most parsimonious cladogram is that shown in Figure 3-2a. The cladogram shown in Figure 3-2a allows all of the characters except M, N, and O to be represented by single lines.

In order to resolve the character conflict shown by M, N, and O, lines of reticulation can be added to the initial cladogram. Nelson suggests that such lines of reticulation should be added one step at a time. In our hypothetical example (Table 3-1), there are characters shared between taxa 1 and 2 and characters shared between taxa 3 and 4. However, there are twice as many characters (M and N) suggesting a relationship between taxa 3 and 4, and thus the first line of reticulation that should be added in this case is a single line that will account for those characters (this is done in Fig. 3-2b). If only one taxon of hybrid origin is represented among taxa 1 through 5, then taxon 4 is most likely the taxon of hybrid origin. If taxa 1 through 5 are species, 4 might be the result of hybridization between species 3 and 5 or the result of hybridization between an ancestor of species 3 that was not also a common ancestor of species 5 and an ancestor of species 5 that was not also a common ancestor of species 3. Accepting one line of reticulation, the only character conflict left unresolved in our hypothetical example is character O. If a second level of reticulation is allowed, then the character conflict shown by character O can be resolved by the addition of one more line to the cladogram (Fig. 3-2c). If there is a second taxon of hybrid origin among taxa 1 through 5, it is most likely taxon 2.

Nelson suggests that when resolving cladograms one should always start at a zero level of reticulation (i.e., no reticulation is initially hypothesized) and then move on to progressively higher levels of reticulation (if so desired) until all character conflicts are resolved. The taxa that are most likely to be of hybrid origin are those that show the most intermediacy in characters between two other taxa. In our hypothetical example (Table 3-1 and Fig. 3-2) taxon 4 is the most intermediate taxon; taxon 4 shares two derived character states with taxon 3 and three derived character states with taxon 5. For a taxon that is truly intermediate between two other taxa, the character conflict seen in a nonreticulate cladogram for that taxon can be resolved by adding a single line of reticulation.

Nelson (1983a) has pointed out that there can also be character conflict in a broad sense where more than two lines leading to a single taxon are

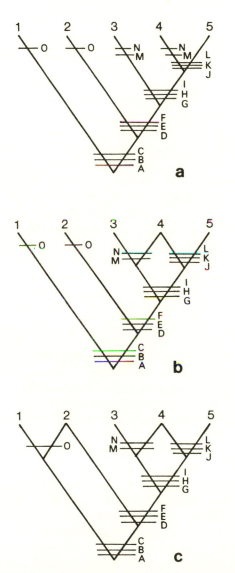

Figure 3-2. Cladograms resolving the phylogenetic relationships among the hypothetical taxa listed in Table 3-1.

Table 3-2. Hypothetical Data Matrix Illustrating Reticulate Evolution

						Character						
	A	B	C	D	E	F	G	H	I	J	K	L
Taxon												
1	+	+	+	+	+	+	+	+	+	−	−	−
2	+	+	+	+	+	+	+	+	+	+	+	+
3	+	+	+	+	+	+	−	−	−	+	+	−
4	+	+	+	−	−	−	−	−	−	−	−	+

needed in order to resolve all of the character conflict. This broad character conflict may suggest a complex hybrid origin for a taxon that involved the union of more than two taxa for its formation. For example, given the data in Table 3-2, the most parsimonious cladogram without reticulation is that shown in Figure 3-3a. The characters whose distribution is not accounted for are J, K, and L. By parsimony, at the first level of reticulation the single line drawn between taxa 2 and 3 shown in Figure 3-3b should be added; this resolves the conflicting distributions of characters J and K. At a second level of reticulation a line joining taxon 2 and 4 will resolve the character conflict of character L (Fig. 3-3c). If taxon 2 has a simple hybrid origin, then it probably involves a union of taxons 1 and 3. However, if a second level of reticulation is allowed, then taxon 2 may have a hybrid origin involving all of the three other taxa. Nelson suggests that in general those taxa displaying high intermediacy in the narrow sense—that is, they are intermediate between two other taxa—are more likely to be of hybrid origin than those taxa that display broad or complex character conflicts with numerous other taxa.

The above discussion has used the convention of utilizing lines of reticulation to represent hybridization in cladograms. Hybridization may also be represented by multichotomies on a cladogram. Thus the cladogram with reticulation shown in Figure 3-4a is equivalent to the multichotomous cladogram in Figure 3-4b, and the cladogram with reticulation in Figure 3-4c is equivalent to the multichotomous cladogram in Figure 3-4d (Nelson and Platnick, 1981, p. 264).

Given character conflict in a set of data, it is thus possible to hypothesize a hybrid origin for one or more of the taxa under consideration. However, it may be equally valid to deal with the character conflict in some other manner (such as hypothesizing homoplasy, and so on). It is currently not possible to reach a conclusion from a set of raw data in isolation as to whether hybridization or homoplasy, for example, would be the best explanation for the observed character conflict. At present the investigator must still depend on other, external (to the data set) knowledge of the organisms involved to try to reach some decision about whether hybridization should be hypothesized for the origin of some of the taxa under consideration.

ANCESTOR-DESCENDANT RELATIONSHIPS

As was alluded to in the Introduction, a major controversy among phylogeneticists concerns the proper place of ancestors and descendants in
phylogenetic reconstruction. One school (the classical evolutionary school)
holds that attempting to discover and identify actual ancestors via paleon-

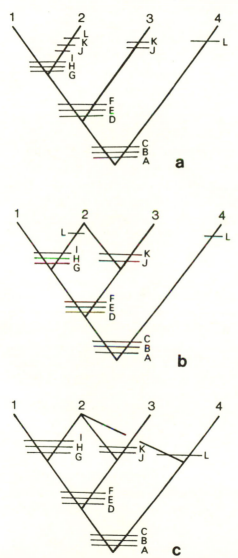

Figure 3-3. Cladograms resolving the phylogenetic relationships among the hypothetical taxa
listed in Table 3-2.

tological data is an integral part of phylogeny reconstruction; ancestors are potentially recognizable (S. S. Bretsky, 1979; Gingerich, 1979). Or at the least, this school holds that genealogy/phylogeny is more than a series of sister-group relationships; it also includes ancestor-descendant relationships (Szalay, 1977a, 1977b, 1977c; of course, such a statement is dependent on one's definition of phylogeny). In this context it is often suggested that phylogenetic trees (including real ancestors and descendants) are more restrictive, bear a higher information content, and are therefore more easily tested and falsified (using biostratigraphic evidence of the sequence of actual organisms) than are hypotheses composed purely of sister-group relationships (Szalay, 1977a, 1977b, 1977c). In contrast, the opposing school (cladistics) variously holds that ancestors are unrecognizable (either theoretically or practically), unprovable, or untestable; at most, ancestors are abstractions (morphotypes: Nelson, 1970; Zangerl, 1948) that cannot actually be recognized (Bonde, 1977, p. 746). The stratigraphic sequence of fossil organisms is considered of limited value in testing proposed ancestor-descendant relationships (or morphocline polarities) because the evidence of the fossil record can always be dismissed by the ad hoc (but possibly true and not necessarily testable) alternative hypothesis that the fossil record is inadequate (not complete enough for the purposes of a particular investigation: cf. Nelson and Platnick, 1981). There are a number of arguments, from both philosophical and practical points of view, for and against the search for and recognition of ancestors. The major forms of these arguments are reviewed in this section.

As Engelmann and Wiley (1977) have pointed out, the term *ancestor* can be used in three ways. First, there is the concept of the individual organism as an ancestor. Second, there is the concept of a population or species-level taxon as an ancestor. Third, there is the concept of the supraspecific taxon as an ancestor.

That an individual organism is an ancestor may simply be an empirical observation in studies of certain extant organisms such as fruit flies, *Drosophila,*

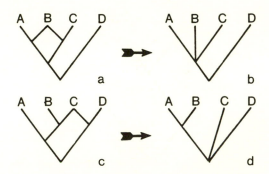

Figure 3-4. Diagram illustrating the equivalence of cladograms with reticulation *(a, c)* and multichotomous cladograms *(b, d).*

bred under laboratory conditions, or of human genealogies traced by the anthropologist. However, when dealing with paleontological data Engelmann and Wiley (1977) could not cite a single example where a paleontologist has or would claim to recognize an individual organism as an ancestor. Furthermore, there is no conceivable way one could possibly hope to accomplish such a task. Lacking historical records, even when dealing with only a couple of generations of extant organisms it is usually virtually impossible to identify ancestor-descendant relationships on the level of the individual organisms. Thus it seems safe to conclude with Engelmann and Wiley (1977) that we can dismiss the concept of the individual organism as an ancestor from further discussion.

As far as ancestors at the species level and above are concerned, Patterson (1983b) has pointed out that there are six basic conventions that can be used to denote ancestry in the context of a phylogenetic tree. Five of these conventions, those dealing with supraspecific ancestors or their equivalents (all but the convention of actual ancestral species), were utilized by Ernst Haeckel (1866, 1868) in his early phylogenetic trees. The conventions are: (1) Taxa of higher rank can give rise to taxa of lower rank; in other words, a certain group gives rise to its included subgroups (cf. Løvtrup's concept that superior taxa have arisen before the inferior taxa that they include). (2) A group may give rise to a group (taxon) of equal or higher rank. (3) Unknown or hypothetical forms (at the species level, i.e., hypothetical stem species) may be seen as giving rise to taxa. As Patterson (1983b, see also E. S. Russell, 1916) notes, such hypothetical forms are logically equivalent to the groups to which they belong or give rise. (4) Extinct groups of low rank may be viewed as giving rise to groups of higher rank. (5) Haeckel inferred certain groups entirely from the ontogeny of the presumed descendant groups. (6) An actual species (either extant or extinct) named on the basis of physical evidence can be considered the ancestor of another taxon.

Patterson (1983b, p. 12) divides these conventions into two sets, which he dubs von Baerian and Haeckelian ancestral groups. As discussed earlier in this chapter, according to von Baer (1828), in the ontogeny of an individual special characters (characters of lower, less-inclusive taxa) develop from more general characters; thus von Baer viewed ontogeny as a guide to common ancestry. In contrast, Haeckel considered the ontogeny of a descendant organism to mimic the adult stages of a series of direct ancestors of the particular organism in question; thus Haeckel viewed ontogeny as a guide to direct ancestry. In this context, von Baerian ancestor conventions are those where a more inclusive group is seen as giving rise to its included subgroups. Von Baerian ancestral groups are taxa named at nodes of cladograms (Patterson, 1983b); the ancestral group is distinguished by certain characters, and the included descendant subgroups are distinguished by having all the characters of the ancestral group plus further additional characters (Løvtrup, 1977a; Patterson, 1983*b*). Von Baerian ancestors and descendants are merely groups characterized by more or less general characters, respectively. The von Baerian conventions fit the concept that for any group the oldest (first) members of the group are just members of the group, not members of any subgroup. Haeckelian ancestral groups are those that give rise to other taxa, which are not included within the ancestral taxon. Haeckelian ancestral groups are paraphyletic and distinguished solely by the absence of characters that distinguish the descendant group or groups. Haeckelian conventions are concerned with the actual, historical descent of particular groups of organisms.

Patterson suggests that von Baerian conventions for ancestors are based on a pre-Darwinian, pre-evolutionary tradition, one in which investigators "attempt to discover and represent natural order" (Patterson, 1983*b*, p. 15). The Haeckelian conventions, in contrast, are formulated with the acceptance of evolution; they are based on the concept that life is "known to be a historical continuum on which biologists must impose order" (Patterson, 1983*b*, p. 15). "Discovery of the natural hierarchy was replaced by the imposition of order, more or less arbitrarily, on a continuum" (Patterson, 1983*b*, p. 16). The continuum was a phylogenetic tree (specifically as opposed to a hierarchical scheme of phylogenetic relationships). At first the phylogenetic trees were speculative genealogies based on embryology and hypothetical ancestors. When these fell out of favor, phylogenetic trees were built up on the basis of the stratigraphic succession of fossils, perhaps an equally speculative approach (Patterson, 1983*b*; cf. various discussions in this book, especially chap. 4. Essentially, it can be argued that ordering fossil specimens stratigraphically [temporally] does not alone yield a phylogeny; the specimens must be ordered or organized on some other basis as well). In recent years there has been a trend of returning to the search for hierarchy and pattern among biological forms (e.g., Eldredge and Cracraft, 1980; Nelson and Platnick, 1981; Patterson, 1983*b*; and numerous other references cited in this book).

It is common, especially in the older paleontological and evolutionary literature, to see references to the various conventions of supraspecific taxa as ancestors (discussed above). In order to dismiss all supraspecific ancestors, certain workers have argued that for a supraspecific taxon to be considered an ancestor of another taxon, certain requirements must be met (Engelmann and Wiley, 1977; Harper, 1976): (1) The ancestral taxon cannot logically include the descendant taxon within it. (Thus, it can be argued, the Vertebrata did not give rise to the Mammalia, although some organism or species that was a vertebrate gave rise to the Mammalia and all mammals are vertebrates. Such a line of argumentation, however, does not recognize the fact that the stem species of any monophyletic taxon is equivalent to the entire taxon. When dealing with strictly monophyletic taxa at the supraspecific level, no taxon gives rise to another distinct and separate taxon, but one monophyletic taxon may arise within another, more inclusive monophyletic taxon.) (2) The species composing the descendant taxon must be descendants of species composing (or if unknown, that would be included in) the ancestral taxon. By the first requirement, von Baerian or Løvtrupian ancestral groups are dismissed by definition (apparently the concept is that such ancestral groups amount to tautologies). As should be immediately obvious, by requirement two above, any supraspecific ancestral taxon must be paraphyletic; it cannot be monophyletic because it does not include all of its descendants. Thus supraspecific ancestors are merely an artifact of an arbitrary and nonobjective taxonomic scheme that recognizes nonmonophyletic taxa (see section on monophyly, polyphyly, and paraphyly, and Ball, 1975, for a discussion of the artifice of paraphyletic groups and Englemann and Wiley, 1977, for general discussion of this point). The descendant taxon of a supraspecific ancestral taxon may be, but need not necessarily be, monophyletic. The descendant taxon may itself be paraphyletic if it gives rise to another taxon. While the descendant taxon may fit the requirements of minimal monophyly (Simpson, 1945, 1961), in a strict sense it may be polyphyletic. For example, if family A is viewed as having given rise to class B, members of family A may be viewed as crossing the threshold from family A to class B several times independently of one another.

Having discussed the possibilities of individuals and supraspecific taxa as ancestors, we are left with only one conception of ancestor-descendant relationships that requires real consideration, that of the population or species-level taxon as an ancestor.

Given the three-taxon relationship (where the taxa are species-level taxa or lower) (A (B, C)), B and C share a more recent common ancestor than either does with A, and in the absence of any other information, B might be the actual ancestor of C or vice versa. If B were indeed ancestral to C, then for *every* character state in which B differs from C, B must bear the plesiomorphous character state. But any taxon that is entirely plesiomorphous in known character states relative to C is not necessarily ancestral to C. And further, if some derived character state should be found in B that is not shared with C (i.e., an autapomorphy of B), then most workers would agree that B could not be ancestral to C. Thus the hypothesis that B is ancestral to C could be falsified by the discovery of autapomorphies of B that would exclude it from the ancestry of C (it goes without saying that if B were found to share synapomorphies with some other taxon not found in C, it would be excluded from the ancestry of C); only the lack of such character states would serve to corroborate the hypothesis that B is ancestral to C. That is, only a lack of evidence (i.e., negative evidence) would support the hypothesis of direct ancestry; no positive proof, analogous to synapomorphy supporting a hypothesis of sister-group relationship, could be cited to support a hypothesis of ancestry.

This lack of positive supportive evidence has been viewed as a major weakness of ancestral-descendant hypotheses by some workers (e.g., Nelson, 1969; see Engelmann and Wiley, 1977).

From the discussion above it should be clear that, by definition, ancestors at the species level are not recognized by unique positive occurrences of characters; they do not have unique identities, individualities, or histories (Forey, 1982). At best, an ancestor can be recognized by what it is not: it is not the descendant. In other words, even species-level ancestors are paraphyletic, not monophyletic, and suffer from all the problems inherent to nonmonophyletic taxa (see section on monophyly, polyphyly, and paraphyly). This problem is not solved by adopting a convention that all species (whether ancestors or not) will, by definition, be considered individuals and monophyletic. Even at the species level, ancestors may be taxonomic artifacts or arbitrary conventions that do not correspond to real groups in nature.

However, it can be argued that a hypothesis of an ancestor-descendant relationship is bolder, more parsimonious, and more readily falsifiable than a simple sister-group relationship (Bonde, 1977, p. 772) and in turn could be considered more highly corroborated if left unfalsified. That is, given the relationship (A (B, C)), an autapomorphy of B would exclude B from the direct ancestry of C, but would not exclude the sister-group relationship between B and C. A hypothesis that is more readily falsifiable and could in theory be more highly corroborated is to be preferred in the context of a Popperian philosophy of science (Engelmann and Wiley, 1977), and thus some phylogeneticists would consider hypotheses involving ancestors and descendants at least theoretically valid and perhaps heuristically useful. In this context, there is one point in particular that must be clarified: the status of sister-group relationships relative to ancestor-descendant relationships. As used in this book (and probably by most cladists), if a sister group is a single species without autapomorphies (a totally plesiomorphic sister group), then such a sister group can conceivably be a true ancestor (even if this is unprovable or unknowable). In other words, a single species may be a sister group and an ancestor. Some workers, however, believe that a species is either a sister group or an ancestor (e.g., Bonde, 1977; Wiley, 1981, p. 106), but not both. According to this latter use, a set of sister groups (or twin taxa) is always the result of the splitting of an ancestral species into distinct daughter species, none of which are actually ancestral to each other. Bonde (1977, p. 776) also suggests that "age overlap or synchronism of two (or more) closely related clades naturally means that the two cannot be ancestral to each other." However, many consider it possible for an ancestral species to persist after giving rise to a daughter species (see section on species concepts); thus, the known age of two taxa need not exclude the taxa from an ancestral-descendant relationship. A 1-million-year-old fossil organism could not be an actual ancestor of a 2-million-year-old organism, but the species represented by the 1-million-year-old organism may have been ancestral to the species represented by the 2-million-year-old organism.

Out of philosophical considerations, Engelmann and Wiley (1977) have invoked a strong argument against the objectivity of ancestor-descendant recognition. They suggest that the designation of the polarity of character states (as apomorphic or plesiomorphic) within a certain group of taxa being analyzed must be done, usually utilizing out-group comparison, within the context of the assumption of a higher-level phylogeny that is accepted in order to deal with the problem at hand. (The higher-level phylogeny itself can in turn be tested at a different level of analysis.) The designation of character states as apomorphic or plesiomorphic merely summarizes the assumed higher-level phylogeny. To quote Engelmann and Wiley (1977, p. 7, italics in the original):

Under these conditions, synapomorphic characters (shared derived characters) are those characters that remain valid tests of the hypothesis examined when placed in the context of the higher-level phylogeny. Symplesiomorphic characters (shared primitive characters) are those that do not provide valid tests of the immediate problem if overall parsimony is to be achieved. In other words, the designation of a character as plesiomorphic at the level of the immediate problem represents an untestable *ad hoc* hypothesis that must be invoked if the most parsimonious overall phylogeny is accepted. . . . We conclude that plesiomorphic characters can neither corroborate nor refute phylogenetic hypotheses.

Since it is only plesiomorphic character states—that is, ad hoc hypotheses—that supposedly corroborate an ancestor-descendant relationship, then ancestor-descendant relationships cannot be corroborated in the fossil record.

But the question remains whether ancestor-descendant relationships can be falsified. As discussed above, it would seem that they could be by the discovery of an autapomorphy in a supposed ancestor, even granted that an autapomorphic character (a character unique to a single taxon) cannot test sister-group relationships between taxa (an autapomorphy will not preclude any possible sister-group relationships). Engelmann and Wiley (1977) also address this point. These authors state (1977, pp. 8–9) that

the concept of autapomorphy, however, is more than a simple statement of uniqueness. Implicit in this concept is the assumption that the taxon that has an autapomorphic character differs in this character from the common ancestor of that taxon and its nearest relative (sister group). . . . The ancestral "morphotype" may be the result of a statement of phylogeny that is testable and that may be highly corroborated. As such, the concept of the hypothetical morphotype might be a very compelling summary statement, but unless the phylogeny on which it is based is assumed to be true, it cannot be considered objective.

Thus, another argument against the falsifiability of ancestor-descendant relationships suggests that the designation of a unique character state as autapomorphous in a particular taxon is actually an ad hoc hypothesis of plesiomorphy for corresponding character states in other taxa; thus, statements of autapomorphy themselves are no more than ad hoc hypotheses that summarize an assumed higher-level phylogeny. Engelmann and Wiley (1977) agree that ancestor-descendant relationships could be testable hypotheses within the context of the assumption of a specific cladistic (i.e., sister-group) relationship between the taxa involved (that is, a particular phylogeny is assumed to be true), but they argue against such a move because it "would place ancestor-descendant relationships in a different system of testing" (Engelmann and Wiley, 1977, p. 9) from that of the assumed phylogeny. However, it would seem that almost invariably (with perhaps the exception of transformed cladists—but then they are no longer worried about phylogeny per se) some higher-level hypothesis of relationships is temporarily assumed in order to test relationships, be they sister-group or ancestor-descendant relationships, on some lower level. Such methodology is advocated by Engelmann and Wiley (1977) in testing sister-group relationships at various levels of analysis. Engelmann and Wiley (1977, p. 1) conclude that "ancestor-descendant relationships based on morphology are not objective statements when applied to fossil populations or species." By the same criteria, statements of sister-group relationships are not objective statements.

Even if we were to accept on a theoretical basis the concept of ancestors and descendants, there are many practical considerations that must be overcome if we are going to search for ancestors in the fossil record. Any ancestor will not be recognizable except by the relative lack of derived character states. As has been often pointed out, specimens and taxa known only from fossils may always (indeed will always) have borne character states unknown to the paleontologist (Carleton and Eshelman, 1979; Cracraft, 1974a; Farris, 1976), and some of those unknown character states may be autapomorphies that if known would exclude the taxon as a direct ancestor. It has been suggested that in most cases, given a sufficiently extensive sampling of characters (although such may not be possible with incomplete materials), it is reasonable to expect that the majority of species will bear apomorphies that exclude them from being direct ancestors (Nelson and Platnick, 1981, p. 205). This is to say that many fossil-based taxa may be considered ancestors only by default because of their poor and incomplete preservation. This is hardly a satisfactory basis for recognizing a taxon as a true ancestor, and a strong argument can be made for leaving hypotheses of relationship only as precise as the sister-group level, sister-group relationships being corroborated by known morphology, not a lack of morphology. Moreover, even if we could have complete and true knowledge of a taxon and it did lack any apomorphy relative to another taxon, this does not necessarily mean that it must be the ancestor of the second taxon.

Another important point to be made is that the suggestion that a particular known species is in the direct line of ancestry of another known species is inherently unlikely to be true. Vast numbers of species must have existed in the past, and only a small percentage have left traces as fossils (and have also been discovered and interpreted by paleontologists: Hull, 1979; Schoch, in press; and see chap. 4). Besides the problems involved in recognizing true ancestors, it appears very unlikely that many true ancestors even exist in our collections of fossils.

It is more out of practical and pragmatic considerations, especially that ancestors are usually recognized only by a lack of derived character states, that one should usually avoid attempting to recognize absolutely ancestors in the fossil record. In a few exceptional cases it may actually be possible to identify (or at least the investigator feels that he or she can identify) with a high degree of probability a population (i.e., species-level taxon) that is ancestral to a later population (species). Nelson and Platnick (1981, p. 202) point out that if one decides to deal with trees (rather than restrict one's investigation to the information contained in cladograms), the most parsimonious trees will be those with the fewest lines, given that the character distributions allow a particular tree in the first place. That is, if the available evidence does not exclude an ancestor-descendant relationship between two taxa (for example, A is completely primitive relative to B and occurs earlier than B), and one decides to construct an evolutionary tree, it is more parsimonious to consider one taxon (in this case A) the ancestor of the second (B) and thus connect them along a single line than to hypothesize that there was an unknown ancestor that gave rise to both taxa separately. Prothero and Lazarus (1980) have outlined the conditions under which this may be possible: (1) the putative ancestor must be older—occur earlier—than its supposed descendant and (2) all potentially ancestral populations must be sampled. As Patterson (1983b, p. 16) has stated, "By assuming that the fossil record is complete, one may treat all potential ancestors (fossil species without apparent autapomorphies) as actual ancestors." The problem is determining that the fossil record is complete or at least sufficiently complete. As Prothero and Lazarus (1980, p. 120; see also Lazarus and Prothero, 1984) note:

in most cases the fossil record cannot meet the criteria outlined above. Preservation is notoriously spotty, time control is often difficult, and most importantly, most fossils sample only a very small part of once extensive and heterogeneous geographic ranges. Thus most fossil collections are a very minute and possibly unrepresentative sample of life at that time. Statements about ancestor-descendant relationships in most fossilized groups are indeed untestable.

Such is surely the case for the local stratigraphic patterns of fossils (primarily mammals in the Bighorn Basin of Wyoming) documented by Gingerich (e.g., 1976a, 1976b, 1979, 1980) and interpreted by him as true evolutionary sequences. Schankler (1980, 1981) has shown that the same stratigraphic patterns do not necessarily hold true as close as 50 km to Gingerich's original sections and that there is "a strong lateral (biogeographical and ecological) component influencing the sequencing of species [and also various morphs of species] in a single basin" (Schankler, 1981, p. 135). In these particular sections, Schankler (1981) has apparently documented local extinctions and ecological reentries, thus demonstrating that separate populations, whose genetic relationships to one another are not known, are sampled in successive layers. It is unwise and unjustified to assume that the particular populations that happen to be sampled in one particular section actually gave rise to each other (Schoch, 1982).

But Prothero and Lazarus (1980) suggest that there is at least one special instance—that of the microplankton of the marine realm—where the requirements to potentially recognize ancestors are fulfilled. Microplankton biotas of the marine realm are relatively homogeneous and widespread, the deep-sea record of fossil forms is of extremely high quality (sedimentation may be in the form of a continuous rain of fine particles, including the microplankton), and it is possible to sample all of the demes that make up an extant or extinct fossil microplankton species with only a few well-chosen samples (either plankton tows for extant forms or sediment cores for fossil forms: Prothero and Lazarus, 1980). Approximately 20,000 to 21,000 piston cores and Deep Sea Drilling Project cores have been made of the sediments of the oceans, and all biotic provinces have been adequately covered (Prothero and Lazarus, 1980). These cores can be correlated using magnetic stratigraphy, biostratigraphy, and isotope stratigraphy. Given such complete data, in these special instances it may be possible to state with a high degree of certainty that a particular species-level taxon gave rise to a succeeding species-level taxon. Prothero and Lazarus (1980) cite as an example a study by J. D. Hay (1970) that recognized the radiolarian *Eucyrtidium calvertense* as the direct ancestor of *E. matuyamai*. Of course, for most taxa, especially macrofossil taxa, the data needed to attempt such studies are lacking. Further, even in the best cases it will usually still be beyond the limits of resolution to specify what part of the putative ancestral population actually gave rise to the descendant population.

TREES FROM PHYLOGENETIC RELATIONSHIPS

From the distribution of characters (which are themselves theory laden) among a group of organisms or taxa, one can postulate a set of phylogenetic relationships among the organisms or taxa. This is a first-level inference and abstraction. From a hypothesis of phylogenetic relationships (often expressed in the form of a cladogram) one can then proceed to construct an evolutionary tree (here evolutionary tree means a diagram that specifies ancestor-descendant relationships, whether hypothetical or real) of the organisms or taxa involved. This is a second level of inference. Finally, if so desired, an evolutionary scenario (a historical-narrative story) can be constructed by adding postulates concerning adaptation, selection, function, and so on to an evolutionary tree (Eldredge, 1979a; Forey, 1982; Hull, 1979; Tattersall and Eldredge, 1977). In general, such a scenario is quite far removed from the data base and may consist of little more than an untestable, nonscientific fairy tale or "just-so story." In the course of an actual scientific investigation, one may choose to begin at the level of the scenario, phylogenetic tree, or cladogram, and there may be feedback between the different levels (Hull, 1979; Tattersall and Eldredge, 1977). However, logically and epistemologically cladograms should precede phylogenetic trees, and phylogenetic trees should precede scenarios. In this section the possibilities of and constraints on the development of phylogenetic trees from cladograms are discussed. Once a phylogenetic tree is created, it is up to the imagination of the investigator to suggest an evolutionary scenario that fits the tree.

Given any set of putative phylogenetic relationships (i.e., a cladogram), there may or may not be several evolutionary trees that are consistent with that set of relationships. If the terminal taxa are supraspecific taxa, then the cladogram cannot be validly translated into any evolutionary tree. Supraspecific taxa cannot be ancestors, unless one is willing to allow nonmonophyletic (specifically paraphyletic) taxa, which, as is discussed elsewhere (see section on monophyly, polyphyly, and paraphyly), are unnatural and biologically meaningless. If one were to allow supraspecific taxa as ancestors, they would of necessity be paraphyletic taxa, which by definition are ancestors. Stating that a paraphyletic taxon is the ancestor of another taxon seems to be the same as stating that the ancestor of the taxon is the ancestor of the taxon. For instance, if the Reptilia (as commonly used) is recognized as including some forms that gave rise to the birds, then stating that the Reptilia gave rise to the Aves is really not very informative. What does this statement say? Certainly not that any extant reptiles, or most other reptiles, gave rise to birds. Perhaps that birds, before they became sufficiently "birdy" (through the evolution of feathers, and so on) looked somewhat like some kinds of reptiles? If so, it is simple enough to state that some early birds looked like some reptiles without saying that reptiles gave rise to birds. Does the statement mean that birds did, and still do, share some traits in common with reptiles? Or that present-day reptiles and present-day birds converge back to a common ancestor? If so, they can both be included in a more inclusive monophyletic group.

If we recognize only monophyletic groups, we can at best state that an organism that is a member of a more-inclusive taxon must always give rise to an organism that is a member of a more-inclusive taxon. Thus, by monophyly, vertebrates always beget vertebrates and hominids always give rise to hominids. Somewhere along the line, it might be argued, a truly ancestral organism (or species-level taxon) gave rise to two different species and lineages that are today recognized as separate monophyletic clades. For example, given the evolutionary tree in Figure 3-5, if species A gave rise, in a dichotomous speciation event, to species B and C, each of which turned out to be (in retrospect) the stem species of a distinct, monophyletic taxon currently ranked as a subfamily, then it could be argued that species A must be classified as belonging either to the subfamily of B or the subfamily of C, or as belonging to a distinct subfamily. In any case, whatever subfamily A is assigned to will automatically be a supraspecific ancestor and furthermore will be paraphyletic. However, if we are to recognize only monophyletic taxa (at least above the species level), then A (if it is a true last common ancestor of A and B) is not assignable to a subfamily. B and its descendants would constitute one subfamily, C and its descendants would constitute the other subfamily, and A and its descendants would constitute some monophyletic group, perhaps ranked as a family; but in such a case A would not be validly assignable to any superspecific taxon below that of the family (for further discussion of the classification of ancestors, see chap. 6).

Given three species-level taxa (= unspecified morphotypes of Wiley, 1979a, p. 222: "unspecified morphotypes which may be species or populations of a single species") A, B, and C, where B and C are sister taxa relative to A (Figure 3-6a), then if we are not given any other information, there are a number of possible evolutionary trees that are consistent with this set of phylogenetic relationships. It is important to note that the terminal taxa must be adequately differentiated, species-level taxa (perhaps distinct subspecies) if we wish to attempt the construction of evolutionary trees from the cladogram. As Wiley (1981, pp. 98–104) has demonstrated, cladistic analysis cannot deal adequately with microevolutionary, population-level evolution that involves repeated intermediate hybridization between semi-distinct populations before differentiating characteristics have become fixed or stabilized in the separate populations. Population-level microevolution may also commonly include evolutionary reversals (for instance, if a characteristic is introduced into a population at low levels and then lost). In other words, if the members of two

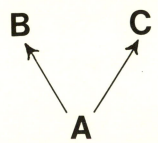

Figure 3-5. A hypothetical evolutionary tree in which species A gives rise directly to both species B and species C (see text for further explanation).

populations that are distinct in ecological time and space do not bear adequate, morphologically based (and inherited), distinguishing features, then all members of both populations will be lumped as one taxon in a cladistic analysis.

Given that we are referring to species-level taxa, A may have given rise to B, which may have given rise to C (Fig. 3-6b) or A may have given rise to C, which then gave rise to B (Fig. 3-6c). A may have given rise to an unknown taxon, X, which then gave rise to B and C (thus X was the last common ancestor of B and C: Fig. 3-6d). Or the unknown taxon X may have given rise to A and the unknown taxon Y; Y in turn then gave rise to B and C (Fig. 3-6e). An unknown taxon X may have given rise to A and B, and B subsequently gave rise to C (Fig. 3-6f), or an unknown taxon X may have given rise to A and C, and C subsequently gave rise to B. Note that if we do not assume that we are dealing with taxa that are immediately and directly ancestral to one another, there could also be any number of unknown

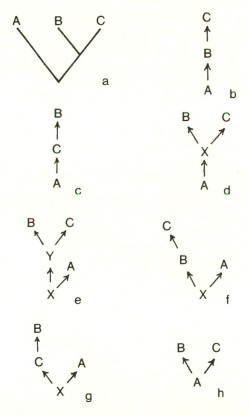

Figure 3-6. A cladogram *(a)* and possible evolutionary trees *(b–g)* that are consistent with the cladogram, plus a tree *(h)* that is not consistent with the cladogram.

ancestors between any two taxa on any branch shown on the trees. For example, in Figure 3-6b if A was not immediately ancestral to B, A may have first given rise to X, which then gave rise to Y, which then gave rise to Z, which then gave rise to B. Also note that A could not have given rise immediately to B and C (Fig. 3-6h); if this had been the case, there would be no synapomorphies to group B and C together relative to A (as shown in Fig. 3-6a). Rather, any synapomorphies shared between B and C would also be shared in common with A (such a situation would give rise to an unresolved trichotomy). If we are given some information concerning the distribution of character states among the taxa under consideration, we may be able to narrow the range of possible evolutionary trees that are consistent with the cladogram. Any presumed autapomorphies would eliminate a taxon from consideration as an immediate, actual, or direct (even if not immediate) ancestor; thus, if the distribution of synapomorphies were as shown in Figure 3-7 (all of the known taxa show autapomorphies), then the only possible topology for an evolutionary tree would be that shown in Figure 3-6e. Such a tree has the same topology as the corresponding cladogram. Wiley (1979a, p. 220) has suggested that when one constructs cladograms using his evolutionary species concept (the terminal nodes of the cladogram represent evolutionary species), then for simple, totally dichotomous cladograms the only possible phylogenetic tree will be that which is "an exact duplicate of the cladogram." This seems to be true as long as (and only when) all of the evolutionary species are characterized by autapomorphies. Yet there seems to be something wrong with this argument, for Wiley claims that such will always be the case, yet he states that species (his evolutionary species)

are *a priori* monophyletic by their very nature. . . . An investigator does not have to justify a species as monophyletic via synapomorphies [i.e., synapomorphies of the subgroups making up the species, autapomorphies of the species] as long as it is accepted that (a) species are lineages . . . with a real existence in nature . . . and (b) species are individuals. (Wiley, 1979a, p. 214 italics in the original)

Just how Wiley proposes to recognize his evolutionary species without the aid of synapomorphies and to distinguish them from what he calls unspecified

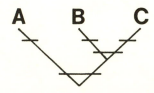

Figure 3-7. Cladogram in which all of the terminal taxa are characterized by autapomorphies (see text for further explanation).

morphotypes is left unanswered (see further discussion of Wiley's evolutionary species concept in the section on species concepts).

Given the cladogram shown in Figure 3-8a for species-level taxa A, B, C, D with a genuinely unresolved trichotomy between B, C, and D, there are several possible evolutionary trees that are consistent with these relationships. A may have been ancestral to an unknown taxon X, which then gave rise to B, C, and D (Fig. 3-8b). An unknown taxon X may have given rise to A and an unknown taxon Y, which then gave rise to B, C, and D (Fig. 3-8c). Or A may have given rise to B, C, or D, which in turn gave rise to the last two taxa (Fig. 3-8d, e, f). Or an unknown taxon X may have given rise to A and B, C or D, the latter of which then gave rise to the remaining two taxa (Fig. 3-8g, h, i). Or any two of the three taxa B, C, D may have hybridized and given rise, by reticulation, to the third (for example, in Fig. 3-9a, B and D hybridized to form C; see also section on reticulation). Again, any presumed autapomorphies of any of the taxa involved would eliminate them from the immediate ancestry of any of the other taxa. If all of the taxa possessed auta-pomorphies, then the sole consistent evolutionary tree would have the same topology as the cladogram (Figs. 3-8a and 3-8c).

Given a cladogram with one level of reticulation, such as that in Figure 3-9a, there are also several evolutionary trees with which it is consistent. Taxon

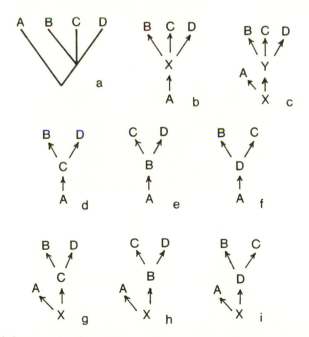

Figure 3-8. A cladogram *(a)* and possible evolutionary trees *(b–i)* that are consistent with the cladogram.

A may have given rise to an unknown taxon Y, which in turn gave rise to taxa B and D; all or part of taxa B and D then hybridized to form taxon C (Fig. 3-9*b*). An unknown taxon X may have given rise to taxon A and an unknown taxon Y, and Y in turn gave rise to B and D, which then hybridized to form taxon C (Fig. 3-9*c*). Taxon A may have given rise to an unknown taxon Y, which then gave rise to the unknown taxa V and Z, which hybridized to form taxon C, and V gave rise to taxon B, while Z gave rise to taxon D (Fig. 3-9*d*). An unknown taxon X may have given rise to taxon A and the unknown taxon Y, Y in turn gave rise to the unknown taxa V and Z, V and Z hybridized to form taxon C; V alone gave rise to taxon B, and Z alone gave rise to taxon D (Fig. 3-9*e*). Note that C cannot be ancestral to D and C cannot be ancestral to B. The cladogram indicates that there are some synapomorphies shared between B and C that are not found in D (if C had been ancestral to D, then these synapomorphies would be found in D as well), and likewise for C and D relative to B. If all four taxa, A through D, are distinguished by autapomorphies, then the only possible evolutionary tree is that shown in

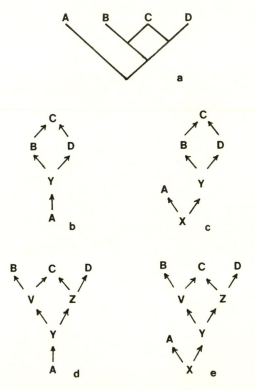

Figure 3-9. A cladogram with one level of reticulation *(a)* and possible evolutionary trees that are consistent with the cladogram *(b–e)*.

Figure 3-9e; this tree has the same topology as the original cladogram. If B is completely plesiomorphous relative to D—that is, it shares synapomorphies only with D but does not bear any autapomorphies—and taxon C is the product of hybridization between B and D, then the topology of the resultant cladogram will be as in Figure 3-10. This cladogram is indistinguishable from one in which B is a possible ancestor (or at least the plesiomorphous sister taxon) of the clade composed of C and D. In both cases B is primitive relative to C plus D, and D will share some synapomorphies with C that are not seen in B; D will not share any synapomorphies with B that are not also seen in C since B does not have any apomorphous features apart from those shared with C. Thus, even on a nonreticulated cladogram, unless there are auta-pomorphies distinguishing all of the terminal taxa on the cladogram, there is the possibility that hybridization may have been involved in the formation of some of the taxa.

Given an evolutionary tree (similar to those described above) that specifies actual ancestor-descendant relationships, one should be able to derive a unique cladogram that corresponds to the tree. However, depending on how one defines an evolutionary or phylogenetic tree, it may not even be possible to arrive at a unique cladogram for a given tree. For example, Estabrook (1984, p. 135) states:

A phylogenetic tree for some collection S of species under study is a tree diagram representing the genealogical continuity of species through time. Each branch point in the tree corresponds to a speciation event, and other points may correspond to speciation events as well. The number of lines at any given time is the number of species at that time. No species is represented at any single time by more than one line.

An example of a phylogenetic tree *sensu* Estabrook (1984) is given in Figure 3-11a. Without more information, specifically what species gave rise to what species, it is impossible to construct a unique cladogram for an Estabrook phylogenetic tree. Assuming that all ancestors of species a through g are already represented among them, two of the many possible sets of ancestor-descen-

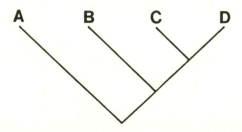

Figure 3-10. Apparently nonreticulated cladogram that may have involved hybridization; see text for explanation.

dant relationships and cladograms that fit this single phylogenetic tree are shown in Figure 3-11*b, c, d,* and *e.*

THE INTERPRETATION OF
UNRESOLVED CLADOGRAMS

Multiple branchings on cladograms (polychotomies or multichotomies; cladograms that are nondichotomous in part or whole, i.e., secondary and tertiary cladograms, Fig. 3-12*a, b*) represent character distributions whose interpretation may be open to question. If all unresolved cladograms were theoretically resolvable into dichotomous primary cladograms, given enough information, then the cladogram in Figure 3-12*b* might be resolvable into any

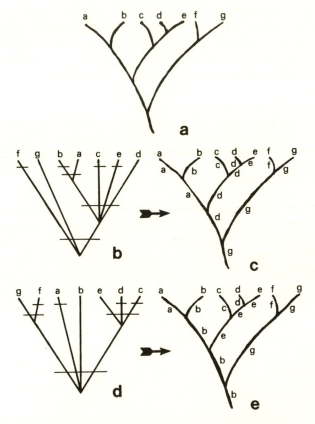

Figure 3-11. A phylogenetic tree *(a) sensu* Estabrook (1984, pp. 135–151) and two possible cladograms *(b, d)* and the respective sets of ancestor-descendant relationships *(c, e)* that are consistent with the phylogenetic tree *(a);* see text for further explanation.

of three primary cladograms. However, in some cases unresolved multiple branchings may represent real phenomena. In such cases, multiple branchings, perhaps along with lines of reticulation, may be the only way to represent certain true character distributions.

Nelson and Platnick (1981, p. 257) suggest several ways of interpreting multiple branchings in cladograms. A multiple branching may reflect our ignorance of certain character distributions and our misinterpretation of putative synapomorphies. In such cases, further analysis or more data should resolve the cladogram. On the other hand, a multiple branching may reflect a real phenomenon. If simultaneous multiple speciation occurred, if hybridization were represented among the taxa analyzed, or if one species under consideration were ancestral to two or more of the remaining species under consideration (perhaps we are dealing with fossil taxa that represent true ancestors,

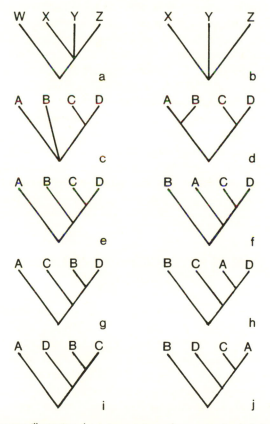

Figure 3-12. Diagram illustrating the interpretation of an unresolved cladogram; see text for explanation.

or an ancestral species underwent sequential speciation via peripheral isolates without change to the central mother population), the distributions of characters could be represented most parsimoniously by a cladogram with multiple branchings. In such cases, the multiple branchings on the cladogram would not be a result of our ignorance—would not be unresolved in the sense of potentially resolvable—but would represent real information.

Thus, from what has been said in the last paragraph, given a cladogram such as that in Figure 3-12c, there are two possible interpretations of the cladogram (see Nelson and Platnick, 1981). The multichotomy between taxa A, B, and (C, D) may represent a real phenomenon or it may be the result of our ignorance. If it represents a real phenomenon, then the cladogram cannot be further resolved. If it results from our ignorance, then there are two further ways to interpret this cladogram. One way to view such a cladogram is that both of the three-taxon problems (A, C, D) and (B, C, D) have been solved, with the solutions (A (C, D)) and (B (C, D)), respectively, and the unresolved aspect of the cladogram is to solve the three-taxon problem (A, B, (C, D)). Thus there are three possible fully resolved cladograms (Fig. 3-12d, e, f) that are compatible with the cladogram of Figure 3-12c. A second way to view such a cladogram is that only one of the above three-taxon problems—that is, either (A, C, D) or (B, C, D)—has been solved. In this case there are seven possible fully resolved cladograms (Fig. 3-12d–j) that are compatible with the cladogram of Figure 3-12c.

From the above discussion, it is obviously important to know what interpretation is ascribed to an unresolved cladogram. If an unresolved cladogram is considered to represent a real phenomenon, it is not expected that it can be further resolved. If an unresolved cladogram is due at least in part to our ignorance, there still may be at least two different ways to interpret it, one of which is much more restrictive (thus giving us more information) than the second.

WHEN PHYLOGENETIC ANALYSIS
IS NOT SUCCESSFUL

For certain organisms and groups it may be impossible to reconstruct convincingly the phylogenetic relationships of the organisms concerned. In many cases this failure may be due to a lack of useful characters; the characters that are available may not form a hierarchical pattern of nested sets, but may simply vary randomly relative to one another. For fossil organisms in particular, many potentially useful characters may be unknown (not preserved). An extreme example of this is the case of ichnofossils (trace fossils—tracks, trails, borings, footprints, and other behavioral marks of extinct organisms: Clarkson, 1979; M. F. Miller, Ekdale, and Picard, 1984; Sarjeant, 1983; Seilacher, 1977) in isolation. When ichnofossils cannot be associated with body fossils, there may be little hope of figuring out the relationships of the organisms represented; for instance, identical (as preserved in the fossil record) borings or burrows might be produced by organisms belonging to several different phyla (many current ichnotaxa may be polyphyletic assemblages). But even extant forms may suffer a loss of characters during their evolution (the loss itself would be an apomorphy, but perhaps of little use in determining relationships at higher levels). In certain organisms whole organs may be lost; in some plants a whole generation, along with all of the characters that that generation may bear, can be lost. For example, in one fern the entire sporophyte generation has apparently been lost (Wagner, 1980).

Related to the lack or loss of characters in known taxa is the absence of obvious intermediates between certain taxa and any other known taxa. There may be such overwhelmingly large morphological gaps between certain enigmatic taxa and any other known taxa that it is virtually impossible to associate or homologize characters of the enigmatic taxa with those of known taxa at certain (lower) levels. This can be a very real problem for paleontologists in particular. Thus one may be able to recognize as monophyletic an extinct group of animals, allocate the group in question to the Mammalia on the basis of synapomorphies shared among all mammals, and reconstruct the internal phylogenetic relationships of the members of the group, but not be able to place the group convincingly within a phylogenetic scheme of all major groups of mammals. Such a group may be left in a classification as Mammalia *incertae sedis.*

Another problem that must always be kept in mind is the possibility of close similarities in gross morphology between only distantly related taxa. Such similarities may result from functional convergence or are traditionally attributed to parallel trends in genetically similar taxa that display similar adaptive responses to similar environments (Lazarus and Prothero, 1984; Newell and Boyd, 1975; but see discussion of parallelism above). Extremely similar taxa that are considered to be not particularly closely related (i.e., homeomorphs) are usually sorted out on the basis of supposedly "clear evidence from the stratigraphic record that earlier homeomorphs became extinct before the evolution of later ones" (Lazarus and Prothero, 1984, p. 164). Of course, not all workers agree on what constitutes "clear evidence from the stratigraphic record" (see discussion in chap. 4 on the nature and completeness of the fossil record).

It may also happen that in the course of evolution phylogenetic patterns become confused, obscured (such as to be undetectable), or lost. This may happen in a group where there has been extensive hybridization and intermixing of different closely related species and lineages, some of which may have hardly diverged from one another (this scenario may be the case for some groups of plants). In the extreme case, using conventional methods discussed in this book, one could not hope to reconstruct the phylogenetic relationships of a number of subspecies that had never fully differentiated and then began interbreeding again. The other situation that immediately comes to mind is when a group undergoes a rapid succession of evolutionary splitting and divergence. An ancestral species may be simultaneously broken up into numerous daughter species, and after only a short while (before many apomorphies have formed and stabilized—become fixed in the daughter species) the daughter species may undergo splitting events. Such a scenario may take place when an ancestral species range is split up into an extensive archipelago. Phylogenetic analysis may be unable to resolve the detailed phylogenetic relationships between the resulting taxa. Rather, an unresolved multichotomy may be indicated.

A Note Concerning Concocted Examples

Many authors have used concocted, hypothetical examples to supposedly demonstrate that certain methodologies are theoretically inoperative. Such examples can be very frustrating in that they are often totally unrealistic and therefore, if we profess to be concerned with the reconstruction of real phylogenies of real organisms, virtually meaningless. To cite a typical example from the literature, Cartmill (1981, p. 85) proposed the following example:

Suppose, for example, that we have 4 taxa, each with 9 two-state characters, as follows:
   Taxon i: ABCDEFGHI
   Taxon ii: abcdeFGHI
   Taxon iii: abCDEfghi
   Taxon iv: ABCDEfghi

Cartmill (1981) then goes on to marvel that the three-taxon statements derived from this data are not mutually compatible and a phylogeny for these four hypothetical taxa cannot be satisfactorily determined. But looking at Cartmill's data, it is at once obvious that the data does not correspond to data for real biological taxa unless there is rampant, and unrecognized, homoplasy and/or hybridization among these taxa. From Cartmill's discussion it appears that he intends, and we must assume, that character state A in taxon i is the same character state as character state A in taxon iv, and so on. If initially we assume that the states symbolized by uppercase letters are the derived conditions relative to the character states symbolized by lowercase letters, then taxon i shares the synapomorphies FGHI with taxon ii as well as sharing the synapomorphies ABCDE with taxon iv. But if these are real biological taxa that are a product of evolution, then two of these taxa must be more closely related to each other than either is to the third (barring that they all arose simultaneously from a fourth, ancestral taxon: this was clearly not Cartmill's intention).

And those two taxa that are more closely related to each other relative to the third will share synapomorphies, but neither of the pair will share true synapomorphies with the third taxon that are not also shared with the other taxon of the initial pair. Thus, Cartmill's hypothetical taxa are unrealistic. Likewise, if Cartmill's data is interpreted such that the lowercase letters represent the derived conditions of the character states, or if any mixture of upper- and lowercase letters is considered the derived conditions for the various characters, the same result ensues. There is little wonder that the phylogenetic relationships of Cartmill's (1981) hypothetical taxa cannot be resolved.

MONOPHYLY, POLYPHYLY, AND PARAPHYLY

There has been much discussion in the literature about the meaning and uses of the terms *monophyly, polyphyly,* and *paraphyly* (for a brief review, see Ashlock, 1984). Much of the disagreement is due to differing definitions being proposed, and considered useful, for these terms. The following are useful definitions of these terms, followed by some alternative opinions.

The majority of persons concerned with this subject would now accept Farris's (1974; as quoted by Wiley, 1981, p. 84) definitions:

1. Monophyletic group: "A group that includes a common ancestor and all of its descendants." (Wiley [1981. p. 85] has demonstrated that this strict, cladistic definition of monophyly has historical precedent and agrees well with definitions that predate the redefinitions of Mayr, Simpson, and other classical evolutionary systematists [discussed below].)

2. Paraphyletic group: "A group that includes a common ancestor and some but not all of its descendants."

3. Polyphyletic group: "A group in which the most recent common ancestor [of members of the group] is assigned to some other group and not to the group itself."

The above definitions define these concepts in an evolutionary sense (that is, in terms of ancestors, descendants, evolution) and agree in spirit with the way Hennig (1965, 1966a) refined, redefined, and used these terms. These might be considered the traditional neo-Hennigian definitions of the terms. (Of course, these terms are relatively meaningless if one is not initially concerned with evolution in some sense, as some transformed cladists are not. Not taking an initial interest in evolution per se is not a denial of evolution.)

Classical evolutionary systematists have defined the term *monophyly* in a very different way, and this has led to seemingly endless discussion and argumentation. A typical classical evolutionary definition reads: A monophyletic group is "a taxonomic group whose members are descended from a common ancestor included in that group" (Szalay and Delson, 1979, p. 563). By this conception, the members of a monophyletic group must all be descended from a common ancestor included in the group, but all organisms descended from that common ancestor need not be included in the group for it to be considered monophyletic. Classical evolutionary systematists define paraphyletic groups in essentially the same way as do the cladists, but the classical evolutionary systematists consider paraphyletic groups to be a type of monophyletic grouping. Many classical evolutionary systematists have adopted the term *holophyletic* (Ashlock, 1971) for the restricted concept of monophyletic as used by cladists. Thus a holophyletic group is "a monophyletic group which includes all of the descendants of any member of that group" (Szalay and Delson, 1979, p. 561).

An evolutionary taxonomist's definition of a polyphyletic group is "a composite taxon derived from two or more ancestral sources; not of a single immediate line of descent" (Mayr, 1969, p. 409). This essentially agrees with the cladistic concept of polyphyly given above if the ancestral source and single immediate line of descent are held to be a single ancestral species. However, this is not always the case. Simpson (1961, p. 121) states: "Parallelism is a widespread phenomenon in evolution, and it is not uncommon to find that some generally recognized taxon arose by parallel evolution through two or more lineages from different ancestral taxa." Simpson believed that in an evolutionary classification, taxa should (if possible) be monophyletic, and in the next few pages of his book he considers how to deal with taxa that arose by some form of parallel evolution. Simpson (1961, p. 123) considers basing a definition of monophyly on descent from a single species, but decides against doing so as "undesirable in principle and usually inapplicable in practice"; he states that a "definition of monophyly as descent from a single species or any other single taxon at one rank fixed by definition is impractical in classification and unnecessary for consistency with taxonomic principles" (Simpson, 1961, p. 124). (It should be carefully noted that here Simpson's primary concern is the pursuit of the art of taxonomy and the building of a stable, practical classification. On page 122 of his book he refers to the "artistic judgment" necessary in classifying organisms. Simpson is not interested in necessarily delimiting evolutionarily meaningful or absolutely natural taxa.) Simpson (1961, p. 121) proposes that to save many strictly nonmonophyletic (here strict monophyly means descent from a single ancestral species) taxa one could frame a definition "of monophyly that would make it relative to the ranks of the taxa involved and that would in some instances, not in all, make taxa that arise from more than one lineage still monophyletic by definition." This Simpson (1961, p. 124) does by redefining monophyly as follows: "Monophyly [= Simpsonian 'minimal monophyly'] is the derivation of a taxon through one or more lineages (temporal successions of ancestral-descendant populations) from one immediately ancestral taxon of the same or lower rank."

Unfortunately, as Bonde (1977, p. 758) has pointed out, a major flaw in Simpson's concept of minimal monophyly is that the monophyly of a group is dependent upon predetermined ranks for both the group under consideration and the putative ancestral group from which it arose. Thus any group can be considered monophyletic if it is given a higher Linnaean rank than the paraphyletic group from which its members arose. To use Bonde's (1977, p. 758) example, the superclass composed of birds and mammals would have to be considered monophyletic relative to the class Reptilia (where all birds and mammals are assumed to have had reptilian ancestors), but the superclass birds plus mammals would not be monophyletic relative to the phylum Reptilia (unless, of course, the group composed of birds and mammals happens to be monophyletic in the strict sense, *sensu* Hennig: see Gardiner, 1982, for a discussion of the latter possibility). Ranks are assigned to the groups arbitrarily; among evolutionary systematists there are no precise rules for determining either group membership or ranks. Rather, certain groups at certain ranks are recognized a priori without evidence that the particular groups are real and should have a particular rank (Bonde, 1977). Obviously, such reasoning is fallacious.

Mayr (1969, p. 75) agreed with Simpson that "monophyly must be required of all taxa." In order to get around the same troublesome problems, Mayr (1969, p. 75, italics in the original) embraced Simpson's (1961) redefinition of monophyly, explaining his position as follows:

It sometimes happens that a certain *grade* of morphological change is reached independently in several lines derived from a single ancestral group. The group showing this grade of development is of course monophyletic. As always in evolution, one must distinguish between what happens to the phenotype and what happens to the genotype. The diagnostic mammalian structure of the jaw-ear region, for instance, is believed to have evolved several times from ancestral theraspid reptiles, which had the needed genetic program to predispose them toward evolving the mammalian grade when exposed to the same selection pressures [here Mayr is referring to Simpson, 1959]. This is not polyphyly, because the genotype permitting these parallel evolutionary changes goes back to the same ancestral program. We classify taxa (= genotypes) and not characters = phenotypes). [Note that Mayr may classify by genotypes, but since he rejects characters as being directly indicative of genotypes he never has direct knowledge of genotypes, so, within certain constraints, it seems he can classify any way he wants and claim that he is classifying by genotypes. In this case why are not the theraspids, "with the needed genetic program," included in the Mammalia?] The usual phrasing of the principle of monophyly ("a taxon is monophyletic if its members are descendants of a common ancestor") is too vague to be helpful in more complicated cases, such as that of the mammals. Simpson (1961) has therefore given a more concrete definition. [quotes Simpson's definition, given above]

Certain evolutionary systematists have been extremely ardent in their opposition to the strict cladistic concept of monophyly, insisting that paraphyly is a type of monophyly. In reference to the cladistic concept that descendants of the ancestral species of a monophyletic group should not be placed in another taxon, Mayr (1969, p. 75, italics in the original) states:

The latter postulate, of course, is completely contradicted by common sense and is in opposition to the phenomena of evolutionary divergence. If a descendant group, such as the birds among the archosaurian reptiles, evolves more rapidly than the other collateral lines, it not only can but it *must* be ranked in a higher category than its sister groups. This does not violate the principle of monophyly, retrospectively defined [i.e., as defined by Simpson and Mayr as not necessarily including all descendants].

(Concerning Mayr's example of birds, archosaurs, and nonarchosaurian reptiles in particular, in a recent classification of the vertebrates [Schoch, 1984*b*] the birds are classified as a subgroup of the Archosauria and the Archosauria are a taxon coordinate with the Reptilia.)

Returning to the cladistic concepts of monophyly, paraphyly, and polyphyly, it should be evident that in general, a monophyletic group is a group that is characterized by the unique possession of synapomorphies, whereas in many cases a paraphyletic taxon will be united only by the possession of symple siomorphies (relative to the total group of organisms under consideration) and a polyphyletic taxon need not be characterized by either. In many cases however, a polyphyletic taxon will be composed of members that have been grouped together by a particular taxonomist on the basis of homoplasy. Following this line of reasoning, Hennig (1975) suggested that the terms *para phyletic* and *polyphyletic* might be useful in characterizing the nonmono phyletic groupings made by other authors. Thus a taxon justified by plesiomorphies would be called paraphyletic and a taxon justified by homo plasies would be called polyphyletic. Wiley (1981, p. 87) has shown that these concepts do not necessarily agree with Farris's (1974) definitions of the terms given above, and in some cases Hennig's (1975) definitions "may result in identically circumscribed taxa being either paraphyletic or polyphyletic and yet have exactly the same genealogical tree topology" (Wiley, 1981, p. 87). For example, given a cladogram as in Figure 3-13, one investigator might justify erecting a taxon that includes only G, A, C, and E of the known terminal

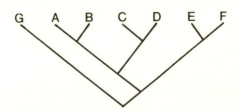

Figure 3-13. Diagram illustrating Hennig's (1975, pp. 244–256) concepts of paraphyletic and polyphyletic groups; see text for explanation.

taxa on the basis of only plesiomorphous character states. Another investigator, looking at different characters or with a different philosophy, may erect the same taxon on the basis of homoplasies.

In the first case the taxon would be considered paraphyletic; in the second case it would be considered polyphyletic. Using Farris's (1974) definitions of these terms, it is clear that given the cladogram is correct, the taxon composed of G, A, C, E is nonmonophyletic because some sister groups are missing. But knowing only the topology of the cladogram and without knowledge of either character-state distributions or common ancestors of the taxa A through G and how these ancestors would be classified, it is impossible to state that the taxon composed of G, A, C, E is either paraphyletic or polyphyletic. As shown in the cladogram, A and B are sister groups, C and D are sister groups, E and F are sister groups, A, B, C, D is the sister group of E plus F, and G is the sister group of taxa A through F. If we assume that A through G are species, then we might hypothesize ancestor-descendant relationships between certain of the terminal nodes. Thus, if A was totally plesiomorphous relative to B, C was totally plesiomorphous relative to D, E was totally plesiomorphous relative to F, and G was totally plesiomorphous relative to all of the other taxa, then A may be the ancestor of B, C the ancestor of D, E the ancestor of F, and G the ancestor of all of the other forms. If such is true and we circumscribe a taxon that includes only A, C, E, and G, then this taxon would be a group that includes a common ancestor and some, but not all, of its descendants—that is, paraphyletic. If, on the other hand, we have a situation where G, A, C, and E all bear autapomorphies relative to their totally plesiomorphous sister taxa (i.e., B might be ancestral to A, D might be ancestral to C, and F might be ancestral to E), then a taxon composed of G, A, C, and E (perhaps G, A, C, E bear homoplasies) among known forms would be a group in which most recent common ancestors would be assigned to some other group than the group under study and would therefore be polyphyletic. Thus, given only the topology of the cladogram, it seems that we can specify only whether a given grouping is monophyletic or nonmonophyletic; this is the first level of analysis. If a group is found to be nonmonophyletic, then one must have some knowledge of the distribution of plesiomorphous versus apomorphous character states or some knowledge of ancestors and descendants to decide whether the nonmonophyletic taxon is paraphyletic or polyphyletic. Thus, returning to Hennig's (1975) characterizations of paraphyletic and polyphyletic, if an investigator does have a good knowledge of the character-state distributions among taxa of a monophyletic group and then, in good faith, distinguishes taxa on the basis of the character-state distributions (this is to say, he does not recognize taxa irrelevant to the character-state distributions), then most likely where the distinguishing characteristics of a taxon are shared plesiomorphies or the lack of apomorphies seen within more derived members of the group (another way to express this is to label the group a *grade*) in question, the taxon will be paraphyletic. Where the distinguishing characteristics of a taxon are homoplasies (the homoplasies may have been misidentified by the investigator as synapomorphies), the taxon will probably be polyphyletic.

Farris (1974) seems to have come to conclusions similar to those discussed in the last paragraph when he gave algorithm definitions of monophyletic, paraphyletic, and polyphyletic groups based on the distribution of characters as follows (quoted from Wiley, 1981, p. 84):

1. Monophyletic group: "A group with unique and unreversed group membership characters." Such a group would be one united by synapomorphies—derived character states.

2. Paraphyletic group: "A group with unique but reversed group membership characters." The plesiomorphies uniting a paraphyletic group may be unique to that group (with a unique evolutionary origin—all plesiomorphies are apomorphies at a higher level), but they have been further derived in the apomorphous forms, thus reversed, and apomorphous forms no longer have the necessary characters so as to be included in the paraphyletic group.

3. Polyphyletic group: "A group whose membership characters are not uniquely derived." If the characters are not uniquely derived, then they are homoplasies.

As has been suggested eleswhere (see section on cladistic methodology), cladistically homology equals synapomorphy and it is only homologies that characterize monophyletic groups; thus monophyly can be defined in terms of "uniquely derived and unreversed group membership" (Patterson, 1982a, p. 28; see also Farris, 1974; Platnick, 1977a, 1977b). Such a definition need not have any, or only minimal, evolutionary/phylogenetic overtones. In this context Patterson (1982a) has pointed out that recognizing a homology is the same as characterizing a monophyletic group; "recognizing a homology is directly comparable to discovering a new species" (Patterson, 1982a, p. 59). A newly recognized monophyletic group is usually named, either with an informal name or by christening it with a proper name (e.g., Aves, Mammalia, Vertebrata). Monophyletic groups are individuals with proper names (if formally named), whereas the names applied to nonmonophyletic groups (e.g., Reptilia, if reptiles gave rise to birds and mammals) are "improper names of universals" (Patterson, 1982a, p. 59). Patterson (1982a, p. 59) summarizes his position as follows:

In other words, monophyletic groups, like species, are real: they exist outside the human mind and may be discovered. Homology, which exists only in the human mind, is the relation through which we discover them. Non-monophyletic groups, unlike species, have no existence in nature They have to be invented, are not recognized by homology, and exist only in the human mind

Taxa (in a strictly monophyletic, phylogenetic system) cannot be defined rather, they are discovered, described, diagnosed, and related (Bonde, 1977).

Nonmonophyletic groups (both paraphyletic and polyphyletic groups) do not have ancestors unique unto themselves, and therefore they do not have unique, individual histories (Hennig, 1965). As the members of a polyphyletic group are derived from various ancestors that are not included within the group, there is no unique ancestor for the group. The ancestor of a paraphyletic group is also the ancestor of all groups derived from the paraphyletic group; thus the ancestor is not unique to the paraphyletic group. The extinction of a paraphyletic group is not comparable to the extinction of a monophyletic group. When one speaks of the extinction of a monophyletic group, one refers to the fact that at some point in time all descendants of the stem-species of the monophyletic group cease to exist (live). In contrast, when one speaks of the extinction (pseudoextinction) of a paraphyletic group, one refers to the fact that at some point in time the suite of characters that, solely in the investigator's mind, define the particular paraphyletic group no longer exist in any living organism (Hennig, 1965). Progeny of the members of the paraphyletic group continue to live. Paraphyletic groups are nothing more than groups of organisms (or subtaxa) that are characterized by the possession of primitive features and the lack of derived features (e.g., Ball, 1975; Eldredge and Cracraft, 1980; Løvtrup, 1973; Nelson and Platnick, 1981). Such taxa have no existence in their own right; they can be defined only in reference to another taxon (or taxa) within a more-inclusive taxon. Within a more-inclusive monophyletic taxon, a paraphyletic group is what remains after subordinate monophyletic taxa are distinguished. In this context, a paraphyletic group may be constructed that consists of any conceivable combination of species (Platnick, 1979). Such a group can be defined by listing the synapomorphies of the inclusive monophyletic group that includes the taxa of interest along with absences of the autapomorphies of taxa to be excluded from the group. As an example, Platnick (1979, p. 544) notes that we could define the group comprised of only man and the platypus by listing the synapomorphies of mammals and the absence of the apomorphies of all species of mammals other than man and the platypus. Just because a group can be easily defined does not mean that it has any real existence in the natural world.

Since nonmonophyletic groups are not real (are not natural—they have no independent existence in nature), they should be abandoned. As noted by Wyss and de Queiroz (1984, p. 608), there is an insidious problem in recognizing paraphyletic grade taxa,

namely that by creating such taxa the systematist invites misuse of classifications by biologists untrained in systematics (although even trained systematists are guilty of these errors). These biologists can now talk about such things as variation, geographic and temporal distribution, and extinction of taxa whose fates have been determined as much by human predilections as by biological processes. Can we honestly say that after 35 million years the "Protorothyridae" became extinct even though some of them may have been the direct ancestors of the vast array of living amniotes? What does it mean to label this group "conservative" (Carroll, 1982:103) if some of its members eventually gave rise to organisms as different as bats, whales, turtles, snakes, and birds?

Yet evolutionary sytematists insist that nonmonophyletic, specifically paraphyletic, taxa are useful and even necessary (see Mayr, quoted earlier). Why?

There is the argument that paraphyletic groups (grades) are adaptively unified and therefore natural and justified (Van Valen, 1978). Paraphyletic groups can represent grades of evolution. They tell us something about anagenesis—the degree and rate of adaptive change or morphological diversification. But it is obvious that what adaptively unified means is in the mind of the investigator. Furthermore, a systematic group is not recognized as natural just because it is adaptively unified. There has to be some component of phylogeny; where phylogeny leaves off and adaptive unification takes over is arbitrary and left up to the artistic judgment of the systematist. Also, it is not necessary to utilize paraphyletic groups to measure anagenetic change;

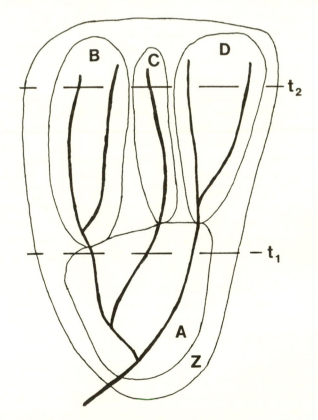

Figure 3-14. Diagram illustrating the concept of monophyly; see text for explanation.

anagenetic change can be measured by the number of homologies that distinguish a particular monophyletic group.

It has also been argued that paraphyletic groups are real, natural groups because they *once* formed real, natural, monophyletic groups. For example, given the phylogenetic tree in Figure 3-14, at time t_1 the group A is monophyletic (it includes an ancestor and all of its descendants), while at time t_2 it is paraphyletic (it is composed of the organisms of which it was composed at time t_1 and only those organisms; the descendants are not included in group A). It can be argued that the mere passage of time should not change the natural status of a group. However, as Wyss and de Queiroz (1984, p. 607) point out, in such cases "it is not the mere passage of time that has changed the status of the group, but the systematist's (not nature's) decision to remove certain direct descendants from the group." At time t_1 group A was synonymous with group Z, and both were monophyletic. At time t_2 group Z is still monophyletic, but group A is no longer considered synonymous with group Z and is paraphyletic because direct descendants of species in group A at time t_1, while still included in group Z, are excluded from group A.

More important, evolutionary systematists appear to be ardent in believing that phylogeny is something more than nested sets or a hierarchy of groupings, something more than systematics per se. Gould (1977, p. 484) defines phylogeny as "the evolutionary history of a lineage," and Mayr (1969, p. 409) defines it as "the study of the history of the lines of evolution in a group of organisms; the origin and evolution of higher taxa." For evolutionists phylogeny necessarily involves statements or hypotheses of ancestry and descent; to deal in such terms one must necessarily deal with hypotheses of paraphyletic groups giving rise to other groups (Patterson, 1982a). To quote a self-proclaimed evolutionary systematist (Carroll, 1984, p. 611): "It is exactly these paraphyletic groups that are of the greatest importance in establishing evolutionary relationships." It makes no difference if the paraphyletic groups that are seen as giving rise to other groups are mere inventions of the investigator's mind (of course, the organisms [fossils] making up the groups are real).

One topic that is seldom discussed is the application of the concept of monophyly to groups that include reticulate evolution (e.g., hybridization or endosymbiosis) in their history. There seems to be an inherent problem in applying monophyly to groups with reticulate evolution. The concept of monophyly distinguishes historically independent, hierarchically arranged groupings, but reticulation breaks down the historical independence and strict hierarchical arrangement of such groupings. Given that taxa 1 through 8 are related as diagrammed in Figure 3-15 (and assuming for the moment that all applicable taxa have been considered), groupings A, C, and D are unquestionably monophyletic (each of these groupings represents an ancestor and all of its descendants). But what of grouping B, for example? Should this be considered monophyletic? It is a group composed of an ancestor and all of its descendants (the usual definition of monophyly), but taxon 4 also has an ancestor that it shares with the taxa of group A and not with the other taxa of group B. Thus, grouping B may be considered polyphyletic (the most recent common ancestor of all components of group B is not included within group B, but within group D). If taxon 4 were not known, there would be no question that taxa 5 and 6 would form a monophyletic group relative to the other taxa shown in Figure 3-15. Thus, it appears that a formerly monophyletic group may become polyphyletic if one of the members of the monophyletic group reticulates (e.g., hybridizes) with a member that is not included within the monophyletic group. A monophyletic group must have a wholly independent phylogenetic history—be a unique historical entity unto itself. As long as there is phylogenetic interaction between members of a putative monophyletic group and nonmembers of the putative monophyletic group, the putative mono-phyletic group has not become completely isolated and does not possess a distinct ontological status—that is, it is not truly monophyletic.

ALLOMETRY AND HETEROCHRONY

There are several major distinguishable (though not mutually exclusive) types of evolutionary changes that have occurred: the origination of morphological (including physiological, behavioral, and so on) novelties either de nova or by the modification of preexisting structures, changes in meristic traits (changing numbers of identical or similar parts, such as bristles on an insect or ribs on a brachiopod), changes in size, and changes in shape. All of these changes, if heritable, may serve as synapomorphies.

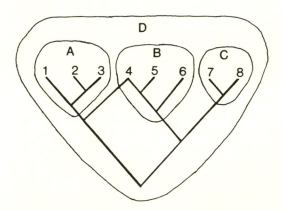

Figure 3-15. Diagram illustrating the application of the concept of monophyly to groups with reticulate evolution; see text for explanation.

Changes in size can be considered in isolation, but in most instances size changes in organisms also involve changes in shape. Allometry is the study of "change of shape correlated with increase or decrease in size" (Gould, 1977, p. 479: for good introductions to allometry, size, and shape in organisms, see Alberch and others, 1979; Benson, 1981, 1982*a*, 1982*b*; Bonner, 1974; Bookstein, 1977; Chapman et al., 1981; Gould, 1966, 1971, 1972, 1974, 1977; J. Huxley, 1932; Lohmann, 1983; Radinsky, 1983; Raff and Kaufman, 1983; Schmidt-Nielson, 1984; Siegel and Benson, 1982; Thompson, 1961). Allometric series may involve changes in relative growth during the development (ontogeny) of a single organism (a series of semaphoronts that comprise a holomorph may be compared), the shape differences between a series of related species of various sizes, or differences in relative growth of organisms in a lineage (such as among ancestors and descendants, or sister taxa). The study of allometric relationships is potentially extremely important for phylogenetic analysis and reconstruction. As Raff and Kaufman (1983, p. 56) state, "Consideration of allometric relationships allows one to sort out which shape changes are caused by growth and which are the result of modifications of the developmental program." Modifications of the developmental program in different lineages and clades may serve as synapomorphies, useful characters in reconstructing phylogeny.

Allometric relationships between two structures can often be expressed using the simple equation (J. Huxley, 1932):

$$y = bx^a$$

where y is a measure of the size (e.g., length or weight) of a particular structure, x is the size of another structure (or perhaps the whole organism bearing the structure), b is a constant scale factor, and a is the ratio of the specific growth rates of y and x. The equation can be written as:

$$\log y = \log b + a \log x$$

and thus when plotted on double logarithmic axes results in a linear plot with a slope of a and an intercept of $\log b$. If $a = 1$, then the relative sizes of the structures in question remain constant and growth is isometric. In most cases a will equal something other than 1, and the relative proportions of the structures will change with size. In comparing several allometric equations of the same basic form, changes in a-values result in differing proportions between the two structures being compared. Changes in b-values affect the proportions between structures by starting the allometric growth of a particular structure at various sizes. For a series of allometric equations of the above form where a is constant and x is a measure of the overall body size of an organism, "when $a<1$, higher b-values are associated with larger sizes; when $a>1$, higher *b*-values characterize smaller sizes" (Gould, 1971, p. 117). Evolutionarily, shifts in *b*-value may occur, corresponding to differing sizes of organisms while preserving overall shape relationships (constant proportions) of particular structures within the organisms.

Given that the basic allometric equation, or some modified form of it, describes the allometric relationships between the members of a monophyletic group (perhaps ancestors and descendants of a lineage), the values of *a* and *b* in the equation can be thought of as characters. Changes in *a* and *b* will result in differing size relationships of structures. As with any other characters, it may be possible to order the values of *a* and *b* in a morphocline and arrive at some judgment as to the polarity of the morphocline. Thus, certain derived values of *a* or *b* may distinguish monophyletic groups at some level. On the basis of presumed ancestor-descendant lineages in the fossil record, Raff and Kaufman (1983, p. 58) argue that evolutionarily allometric relationships may change fairly rapidly, thus distinguishing new lineages or clades. It should also be noted that allometric growth for a single organism during ontogeny may follow a different curve than that relating a group of individuals of closely related adults of different species-level taxa (Gould, 1971). It is possible that the use of allometry may help distinguish between semaphoronts of a single species and individuals of separate but closely related species.

Heterochrony refers to "changes in relative timing of developmental processes" (Raff and Kaufman, p. 173: for a good review of the history of ideas of heterochrony, see Gould, 1977). Heterochronic changes can result in radical morphological differences between closely related species. In an evolutionary sense, through heterochrony an ancestral form may give rise to a descendant of significantly different form without the presence of any intermediates or transitional forms (Broadhead and Waters, 1984). For example, a sexually immature larval or juvenile stage may be significantly different in form, habitat, and ecology from the adult of the same species. If suddenly, through simple heterochronic change via regulator genes, an immature stage undergoes sexual maturity (and the later adult stages are lost), a radically new species could result that might bear most of the previously larval or juvenile characters in the sexually mature adult. Clearly, consideration of heterochronic processes is extremely important in phylogeny reconstruction.

The most obvious heterochronic changes "involve the dissociation of rates of development of somatic traits with the rate of maturation of the gonads" (Raff and Kaufman, 1983, p. 174), and it is on this basis that the classic definitions of the processes and morphological results of heterochrony are based (Gould, 1977). Recapitulation is the result where a formerly adult trait becomes a juvenile trait in the descendant. It may occur either by the acceleration of the appearance of the formerly adult somatic feature such that it is present in the juvenile before gonadal maturation (classic Haeckelian recapitulation via the process of acceleration) or by the retardation of the maturation of reproductive organs such that a formerly adult stage of somatic development becomes preadult (recapitulation via hypermorphosis). In paedomorphosis, characteristics present in an ancestral juvenile are present in an adult descendant. This result may be obtained either by the retardation of the development of adult somatic features with respect to reproductive maturation (neoteny) or by the acceleration of the maturation of the reproductive organs (paedogenesis = progenesis).

Paedomorphosis in particular may occur very quickly as a genetically easy evolutionary response to an environmental pressure (Raff and Kaufman, 1983) that results in very different (perhaps regressive) morphological changes. Superficial comparison of closely related paedomorphic and nonpaedomorphic species, without consideration of the heterochronic changes involved, may be very misleading in terms of the apparent phylogenetic relationships between the species under consideration. In some cases paedomorphosis may be considered as a single derived character that distinguishes a certain taxon; many other derived characters that would unite the particular taxon with other taxa may be secondarily lost due to the paedomorphic condition. For examples of heterochronic changes in the course of evolutionary lineages (reconstructed on the basis of fossil organisms), see Hallam (1982) and McNamara (1982).

EVOLUTIONARY RATES

Evolutionary rates have been, and continue to be, one of the primary concerns of many paleontologists (e.g., Eldredge and Gould, 1972; Gingerich, 1983; Gould, 1984; Gould and Eldredge, 1977; Schopf, 1984; Simpson, 1944, 1953a; Stanley, 1979, 1985). Although evolutionary rates are not directly applicable to phylogeny reconstruction, they do supplement phylogenies and are an essential part of many evolutionary scenarios derived from phylogenetic analysis. Therefore, various common ways of expressing rates of evolution based on the fossil record are briefly considered.

First, however, it should be noted that what rates based on recorded durations of taxa in the stratigraphic (fossil) record actually mean, if anything, is open to debate (Cracraft, 1981a; Novacek and Norell, 1982). Durations of taxa at any level, based on the known fossil record, may be extremely incomplete and deceptive. The known fossil record may at times appear to contradict well-corroborated hypotheses of phylogeny. For instance, a well-corroborated hypothesis may yield the following grouping for the three taxa A, B, and C: (A (B, C)), indicating that taxon A is historically equivalent to taxon (B, C) and diverged from a common ancestor of B and C before B and C diverged from one another. Yet, based on the known fossil record, all three taxa might have gone extinct simultaneously, but the earliest occurrence of taxon C may appear before the earliest occurrence of taxon B, and likewise the earliest occurrence of taxon B might occur prior to the earliest occurrence of taxon A. Taking the fossil record at face value, we might incorrectly interpret C as the sister group of (A, B) and regard C as of longer duration than either taxon A or taxon B. Such misinterpretations will seriously affect our calculations of evolutionary (particularly taxonomic) rates (Novacek and Norell, 1982).

"An evolutionary rate may be defined as a measure of change in organisms relative to elapsed time or to some other independent variable correlated with time" (Simpson, 1953a, pp. 3–4). In paleontology, evolutionary rates are usually divided into two broad categories: taxonomic rates and morphological rates.

Taxonomic rates: Classically, a lineage of sequential ancestral and descendant taxa (at any level) may be established on the basis of fossil material, and from such a lineage one can read off directly rates of taxonomic evolution in terms of number of taxa per unit of time (or conversely, average duration per taxon). Thus, for Kauffman (1977, p. 112) an evolutionary rate is

the number of species added to a lineage per unit time (usually per million year interval). . . . Lineages are narrowly defined here [in Kauffman, 1977] as phylogenetically closely related taxa [i.e., species and subspecies] directly derived from one another, or nearly so, in the evolutionary process. In most Cretaceous examples analyzed here, this relationship can be demonstrated by successive analysis of stratigraphically closely spaced species populations.

In order to avoid the assumption of ancestor-descendant relationships, a form of survivorship curve can be used to determine taxonomic rates of evolution. To do this the durations (expressed in absolute or relative time) of some chosen low-level taxon (e.g., species or genera) within a larger taxonomic group are compiled. Then percentage or proportion of taxa that survive a given amount of time or longer can be plotted on the ordinate against time on the abscissa (Simpson, 1953a). From such plots the average duration of a certain taxon within the group under study and the half-life of the taxon can be found. If the ordinate of such a survivorship curve is logarithmic rather than arithmetic, the slope of the curve at any particular time (age) is proportional to the probability of extinction of a taxon at that age (Van Valen, 1973).

Using taxonomic data curves of taxonomic frequencies and frequency rates can be plotted (Simpson, 1953a). The data for these plots "consists of taxonomic counts, at any one level in the hierarchy, at defined times or for defined spans of time" (Simpson, 1953a, p. 49). Thus total frequency for some taxonomic category for a particular group, such as number of mammalian genera per age, might be tabulated and plotted. The slope of the curve of total frequency at any particular point is simply the rate of change of the frequency at that point. The rate of change of frequency can also be plotted against time. The rate of change of frequency is actually controlled by the rate of origination and the rate of extinction. Correspondingly, first appearances (originations) per unit time, last appearances (extinctions) per unit time, and rates of origination and extinction can be plotted against time. For detailed discussion with examples and analyses of the plotting of taxonomic rates of evolution, the interested reader should consult the work of Simpson (1944, 1953a) and Stanley (1979, 1985).

Van Valen (1973, p. 12) has proposed the *macarthur* as a unit of rate for discrete phenomena, such as origination rates and extinction rates. According to Van Valen (1973, p. 12), if P is the probability per t thousand years, then:

$$\text{ma (macarthurs)} = -\log_2(1 - P^{2t}).$$

One macarthur is a rate of extinction for a particular taxon that gives a half-life of 500 years for the taxon. For origination rates, "one macarthur is the occurrence of one origin per thousand years per potential ancestor" (Van Valen, 1973, p. 12).

Morphological rates: The most common way of measuring morphological evolutionary rates is to compare the size of homologous structures in an ancestor and a descendant and convert this to a fractional size change over a given time span. Taking up this line of reasoning, Haldane (1949) introduced the *darwin* as a unit of evolutionary size change that is equal to a factor of e (2.718, the base of natural logarithms) per million years (1 darwin $= e/10^6$ years). The darwin is derived from the equation:

$$(1/x)\,(dx/dt) = d(\ln x)/dt = (\ln x_2 - \ln x_1)/t$$

where t is time, x_1 is the initial length of some structure in an ancestor, and x_2 is the length of the same structure in a descendant. A darwin is equivalent to a change of about 1/1,000 per 1,000 years, or a doubling of the length of a structure in a million years (Raff and Kaufman, 1983).

In some cases another quantitative assessment of morphological evolution may be the progressive change through a lineage of the factors a and b in the allometric growth equation, $y = bx^a$.

One can also attempt to calculate morphological rates of evolution in terms of the rate of the progressive acquisition of certain character states or character complexes in a putative lineage through time (see, for example, Westoll, 1949, and Simpson, 1953a). Recently, Derstler (1982, p. 131) has formalized this method, which he calls Westoll's Technique, by attempting to estimate rates of morphological evolution "for a fossil lineage by tabulating the number of derived characters which accumulate during the group's history." Derstler's (1982) unit of morphological change is one derived character state. Beginning with a cladogram, Derstler reconstructs a direct lineage from a clade ancestor (stem species of the clade, perhaps a hypothetical reconstruction) to the terminal species under consideration by the use of reconstructed ancestors represented by the nodes of the cladogram. The score for each member (many of which will be hypothetical, reconstructed ancestors) of such a reconstructed lineage is the total number of character-state derivations required to transform the clade's common ancestor into the particular member of the lineage. For a particular lineage leading from the common ancestor of the clade to a particular species, the score of each ancestor can then be plotted against the minimum geological age of the ancestor. The first derivative of the score-versus-age curve taken with respect to age can also be plotted—that is, change in score versus geologic age. The latter plot may point up times or periods of relatively rapid morphological change, and periods of morphological stasis.

Simpson's Classification of Rates of Evolution

Simpson (1953a) proposed a qualitative classification and terminology for organismal rates of evolution that has gained wide currency. Simpson (1953a, p. 318) argues that for any particular group (lineage) of organisms "there is a definite range and pattern" of evolutionary rates. Horotelic rates of evolution are the average or modal rates of evolution for a particular group, "and evolution at rates so distributed is horotely" (Simpson, 1953a, p. 318). Bradytelic/ bradytely refers to very low rates of evolution (as compared to normal rates of evolution within some more inclusive group—perhaps even the group of all organisms, extant and extinct)—that is, what can be termed slow or arrested evolution. Tachytelic/tachytely refers to exceptionally fast rates of evolution.

Horotely represents a sort of normal or average turnover or [evolutionary] metabolism in the evolution of a group of organisms. To the extent that evolution is defined as change, bradytely is a cessation of evolution without extinction. Tachytely is an episodic acceleration, figuratively a brief fever imposed on the normal metabolism. Tachytelic lines arise from horotelic or bradytelic lines, in the latter case usually (perhaps always) by splitting. A tachytelic line must soon become horotelic, bradytelic, or extinct. Horotelic lines usually arise from other horotelic lines, occasionally from tachytelic lines (but these occasional events are unusually important) or from bradytelic lines, in the latter case by splitting and usually through a tachytelic phase. Horotelic lines may remain horotelic, or become bradytelic or tachytelic. Bradytelic lines develop from horotelic or often rather directly from tachytelic lines. The bradytelic lines themselves remain bradytelic and rarely if ever become horotelic or tachytelic again, but it is rather common for them to give rise to tachytelic branches, which in turn become horotelic or extinct. (Simpson, 1953a, p. 337)

Note that in the above quote Simpson considers paraphyletic lines (lineages) to be true or good lineages independent of the lineages (branches) to which they give rise. Thus he can refer to a bradytelic line itself remaining bradytelic while it gives rise to many tachytelic branches. One of the theses of this book is that paraphyletic groups such as Simpson's bradytelic line that gives rise to tachytelic branches are nothing more than artificial constructs of the mind of the investigator (see section on monophyly, polyphyly, and paraphyly).

Besides the various rates of evolution discussed above, Simpson (1953a) also recognizes several different patterns or modes of evolution. In this context, he first recognizes the dichotomy between evolution by splitting (i.e., speciation, cladogenesis) and what he terms *phyletic evolution,* evolution within a single lineage without splitting (i.e., anagenesis). Simpson (1953a, p. 385) takes an extreme adaptationist view with regard to phyletic evolution: "phyletic evolution is usually characterized by progressive and nonrandom change" and "has as its definitive characteristic the continuous maintenance of adaptation in ancestral and descendant populations."

Within the general concept of phyletic evolution, Simpson (1953a, p. 385) distinguishes four basic rates or patterns. (1) Arrested evolution: Organisms in a stable environment (stable, or at least predictable and recurrent as far as the organism is concerned) may be adequately adapted to that environment and thus undergo little or no evolutionary change for long periods of time. (2) Evolutionary trends: A lineage may, in evolutionary time, undergo progressive (here progressive is used as Simpson [1953a, p. 385] uses it, to mean "in a progression," with no implication of progress as commonly used in the vernacular) or unidirectional change as an adaptation to some unidirectional and progressive environmental change. (3) Causal or episodic change: short-term adaptations to nonrecurrent environmental changes. (4) Quantum evolution: "Short-range and relatively rapid shift from the ancestral adaptive zone into another that happens to be available, which may in some cases and not in others involve maintenance of adaptation in the face of radical environmental change" (Simpson, 1953a, p. 385).

Following up on the groundbreaking work of Simpson (1944, 1953a), in the last decade there has been much interest and debate over the relative frequencies of broad patterns, tempos, or modes of evolution in various groups of organisms (e.g., Stanley, 1979). Much of this work has polarized around the question of which of two scenarios, *punctuated equilibria* or *phyletic gradualism,* has more validity as a general description of most macroevolutionary patterns in the fossil record (Eldredge and Gould, 1972; Gingerich, 1976a, 1976b, 1980, 1983; Gould, 1982, 1983b; Gould and Eldredge, 1977; Schopf, 1981; Schopf and Hoffman, 1983; Vrba, 1980, 1982a). To characterize phyletic gradualism, classic Neo-Darwinian evolutionists have assumed a priori that evolution occurs through the accumulation of slow, steady, and gradual adaptive change. Given a complete succession of fossils through time, these will show a continuous gradation, and all morphological gaps will be filled between any two end forms. According to some extreme gradualist scenarios, rates of evolution during cladogenesis (splitting events between species) are of approximately the same order of magnitude as rates of evolution during anagenesis (phyletic evolution, cf. Simpson's horotely). The scenario of punctuated equilibria suggests that rather than being gradual, evolution tends to be episodic in most clades. Extremely high rates of evolution occur in spurts (cf. Simpson's quantum evolution), preceded and followed by evolutionary stasis (cf. Simpson's arrested evolution) during which a species does not appreciably evolve or change, often merely fluctuating mildly around a basic morphology. An important point of the punctuated equilibria model is that high rates of evolution occur during cladogenesis (speciation) but relative stasis is the rule for evolution within a species; the punctuations occur at the splitting (speciation) events.

Trying to decide between the punctuated equilibria and phyletic gradualism scenarios has apparently proved relatively futile thus far (Schoch and Meredith, 1984). Both are scenarios, and one can enumerate putative examples of each (e.g., Gould and Eldredge, 1977). Both rely on the recognition of series of ancestors and descendants, over extended periods of time, in the fossil record. As argued elsewhere (see section on ancestor-descendant relationships), it is highly questionable that these can be recognized with any reliability. Furthermore, punctuated equilibria and phyletic gradualism are only two of several possible scenarios. For example, Turner (1983; cf. the *punctuated gradualism* of Malmgren, Berggren, and Lohman, 1983, 1984, and *punctuated anagenesis* of Gould, 1985) has suggested another scenario, which he calls punctuated phyletic evolution: Nonconstant rates and directions of evolution may occur within a species or series of successional subspecies without cladogenesis.

Simpson (1953a, pp. 282–291) also discussed, and dismissed, the concept of hypertely. The basic notion of hypertely is "that somehow evolutionary momentum carries [evolutionary] trends to inadaptive lengths and causes extinction" (Simpson, 1953a, p. 282). Extinction results because the end forms of a lineage are of decreased fitness. The classic example is the Pleistocene Irish elk, *Megaloceros giganteus* (see Gould, 1974), the males of which grew antlers so large that they must have been inadaptive. (Note that the concept of hypertely is very different from the concept of *success-extinction* [Schoch and Meredith, 1984: see chap. 7]. Success-extinction suggests that superiorly adapted individuals of a particular species may cause the demise of the entire species by being so successful that resources on which the members of the species rely are overutilized.)

CHAPTER 4

The Fossil Record

THE ADEQUACY OF THE FOSSIL RECORD

It has become a kind of dogma among some circles of paleontologists and biologists that the fossil record is inherently incomplete and inadequate for many sorts of evolutionary, phylogenetic, and stratigraphic studies (Shaw, 1964, p. 105). In many ways this stigma and prejudice appears to be a direct outgrowth of Charles Darwin's chapter in *The Origin of Species* (1859) "On the Imperfection of the Geological Record," in which he discusses explanations for the relative lack of direct evolutionary (i.e., ancestor-descendant) sequences in the rock record. But what is really meant when one refers to the inadequacy of the fossil record? There seem to be several differing concepts being addressed. Ager (1976, p. 131) has given one impressionistic view, with which many workers probably agree: "The fossil record, like the stratigraphic record [see Ager, 1973], is thought to be episodic with long periods of explosive evolution, expropriations and extinctions." This view may be due in part to the possibility that evolution, over geological time, is quantized (cf. "punctuated equilibria," described in the section on evolutionary rates; Eldredge and Gould, 1972). Many authors are also impressed by the seemingly haphazard nature, and in some cases absolute rarity, of the fossilization process per se. For any organism to be fossilized and preserved, it must meet certain conditions and requirements. Organisms are rarely preserved that live and die in areas of nondeposition and erosion, and it is the hard parts of organisms that are most readily preserved (such as the bones and teeth of vertebrates, the wood of trees, and the shells of certain invertebrates). Whole groups of soft-bodied organisms, such as jellyfish and worms, have virtually no fossil record. Thus, even if we had an entire knowledge of the organisms preserved in the rock record, we would still not have knowledge of every species that ever inhabited the earth; the fossil record would still be incomplete. And paleontologists do not even know of every organism that has successfully been fossilized. Not only must an organism be preserved in the fossil record, but it must be found and interpreted by a paleontologist; new fossil species are being described all the time. But then too, new species of extant organisms continue to be described; our knowledge of even the species that presently inhabit the earth is far from complete. It should be obvious that our knowledge of life on earth will be less complete if we do not take into consideration fossil organisms, no matter how incomplete the fossil record may be.

In 1940 Efremov (1940, p. 85) defined the field of taphonomy as "the study of the transition (in all its details) of animal [and plant] remains from the biosphere into the lithosphere, i.e., the study of a process in the upshot of which the organisms pass out of the different parts of the biosphere and, being fossilized, become part of the lithosphere," or more simply, "the science of the laws of embedding" (Efremov, 1940, p. 93). This branch of paleontology deals specifically with many of the problems posed in the last paragraph, especially in exposing preservation and collecting biases that affect our knowledge of fossil organisms (Behrensmeyer, 1984: Behrensmeyer's paper provides a concise, recent review of the field of taphonomy by one of its leading authorities; see also Behrensmeyer and Kidwell, 1985). At present, taphonomic considerations appear to be most important for studies involving paleoecology, evolutionary scenarios, and large-scale processes, tempos, and patterns of faunal replacement through time. Taphonomic studies are only marginally important to the study of phylogeny in the strict sense, as dealt with in this book. However, taphonomic considerations are extremely important if one wishes to pursue a stratophenetic program of phylogeny reconstruction; taphonomic studies could potentially indicate to an investigator when a certain set of data is not suited to such a mode of investigation. The study of taphonomy can give a clear indication of what types of organisms will be preserved in certain types of environments. While in many cases fossilization is probably the exception, in some instances (e.g., among certain deep-sea microorganisms) fossilization may actually be the rule, and thus one may be able to read evolution fairly directly from an adequately sampled record.

Once a fossil specimen is found, the study of taphonomy also aids the investigator in determining the information that can be gained (or that has been lost) from the specimen. A fossil may have undergone crushing, plastic deformation, or some other form of distortion. In most instances only selected hard parts will be preserved, and these may be heavily mineralized or abraded, obscuring potentially significant morphological details. On the other hand, postmortem features of a fossil and the context in which it is found may prove significant. Certain marine fossils may be preserved in life assemblages—and this could provide useful information about the living organisms. Or it may be important to know that a fossil has been transported postmortem far from the original environment in which the organism lived. Taphonomic considerations can also be extremely important to the practicing biostratigrapher who depends on the presence (or absence) of fossils in various rock units for correlation purposes. In some cases a species may have occupied a certain area at a certain time but is not preserved in the corresponding rock record due merely to preservational biases against the species. The same species might be preserved in abundance in a different area under slightly different preservational conditions. For the field-oriented paleontologist, taphonomic considerations can be utilized to help predict and locate the best collecting areas for certain kinds of fossils.

Even when an ancient organism is preserved, fossilized, and recovered, it is still often seen by many systematists as a mere ghost of an organism compared to much better preserved neontological specimens. The argument runs that the average fossil specimen preserves a small sample of the total morphology (usually only the gross morphology of hard parts) of the original organism, and that any neontological specimen is usually far superior to any fossil specimen in its completeness and thus in the information bearing on its phylogeny that it contains. Furthermore, with extant organisms one often has access to developmental, ecological, and behavioral data that is virtually nonexistent for all extinct organisms. But it can be argued that for a large number of groups preserved in the fossil record with extant members, the morphological characters used by the practicing museum taxonomist working from preserved neontological specimens are basically the same morphological characters preserved in fossils and utilized by the practicing paleontological systematist. In many cases the same worker may be an authority on both extinct and extant members of a particular taxon.

Many stratigraphers and paleontologists worry that because of the incompleteness of the fossil record it may be impossible to find the stratigraphic ranges and the sequences of first and last appearances of fossil organisms with any degree of accuracy and certainty. The relative stratigraphic order of fossil species (and thus higher taxa) is extremely important for biostratigraphic correlation, for putting constraints on the ages of various clades, and as an independent test of phylogenies constructed on the basis of other sets of data (such as pure morphology). At times, for certain organisms, given that certain assumptions and requirements are fulfilled, it may be that a fossil record is even good enough to attempt to read an evolutionary history directly from the rocks (cf. Lazarus and Prothero, 1984; see the sections on classical paleontological methods and on ancestor-descendant relationships).

Recently, several different methodologies have been proposed to arrive at estimates of the relative completeness of the fossil record. These contributions are reviewed here.

Paul (1982, p. 78) has looked at the problem from the point of view of gaps in the fossil record. As Paul defines it,

A gap occurs when a taxon is known from below and above, but not actually within, a stratigraphic interval. Each such gap represents a situation where at least one species must have existed to carry on the line, but has not been preserved or is yet to be discovered. Unfortunately this is so obvious a fact that very few range charts show any gaps at all. We all make the logical deduction and plot known range irrespective of whether it is based on two, or two million, occurrences.

Paul (1982) then proceeded to analyze the proportion of gap to record for cystoid (Echinodermata) families at the series level. He plotted the known stratigraphic ranges, including gaps, at the series level for all cystoid families that are known from two or more series. For each family of cystoids Paul could then calculate the proportion of gap relative to the total stratigraphic range of the family. For example, a certain family may be first known from a certain series, be unknown from the following series, but be again represented in the series that follows and unknown thereafter. This family would thus have a total range spanning three series, but since it is absent from one of them (the middle series) its "gap index" could be said to be $1/3$, or 33% at the series level. As Paul points out, such estimates of the incompleteness of the fossil record are minimal estimates. There may be additional gaps above and below the known range (i.e., the known range may not correspond to the complete or total range) of the fossil group, but this method detects gaps only within the known range. Of course, it is not applicable at the series level for a group that is known only from a single series, but it can be applied to any temporal level of resolution and to any taxonomic unit (such as to genera or species at a substage level).

The cumulative ratio of gap to range for all of the taxa within a larger grouping can also be found. Thus, Paul found that for all Ordovician through Devonian cystoids at the family/series level, there was a total of 27 gaps out of a total range of 108 series, and thus the group must be at least $27/108$ or 25% incomplete at this level. Using the same matrix of data it is possible to get an idea of how complete the fossil record is for some larger taxonomic group during certain times. From Paul's (1982) data, 12 cystoid families are known from the Ashgillian (the last series of the Ordovician), and only 2 cystoid families are known from both earlier in the Ordovician and from the Silurian, but are not represented in the Ashgillian. Thus we can calculate that for Ashgillian families of cystoids, the ratio of gap to representation is $2/14$, or an incompleteness of 14%. In contrast, for the same group the Llandoverian (first series of the Silurian) has an incompleteness of $7/8$, or 87.5%.

Shaw (1964, p. 107; see also the work of Dennison and W. W. Hay, 1967, and W. W. Hay, 1972) proposed a way to estimate the reliability of the observed ranges of species (or other taxonomic units). In particular, Shaw posed and answered the question, "What is the probability that we shall find one or more specimens of the species we are seeking (which we may designate a) in a sample of N identified specimens?" If we let p represent the proportion of species a in the paleofauna that we are sampling and q equal the proportion of all other species in the fauna, then $p + q = 1$. If we collect and identify N specimens, then the probability, Q, that we will not sample a but will sample only b (specimens that we are not seeking) is:

$$Q = q^N,$$

and the probability of finding at least one specimen of a is:

$$P = 1 - Q.$$

From the above equations it can be readily seen that as we collect and identify more specimens (N) from any single fauna or locality, the probability of finding even a relatively rare species increases dramatically. To take a single example, if the species a occurs in a proportion of 1%, then in 99 cases out of 100 a sample of 459 specimens will sample species a. ($p = .01$, $q = .99$, $P = .99$, $Q = 1 - P = .01$, $Q = q^N$, $.01 = .99^N$, therefore $N = 459$ [$.99^{459} = .00992097$, which is approximately .01]). Both Shaw (1964) and Paul (1982) present selected tables of values for p, Q or P, and N, but with powerful handheld calculators and desk-top computers now readily available, the need for such tables is greatly decreased. Workers can solve the equation for whatever particular values of the variables they are dealing with.

Shaw's method is particularly useful for deciding how large a sample we need to take to insure, at some probability level (P or Q can be specified), that we do not miss a certain species. But to do this we must have some idea of p, the proportion of the species making up the entire fauna. Shaw (1964, pp. 111–115) discusses this point. He suggests that there are three basic ways to obtain estimates of p. (1) *Proportion per collection:* The number of specimens of the species in question can be divided by the total number of specimens for any particular collection (sample) taken. The larger the sample used, the closer the estimate will be to a true value for p for the particular horizon, zone, stratum, or area sampled. (2) *Local average proportion:* The total number of specimens of the species in question can be divided by the total number of fossil specimens collected throughout the range of the species in a single (local) stratigraphic section. (3) *The overall proportion:* The total of all specimens of the species found in all stratigraphic sections under study can be divided by the total number of specimens from all of the sections under study to find a very general estimate of *p*.

In attempting to establish the stratigraphic and temporal limits of the range of a species (or other taxon), it is usually the *proportion per collection* and *local average proportion* that will serve most usefully as estimates of *p*. For many species *p* will vary from section to section and throughout their geographic and stratigraphic range. When establishing the base or top of the range of a species, we wish to know how many specimens we need to collect from the layer just below or just above the presumed base or top of the species' range to test, with reasonable certainty, whether the species occurs in, and can be recovered from, this next layer. Ideally, we would like to know *p* for the new layer that we are testing; but we can never absolutely know *p* for any layer or section, and we cannot estimate *p* until we begin to find specimens of the species in question. Thus it seems that in most cases the next best estimate of *p* will be based on collections from strata immediately adjacent to the stratum in question or on the occurrence of the species in the localized stratigraphic section that is presumed to contain the top or the base of the range of the species.

Using Shaw's methods one can find what Shaw (1964, p. 103) calls the *total stratigraphic range* of a species—that is, the determinable range of a species in the rock column, the upper and lower boundaries of which are marked by such small values of *p* that "there is little or no rational hope of recovering more individuals beyond the limiting occurrences" (Shaw, 1964, p. 115). Shaw (1964, p. 103) is careful to distinguish this concept from the concepts of the biozone or teilzone of a fossil species. The term *biozone* refers to the true (and unknowable) total stratigraphic and temporal range of a species; *teilzone* refers to the local, and determinable, stratigraphic range of a fossil.

Paul (1982, p. 86) has further expanded on Shaw's (1964) method and his own analysis of gaps by looking at the length of intervals of specimens of a single species within a particular measured stratigraphic section. He starts with the basic idealized assumptions that the preservation of individuals of the species is a random process and that deposition of the stratigraphic section under study was characterized by uniform rates of sedimentation. If this is the case, Paul (1982, p. 86) defines the *specific preservation time* (*T*) as the "average interval of time between the preservation of two successive specimens" of the species in question. If the measured section is of uniform width laterally, and given uniform sedimentation rates, then the specific preservation time will correspond to a *specific preservation interval* (*I*), which can be found empirically in the measured section. Paul goes on to show that given the above assumptions and the null hypothesis of random preservation and uniform sedimentation, the frequency distribution of actual measured intervals between specimens of the species in question should ideally form an exponential declining curve (Fig. 4-1) taking the form:

$$y = ke^{-\upsilon x}$$

where y = the frequency of a stratigraphic interval of certain size, x = the size of the stratigraphic interval, k and υ are constants, and e is the base of natural logarithms.

Paul further develops this equation in the following manner. We can let k = the total number of intervals ($k = N - 1$, where N is the total number of specimens of the species in question collected from the stratigraphic section, assuming that no two specimens were collected at precisely the same level), and we can find the size of the median stratigraphic interval, i (which is an approximation of I), such that half of the intervals (x) are smaller and half are larger than i. We can then compare any actual frequency distribution with a theoretical curve (Fig. 4-1) by normalizing k to 1 or 100% and expressing the stratigraphic intervals in terms of i.

As Paul points out, if the actual frequency distribution fits the idealized curve, then half of the measured intervals should be smaller than i, one-quarter should have values between i and $2i$, one-eighth should have values

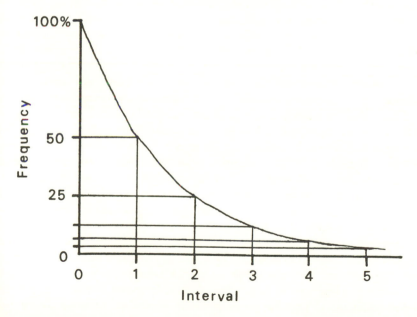

Figure 4-1. Idealized frequency curve of stratigraphic intervals (in terms of the median stratigraphic interval) in a stratigraphic section After Paul, 1982, p. 87, fig. 4; see text for further explanation.

between 2*i* and 3*i*, one-sixteenth should have values between 3*i* and 4*i*, and so on. In any particular case, if the actual frequency curve fits the theoretical curve, then there is little reason to reject the null hypothesis. Although it is theoretically possible that nonuniform sedimentation and nonrandom preservation could combine in just the right combination to produce the expected frequency distribution, the chances of this happening are exceedingly small. Further, if the actual curve fits the expected distribution and we accept the null hypothesis, then we can place confidence limits on the known local stratigraphic range of the species under question if we assume that we have samples located reasonably close to the bottom and top of its range—that is, if below the lowest known occurrence and above the highest known occurrence we never find another specimen of the species in question. Of course, the barren interval above and below the known range of the species must be significantly large in terms of *i*. As Paul (1982, p. 89) states, approximately 93.75% of the intervals (that is, 93.75% of the area under the idealized curve) will lie between $x = 0$ and $x = 4i$, and likewise 99.22% will lie between $x = 0$ and $x = 7i$. We can be approximately 95% confident that the highest local occurrence of the species is no more than 4*i* above the highest sampled specimen and likewise that the lowest local occurrence is no more than 4*i* below the lowest sampled local occurrence. Thus the 95% and 99% confidence limits of the range of the species in question in the local section lie approximately 4*i* and 7*i*, respectively, above and below the known range of a species. As any particular section is collected more throughly, the intervals between the occurrences of collected specimens of the species in question will tend to decrease. Correspondingly, *i* will decrease in absolute size while the absolute size of the confidence intervals (4*i* and 7*i*) will decrease; thus we will be more certain about the limits of the local stratigraphic range of the species.

If the actual frequency does not fit the idealized curve, then at least one of the initial assumptions (random preservation, uniform sedimentation) must be rejected. Sedimentological analysis may be able to detect nonuniform sedimentation (Paul, 1982). A storm or other short-term event might deposit a large amount of sediment very quickly with few or no fossils buried in it. On the other hand, winnowing may remove sediment and form a lag deposit that has an undue representation of fossils. Or, the local section may be composed of a sequence of alternating lithologies, and fossils of the species in question may be more common in one particular lithology. However, if the frequency distribution for one particular species in a certain stratigraphic section does fit the idealized curve, and we accept the assumption that sedimentation was uniform within the local section, then sedimentation must have been uniform relative to all species found in the local section. If the frequency distribution for a second species from the same section does not correspond to the idealized curve, then either our initial acceptance of uniform sedimentation, based on the frequency distribution of the first species, was incorrect or the second species experienced nonrandom preservation throughout the section. For instance, the second species may not have been continuously present in the area where the local section was deposited during the time interval under consideration, or the second species may have been more abundant at certain times and rarer at other times. Or lithologic variations in the local section may have had a nonrandom correlation with preservation of the second species.

In such cases of nonrandom preservation or nonuniform sedimentation rates, the largest measured stratigraphic interval between two specimens of the species in question will seem too large, and there will be too many of the large intervals, according to the idealized curve (remember that the stratigraphic intervals are expressed in terms of i, which is the median value of all of the intervals). Paul (1982, p. 90) has developed a methodology where "the largest recorded interval may be used to test the actual distributions and we can predict exactly the minimum sample size required for a given maximum interval from the theoretical curve."

Due to the form of the curve (Fig. 4-1), as stated above, it is predicted that half of the sampled intervals will fall below i and half will fall above i. Thus, to sample an interval above i and an interval below i we ideally need a minimum of 2^1 samples, one of which will be below i and one of which will be above i—that is, ideally one half of the intervals will be larger than i. Likewise, one-quarter will ideally be larger than $2i$ and one-eighth will be larger than $3i$, so ideally we will need at least 2^2 or 2^3 sampled intervals to sample an interval larger than $2i$ or $3i$, respectively. And in general, we would need at least 2^n sampled intervals to sample a maximum observed interval of size ni or greater. If the actual number of intervals sampled is k, then the expected probability, p, of observing an interval that has a size between ni and $(n + 1)i$ would be less than or equal to $k/2^n$ and greater than or equal to $k / ((2^{n+1}) - 1)$ (Paul, 1982, p. 90). For example, if the largest observed interval were slightly larger than $5i$, then $2^5 = 32$ and $2^6 - 1 = 63$; the predicted size of the population of intervals to include a maximum interval between $5i$ and $6i$ would be between 32 and 63. If the actual number of observed intervals is 10, then the probability, p, of observing a maximum interval between $5i$ and $6i$ should lie between approximately .31 ($= 10/32$) and .16 ($= 10/63$).

If p is unreasonably small for a particular species as distributed throughout a particular stratigraphic section, we should look for some nonrandom explanation to account for the large intervals observed. Perhaps the particular species in question was not continuously present in the local area during the time of deposition of the local stratigraphic section. If p is relatively large, this supports the null hypothesis of random preservation and uniform sedimentation. This is one more way in which one can obtain a semi-quantitative assessment of whether the distribution of fossils in a particular section is due to random processes or if some special explanation for the distribution must be sought.

Another way that has been developed to measure the quality of the fossil record is to calculate completeness estimates for particular stratigraphic sections that are fossil bearing (Dingus, 1984; Dingus and Sadler, 1982; Gingerich, 1982; M. L. McKinney, 1985; M. L. McKinney and Schoch, 1983; Sadler, 1981, 1983; Sadler and Dingus, 1982; Schindel, 1980, 1982a, 1982b, 1982c, 1982d; Schoch, 1983; Tipper, 1983). Completeness can be defined as the portion of a given time span that is represented by sediment or rock (Sadler, 1981); when we say that an interval of time is represented by a sedimentary/rock record in a certain area, we are referring to the fact that sediment was deposited in the area during a certain period of time (M. L. McKinney and Schoch, 1983). Completeness estimates are calculated at various levels of resolution—for instance, at the 10-year, 100-year, 1,000-year, or 10,000-year levels—and are expressed as percentages. The resulting completeness values can be thought of as the percentage of intervals, at the specified level of resolution, that is represented by some net accumulation of sediment, no matter how small, in the stratigraphic section. To calculate completeness, the net rate of accumulation of a sedimentary section (thickness of section divided by length of time over which it was deposited) is divided by a short-term rate of sedimentation (based on modern analogues and compilations of sedimentation rates from various sedimentary environments). As an example, in fluvial environments median sedimentation rates observed over 100-year intervals are approximately 600 mm (mm = millimeters) per 100 years (= 6,000 m/1 m.y. [m = meters; m.y. = million years]) (M. L. McKinney and Schoch, 1983; Sadler, 1981; Schindel, 1980). Let us assume that we are interested in calculating the completeness of a 500-meters-thick fluvial siltstone that was deposited over the course of 1 m.y. The completeness of this section at a resolution of 100 years would be 500 m/1 m.y. divided by 6,000 m/1 m.y. = 8.3 percent; about 830 of the 10,000 100-year intervals constituting the 1 m.y. period saw some deposition.

A major problem with calculating such estimates, however, is that of compaction. In the above example the observed 100-year interval rate would be in terms of thickness of recent, uncompacted sediment, whereas the 500-m stratigraphic section has probably undergone some sort of compaction. In general, sedimentation rates calculated from observations of accumulation over longer spans of time (at coarser levels of resolution) in already lithified strata will already incorporate compaction, whereas rates calculated for short spans of time are generally based on observations of unconsolidated, uncompacted sediment. To help offset this problem, M. L. McKinney and Schoch (1983) doubled their calculated completeness estimates for the finer levels of resolution (10- and 100-year levels), assuming a factor of 50 percent compaction. At coarser levels of resolution (1,000- and 10,000-year levels) they did not make any adjustment for compaction, believing that compaction was already accounted for in the short-term (1,000 and 10,000 year) rates of sedimentation used. Thus, in the example in the last paragraph, they would double the completeness estimate of 8.3% to arrive at a final completeness estimate of approximately 17%. Their method of dealing with the problem of compaction is extremely rough, but it is the best that can be done until compaction data is synthesized so as to give more accurate estimates of compaction in different rock types under different diagenetic environments for different amounts of time.

To estimate completenesses for stratigraphic sections at various levels of resolution, one also needs accurate data on sedimentation rates in different depositional environments and over different periods of observation. Reineck (1960), Sadler (1981), and Schindel (1980) have compiled depositional rates for a variety of environments, and these authors have all noted that there is always an inverse relationship between the rate of sedimentation reported by an investigator and the period of observation from which the rate was derived. For each environment, the rates of sedimentation reported decreased as the time span of observation increased. For fluvial environments, sedimentation rates calculated from observations of 100-year intervals were higher (with a median of roughly 6,000 mm/1,000 years) than those calculated from 1,000-year intervals (1,000 mm/1,000 years). The main reason for this result seems to be that, as the period of observation increases, more and longer intervals of nondeposition and/or erosion are averaged into the rate calculations, thus lowering values. The implication for completeness estimates is that any stratigraphic section will be ever more complete at coarser levels of resolution.

To estimate completenesses of sections one must also know, or be able to determine, the depositional environment under which the sediment was deposited and then apply the appropriate short-term sedimentation rate discussed above. Original depositional environments for stratigraphic sections can usually be determined through sedimentary analysis of the lithologies of the rock bodies (see, for example, Blatt, Middleton, and Murray, 1972; Füchtbauer, 1974; Moorhouse, 1959; Selley, 1976; Williams, Turner, and Gilbert, 1954). Certain types of rock bodies are deposited in certain environments. However, even after the original sedimentary regime is determined as accurately as possible, it may still be difficult to find, or decide on, the appropriate short-term sedimentation rates to use in the calculations.

Finally, in order to calculate completeness estimates, one must know the total interval of time over which the section (or portion of a section) was deposited. This time interval may be very difficult to determine if there is not some kind of absolute dating of the section or some way to tie the section into an absolute (i.e., dated in years) temporal framework. Needless to say, the thickness of the section and estimates of average sedimentation rates must not be used to calculate how much time the section represents and then in turn used to calculate completeness estimates. It may be possible to tie a section into a radiometric time scale, perhaps directly if there are minerals suitable for radiometric dating within the section under consideration, or indirectly through refined bio- or lithostratigraphic correlations. Another possibility, utilized by M. L. McKinney and Schoch (1983), is to use magnetostratigraphy to tie a section into the standard magnetic polarity time scale.

Once completeness estimates for a stratigraphic section are calculated, they are open to interpretation. One important thing to remember is that the completeness estimates are of the lithostratigraphic completeness, not the biostratigraphic completeness (Dingus and Sadler, 1982). In some cases, the biostratigraphic completeness will be much less (that is, there is a fair amount of sediment separating any two fossils). In other cases, the lithostratigraphic and biostratigraphic completeness may be essentially the same (such a case might involve a sedimentary rock composed entirely of organic clasts—the fossils). In exceptional cases it may be that the biostratigraphic completeness is higher than the lithostratigraphic completeness; perhaps fossils accumulated on a surface as sediment was winnowed away (Behrensmeyer, 1982; Kidwell, 1982).

In connection with completeness estimates, Schindel (1982c) has introduced the concept of *microstratigraphic acuity* to refer to the amount of time represented by each individual fossiliferous sediment sample in studies based on sequences of such samples collected up a stratigraphic section. To calculate the maximum stratigraphic acuity it must be assumed that, by careful sampling, all gaps (which are longer than a specified level of resolution) in a stratigraphic section can be avoided. Thus the total, gap-ridden stratigraphic section is broken down into a number of short, complete, gap-free stratigraphic sections. But hiatuses in a sedimentary sequence may be both very subtle and very significant (Schoch, 1983; Van Andel, 1981); many gaps may remain unmarked, and it is unrealistic to suppose that all hiatuses can be avoided when collecting samples in the field. The concept of microstratigraphic acuity requires assumptions that cannot be justified by the ad hoc, probabilistic nature of completeness estimations of sedimentary sections. The strength of completeness estimates lies in the fact that no such assumptions need to be made. Completeness estimates make no predictions as to the distribution of gaps versus temporal intervals represented by sediment in a particular stratigraphic section.

Schindel (1982c) has also presented what he refers to as a *habitat-shift* model in which gaps in the stratigraphic record are considered to correspond to interruptions in within-populational processes of organisms (later preserved as fossils) living in a particular environment. Yet there is no a priori reason why periods of sediment accumulation necessarily have any correlation with hospitable or inhospitable conditions for organisms that later die and are preserved in the sediments (Schoch, 1983). For example, in a fluvial system the mobile terrestrial mammals would have been relatively independent of the exact coordinates of the river system, while for many marine organisms relatively high rates of sedimentation (perhaps burying the organisms in sediment), rather than profound breaks in sedimentation, might represent the most inhospitable conditions.

SEQUENCES OF FOSSILS

It is often argued that because of the incompleteness of the fossil record, the order of first occurrences of fossil taxa is inherently unreliable. This is not to say that the ordering or dating of actual, known fossil specimens is unreliable; a taxon may have existed prior to the age of its first known representative and subsequent to the age of its last known representative. Stratigraphic data cannot or should not be used to determine the historical sequence of first appearances of taxa and morphologies; primitive taxa or morphologies may just as easily occur later than advanced or derived morphologies as vice versa. There seem to be two aspects to this discussion. First, there is the biological concept that even if we had absolute knowledge of the complete sequence of all organisms through time, there would still be some taxa that occur relatively late in time with predominantly plesiomorphic suites of character states (extant taxa that fit this description are often referred to as living fossils: Eldredge and Stanley, 1984; Schopf, 1984). On the other hand, some very ancient taxa might bear many apomorphies. And it is also theoretically possible, and has probably occurred (Van Valen, 1978; Wiley, 1978, 1979a), that an ancestral species can outlive some of its descendant species. Thus, even given a perfect fossil record, there will be some interpretation involved in determining polarities of morphoclines and phylogenetic relationships. Second, there is the geological/stratigraphical concept that the fossil record is not known in its entirety and the total stratigraphic range of many taxa is not known. There is often the possibility that the order of first-known occurrences will be incorrect.

Harper (1980) has explicitly addressed the rationale assumed by many stratigraphers and paleontologists in inferring the relative ages of fossil taxa based on the known stratigraphic record (see also various articles in Cubitt and Reyment, 1982; Kauffman and Hazel, 1977). In his discussion Harper makes seven basic points.

1. The only thing that a stratigraphic paleontologist can document objectively is a succession of fossils in a local stratigraphic section. The relating of sections to one another and implying that successions of fossils in strata correspond to temporal sequences are acts of inference (Harper, 1980, p. 239). The fact that a certain collection of fossil taxa, or fossil assemblages, consistently occurs in a particular (nonrandom) order in space (through stratigraphic sections covering a certain, perhaps large, geographic area) does not necessarily

imply that the taxa or assemblages represented succeeded each other temporally. If a vertical pattern of occurrences is repeated in a single local section (i.e., in a local section taxa are found as such, from bottom to top: A, B, C, A, B, C, A, B, C), the pattern obviously does not imply a unique temporal sequence for the taxa concerned. But even if this is not the case, a certain sequence may or may not have temporal significance. For any particular stratigraphic section one cannot always assume a priori that reworking of fossils (usually resulting in older fossils in younger sediments), and stratigraphic leaks (younger fossils displaced into older rocks) are absent or negligible. In order to clarify this notion, T. H. Huxley (1862) coined the term *homotaxis* to refer to the concept of similarity or identity of the spatial ordering of fossil taxa and assemblages at different localities or in different areas without the notion that they are necessarily time correlative. Homotaxial patterns may have time significance, but they need not. A major endeavor of the biostratigrapher is to determine what homotaxial patterns are temporally significant. Some workers (e.g., Jeletsky, 1956) use the term *homotaxial* to refer only to patterns that are homotaxial in T. H. Huxley's sense, but are not time correlative.

2. First occurrences need not necessarily correspond in time from one area to another (Harper, 1980, p. 240). Rates of migration and geographic dispersal vary for various organisms. Even if rates of migration were extremely fast geologically, there may have been barriers to dispersal that lasted for significant periods of time. The preservation of many fossil organisms is facies controlled, and hiatuses in deposition, stratigraphic mixing, and pure accidents in preservation and collection may pose problems.

3. Likewise, last occurrences need not correspond in time from one area to another. The timing of local extinctions of the same organisms may have varied significantly from one area to another. A last occurrence may also represent a hiatus, or it may represent only the last occurrence of a particular facies in a particular section.

4. "An argument for the time significance of a given homotaxial pattern is greatly strengthened if the taxa involved were likely to be preserved in a very broad range of depositional environments. In the case of marine patterns this means a broad range of bottom depths and/or distances from shore" (Harper, 1980, p. 241). One must be wary of the situation where a "homotaxial pattern reflects a succession of laterally arranged biofacies that succeeded one another at different times at different localities" (Harper, 1980, p. 241).

5. Harper (1980, p. 241) suggests that the assumption of punctuated equilibria would strengthen the argument for the time significance of a homotaxial pattern. Correlations based on species that are static and unchanging throughout their range (both temporally and geographically) should be easier to establish than correlations based on continuously gradating or evolving ancestor-descendant lineages.

6. "In arguing for the time significance of a given homotaxial pattern, the areal extent of the pattern is irrelevant" (Harper, 1980, p. 241). A homotaxial

pattern may be temporally significant within a certain area or biotic province, but perhaps not between areas or biotic provinces (there may have been temporally significant barriers to geographic dispersal).

7. "In arguing for the time significance of a given homotaxial pattern, the phylogeny of the taxa is irrelevant" (Harper, 1980, p. 242). Phylogenies themselves can only be inferred, and first occurrences in a certain area may represent migrations rather than in situ evolution without affecting their utility.

Paul (1982) has addressed the geological concept of the unreliability of the sequences of appearances and events in the fossil record. He has pointed out that for any two species, or other taxa, whose true stratigraphic and temporal ranges did not overlap, there is no possibility that specimens representing the taxa could be preserved in the wrong order of occurrence (first appearance). (Of course, here it is understood that the stratigraphic relationships of the rock bodies containing the fossils are correctly interpreted and that post-depositional mixing of fossils from different temporal intervals has not occurred.)

The important consideration is the probability for two taxa whose ranges overlap, if just one example of each taxon is collected, that they will occur in the correct order. Paul (1982, p. 105) proposes the following analysis of this problem. Assume two species, C and D, whose ranges overlap, but the first appearance of C is earlier than the first appearance of D and the last appearance of D is later than the last appearance of C. Further assume that the chance of discovering either species is constant throughout the range of the species. Let the total range of C be N_C stratigraphic intervals, the total range of D be N_D stratigraphic intervals, and the zone of overlap be N_O intervals where all of the stratigraphic intervals are of equal length. N_O/N_C and N_O/N_D will be the probabilities of finding specimens of C and D, respectively, in the zone of overlap, and $(N_O)^2/(N_C N_D)$ will be the probability of finding one specimen each of both taxa in the zone of overlap.

In order to analyze what happens within the zone of overlap Paul (1982) suggests that we construct a probability matrix of N_O by N_O intervals. If the specimen of C and the specimen of D both occur in the same, N_n, interval of overlap, then they are considered contemporaneous and are in neither the correct nor the incorrect order. The probability of this occurring is $N_O/(N_O)^2$ The other two possibilities are that either C occurs above D or that D occurs above C. The probability of either scenario is equal to that of the other and each is equal to $((N_O^2 - N_O)/2)N_O^2 = (N_O^2 - N_O)/(2(N_O)^2)$, which is always less than 0.5. Therefore, the chances of finding specimens of both C and D in the zone of overlap and also finding C and D in the wrong sequence is $p = N_O^2/(N_C N_D) \times (N_O^2 - N_O)/(2(N_O)^2) = (N_O^2 - N_O)/(2N_C N_D)$. The probability of finding the two specimens in the zone of overlap in the correct order or not in the zone of overlap (in which case they will occur in the correct order) is $q = 1 - p$. As p can approach 0.5 but will never be larger than 0.5, q must always be 0.5 or greater. Thus, as Paul (1982, p. 108) states, "even finding just two examples at random, there is never more than a 50% chance of getting the wrong order."

If we find more than one example of either species, then the chances of observing the correct order increases. We need only to find a single specimen of the older taxon below the zone of overlap, or within the basal zone of overlap, and it will be impossible to get the wrong order of first appearances. The chance of finding a single specimen of C in or below the basal zone of contact is given by: $p = ((N_C - N_O) + 1)/N_C$ (Paul, 1982, p. 109), and the chance of not finding a specimen of C in the basal zone of contact or below is $q = (1 - p)$. After n trials, the chances of not finding a specimen of C in the basal zone of overlap or below will be q^n, and since q is always less than 1, the probability of not finding a specimen of C in the basal zone of overlap or below will decline rapidly with the increase of n, number of specimens of C collected.

Thus far the assumption has been that the chances of finding a specimen of a particular species throughout its range are equal. Of course, this is not true for most species, but Paul (1982) has also addressed this problem. He demonstrates qualitatively that if the earlier species, C, were initially rare whereas the later species, D, were initially extremely common, then the chances of finding one specimen of C and one specimen of D in the wrong order would be increased. But if another later species, D', were initially rare, the chances of finding a single specimen of C and a single specimen of D' in the correct order would be increased. Paul (1982, p. 109) argues that "the maximum distortion of the theoretical probabilities would occur if half of all fossil species had a stratigraphic distribution" as hypothesized for C and half had a stratigraphic distribution as hypothesized for D. But even then, in only approximately 25% of the cases comparing two taxa with overlapping ranges will cases like that hypothesized for C and D above be confronted. In another 25% of the cases the chances of finding the species throughout their ranges will have the effect of increasing the chances of getting the sequence of first appearances in the correct order. Finally, if we are dealing with two species that were both initially rare or both initially common, then the probabilities of getting the sequence of first appearances in the correct order should not be significantly altered.

From such considerations, Paul concludes that, in general, taxa (species or higher taxa) preserved in the fossil record are most likely preserved in the correct order; sequences are meaningful in the paleontological record.

Harper (1980) and Paul (1982) addressed the general question of the meaningfulness of sequences of fossils; empirical biostratigraphers usually take it as a first assumption that certain sequences of fossils in rocks are nonrandom and meaningful. One of the prime concerns of many biostratigraphers is to construct hypotheses as to the ranges and sequences of ranges of certain fossil taxa. In approaching this problem, biostratigraphers tend to follow one or another of two different philosophies (L. E. Edwards, 1982a). Given a series of stratigraphic sections (either the sections in a local study area, or perhaps all known sections containing the fossil taxon under consideration), some biostratigraphers aim to locate the absolute lowest first occurrence and the absolute highest last occurrence of the fossil taxon, thereby hoping to obtain the total, true, or actual range of the taxon. The total range of a particular taxon may be observed very infrequently in any particular local stratigraphic section. Using total ranges to construct biostratigraphic correlations and to date fossil samples tends to result in correct but coarse (nonprecise) results (L. E. Edwards, 1982a). Alternatively, some biostratigraphers tend to discount infrequent occurrences of taxa at the extreme limits of their ranges; they deal primarily with most probable ranges, rather than with total ranges, of taxa. When dealing in terms of most probable ranges, the biostratigrapher will tend to arrive at more precise age determinations and correlations, but the chances of being wrong increase (L. E. Edwards, 1982a)—that is, accuracy is sacrificed in order to increase precision. L. E. Edwards (1978, 1982a, 1982b, 1984; see also Harper, 1981) has reviewed various techniques of arriving at biostratigraphic correlations between local sections, constructing range hypotheses, and reconstructing the sequences of biostratigraphic events. Following is a brief review of two basic techniques that may be of particular interest to stratigraphically inclined phylogeneticists: L. E. Edwards's (1978) no-space technique of producing range charts and sequences of biostratigraphic events and W. W. Hay's (1972) probabilistic stratigraphy. Shaw's (1964; see also F. X. Miller, 1977) "graphic expression of biostratigraphic correlation" (see below) can also be used to construct ranges and sequences of range occurrences. W. W. Hay's techniques yield most probable sequences and ranges of fossil taxa. L. E. Edwards's method and Shaw's graphic expression of correlation yield total stratigraphic ranges, within the universe of sections analyzed, of the taxa under consideration.

L. E. Edwards's (1978) method applies no-space graphs to the problem of constructing range charts and sequences of events (biostratigraphic events, lowest and highest observed occurrences of particular taxa in the study area, which should approach the biologic events, first [e.g., by evolution or immigration] and last [e.g., by extinction or emigration] occurrences of taxa in the study area). L. E. Edwards's method deals only with sequences of events and thus requires only knowledge of the stratigraphic succession of fossil taxa in a series of local sections. Unlike Shaw's (1964) method, L. E. Edwards's method does not require the use of measured sections. However, L. E. Edwards's method yields unscaled results; the relative sequence of biostratigraphic events is reconstructed, but the amount of spacing between the events is left undetermined. In contrast, Shaw's (1964) method can be used to develop a relative chronologic scale (F. X. Miller, 1977). L. E. Edwards's (1978, p. 248) method "seeks to determine the total range in the area under consideration of each taxon studied."

In the presentation of her technique, L. E. Edwards (1978, p. 248) first discusses some basic concepts and assumptions that bear on the no-space graph method. (No-space graphs are graphs of relative position of events only; spacing between the events is not significant.) She notes that biostrat igraphic events may reflect biologic events in an imprecise manner, and thus the placement of biostratigraphic events in a local section other than at the level predicted may be subject to various interpretations. If, in a local strati graphic section, a first occurrence is found lower than was predicted, this may be due to either of two possibilities: the maximum range is actually greate than the predicted range and the initial hypothesis of the range of the taxon must be revised, or the observed putative first occurrence is an error; perhaps it was based on misidentification of a fossil specimen, downward mixing or fossils, or contamination of a sample with younger fossils. If, in a local section a first occurrence is found higher than was predicted, the reason may be that the range of the particular taxon in the particular local section is not as grea as the hypothesized maximum range. The local stratigraphic section may suffe from lack of preservation of the particular taxon under consideration, there may have been delayed dispersal to the local section, the facies of the loca section may have been unsuitable to the taxon under consideration, or fossil may have been misidentified.

If, in a local section, a last-occurrence event occurs higher than was pre dicted, this may indicate that the maximum range is actually greater than the predicted range, and the initial hypothesis must be revised, or the observed putative last occurrence is an error, perhaps the result of upward mixing contamination, or misidentification. If a last occurrence occurs lower than pre dicted, the range of the taxon in the local stratigraphic section may not be as great as the maximum hypothesized range of the particular taxon. Again this situation may be the result of such factors as misidentification of foss remains, lack of preservation, incomplete or inadequate sampling, facies con trol of fossil deposition, or local extinction or emigration in the local section (L. E. Edwards, 1978, p. 248).

In order to distinguish between the different implications of disagreement between observed and predicted occurrences of biostratigraphic events, Ed wards (1978, p. 248) has coined two useful terms: "Unfilled-range (UFR events are first-occurrence events actually observed higher than predicted c last-occurrence events actually observed lower than predicted. Range-revisio (RR) events are first-occurrence events actually observed lower than predicted and last-occurrence events observed higher than predicted" (L. E. Edwards 1978, p. 248). UFR events might be viewed as deficiencies in local section relative to the hypothesized sequence of events, while RR events, if not based on errors, call for revision of the initial hypothesis. UFR events are of no valu in reconstructing the total ranges of taxa and their relative sequences.

Any observed biostratigraphic event in a local stratigraphic section may b in place or out of place relative to a hypothesized sequence of events. An ir place event occurs in the correct position relative to a hypothesized sequence

An out-of-place event may indicate either that the range of the taxon in question in the particular section has an unfilled range (represents an UFR event) or that the hypothesized sequence needs to be revised.

Edwards (1978, pp. 248–249) states that her no-space graph technique of constructing ranges and sequences of events is based on at least seven assumptions; most or all of these assumptions apply to other biostratigraphic techniques as well.

1. The law of superposition is valid. In a stratigraphic section that has not been overturned, younger fossils and strata occur stratigraphically higher than older fossils and strata. Reworking and stratigraphic leaks are assumed to be identifiable and are eliminated prior to analysis.
2. The fossil taxa utilized must be consistently recognizable.
3. A particular taxon evolves only once and goes extinct only once.
4. "A consistent sequence of biostratigraphic events implies a consistency of temporal relationships" (L. E. Edwards, 1978, p. 248); that is, homotaxial patterns (similar orderly successions of fossil taxa at different localities) indicate successions of events (first and last occurrences of taxa) through time (Harper, 1980).
5. The fossilized organisms used in the study are assumed to have been living at the time of sediment deposition and accumulation; that is, it is assumed that the samples are not contaminated, perhaps by reworking, with fossils that did not live contemporaneously with the sediment accumulation. Or if there is contamination of the sample, the relevant fossils can be recognized and discounted.
6. Two or more taxa may occur together at a single stratigraphic horizon or sampling unit. There are two possibilities in such cases; the relevant fossil specimens "may have lived together, or they may have been brought together after death" (L. E. Edwards, 1978, p. 248). For biostratigraphic purposes, in many instances it does not matter whether the specimens literally lived contemporaneously or were brought into association after death. However, if the last occurrence of one taxon occurs in the same stratigraphic horizon as the first occurrence of a second taxon, it may be that the ranges of the two taxa merely touch, or that they overlap by some (perhaps very small) amount. To deal with such cases, Edwards (1978, p. 248) proposed her "rule of overlapping ranges"—"a first-occurrence event and a last-occurrence event which occur at the same stratigraphic horizon indicate an overlap of the ranges of the two taxa involved."
7. Edwards's method is semi-objective.

The best interpretation of the data is determined from the best judgment of the user and takes into account knowledge of the fossils and their sequence of occurrence, the environment of deposition, and the stratigraphy, and causes a minimum net disruption of the most reliable ranges ("economy of fit" as described by Shaw, 1964, pp. 254–257). The best interpretation has the most, reliable filled ranges, where "reliable" is determined by the user's best judgment. (L. E. Edwards, 1978, p. 249)

Edwards's basic technique is to use no-space graphs to compare particular sequences of events, as observed in stratigraphic sections, to a hypothesized sequence of events. The initial hypothesized sequence of events has no significant effect on the final hypothesis, but Edwards (1978, p. 251) found it easiest to use an observed sequence, with events missing from the particular section added by interpolation, as the initial working hypothesis.

The graphic display then is used to revise the hypothesis, according to the best judgment of the stratigrapher. The hypothesized sequence is revised and retested until a final hypothesized sequence is reached. In the final hypothesis, all biostratigraphic events occur in-place in at least one section and all events in all sections may be interpreted to represent in-place events or unfilled-range events. No event may indicate a need for revision of the final hypothesized sequence. (L. E. Edwards, 1978, p. 249)

In the actual construction of the no-space graph (see Fig. 4-2), the hypothesized sequence of events is listed on the vertical axis (later events above earlier events) and the observed sequence of events is shown along the horizontal axis (higher events are placed to the right of lower events). On the no-space graph, spacing between the events listed on the axis is not significant. Edwards suggests arbitrarily setting the spacing at one unit between each pair of events on the vertical axis. On the horizontal axis Edwards suggests listing sequentially numbered levels at which one or more biostratigraphic events occur in the particular section under consideration. The spacing between the listed levels is arbitrary.

At any level, all first-occurrences events are plotted slightly to the left of the tic mark for that level and all last-occurrence events are plotted slightly to the right of the tic mark for that level. These slight offsets in the plotting incorporate the rule of overlapping ranges. (L. E. Edwards, 1978, p. 250)

The next step is to plot each event in a particular local section at the intersection of its observed value and its hypothesized value on a no-space graph. A separate no-space graph is constructed for each local stratigraphic section comparing the observed sequence with the hypothesized sequence. On the no-space graphs first-occurrences are plotted with a symbol distinct ("O" on Fig. 4-2c) from last-occurrences ("X" on Fig. 4-2c). Once the points are plotted, as many of the best (those most significant chronostratigraphically according to the judgment of the investigator) points as possible are connected by line segments into what Edwards (1978, p. 250) terms a *line series*. Each line segment of the line series must have either a positive slope or be vertical; there can be no negative slopes.

Once the best line series is determined and plotted for a particular graph, all points on the graph must fall either on the line series, above the line series, or below the line series. Points that fall on the line series are in place relative to the hypothesized sequence of events, but points above or below the line series are out of place. Points above and to the left of the line series represent events that occur in the local section at lower levels than predicted, while points below and to the right of the line series occur in the local section at higher levels than predicted from the hypothesized sequence. First-occurrence events that occur below and to the right of the line series and last-occurrence events that occur above and to the left of the line series can be interpreted as events in the local section that represent parts of unfilled ranges. First-occurrence events that occur above and to the left of the line series and last-occurrence events that occur below and to the right of the line series are observed events that represent needed range revisions in the hypothesized sequence.

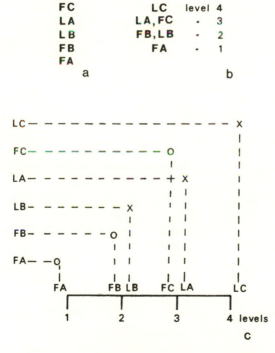

Figure 4-2. Example of a no-space graph. After L. E. Edwards, 1978, p. 249, fig. 1: *(a)* hypothesized sequence of events (F = first occurrence and L = last occurrence of species A, B, or C; sequence listed from bottom to top); *(b)* observed sequence of events in a local stratigraphic section with four fossil-bearing levels; *(c)* no-space graph constructed for the data shown in *a* and *b;* hypothesized sequence on the vertical axis and observed sequence on the horizontal axis.

Vertical line segments may occur on the graph if two or more like events (i.e., two or more first occurrences or two or more last occurrences) are found at the same level in a local section. Such vertical line segments are called terraces by Edwards (1978, p. 250), are considered in place relative to other events on the line series, and neither corroborate nor contradict the hypothesized sequence of the particular events composing the terrace.

Once no-space graphs with the best line series are plotted comparing all of the observed stratigraphic sequences to the hypothesized sequence, the graphs can be visually inspected, particularly for trouble spots (where it is unclear which event is in-place, i.e., part of the line segment, and which is out-of-place) and for needed range revisions. Range revisions can be handled in the following manner. For a first occurrence that entails a range revision, a vertical line can be extended from the out-of-place first occurrence to the line series, and the event to the right of this intersection is the next higher event. The order of these events in the hypothesized sequence can then be reversed. Likewise, for a last occurrence that entails a range revision, a vertical line can be extended from the out-of-place last occurrence to the line series, and the event to the left of this intersection is the next lower in-place event. The order of these events in the hypothesized sequence can then be reversed. Based on such inspection, the hypothesized sequence is revised and the observed sequences are plotted against the revised hypothesized sequence. If necessary, the revised hypothesized sequence can again be revised and re-tested. This process is continued until a final hypothesized sequence is reached.

Edwards (1978, p. 253) suggests the following procedure as a final test of the preferred hypothesized sequence. Successive pairs of events in the hypothesized sequence are compared, and for each pair a final test fraction is constructed. The denominator consists of the "number of sections in which both events are in-place or in trouble spots," and the numerator consists of the "number of times [the] hypothesized relationship is observed" in a local stratigraphic section. As the final test, all pairs of like events must have test fractions equal to or greater than one-half, and all pairs of unlike events must have nonzero numerators. If these criteria are fulfilled, then all like events will be actually observed in the hypothesized order in most local sections (unfilled range zones are excluded), and each predicted range overlap will be actually observed in at least one section. If the conditions of the final test are not satisfied, then the hypothesized sequence of events must be revised.

In certain situations it may not be possible to resolve a sequence of events completely and unambiguously.

If, in the final check, pairs of events show a fraction equal to one-half, never occur at separate levels, or never occur in-place in the same section, there is no preferred order to the two events. The stratigrapher may wish to note that the relationship between the two events seems to be random. (L. E. Edwards, 1978, p. 254)

W. W. Hay (1972; see also Harper, 1981) has proposed a method for determining the probability of discovering the most probable ranges and sequences of events (for our purposes, first and last appearances of fossil taxa) in the stratigraphic record given the information (i.e., ordering of fossil occurrences) contained in a series of local stratigraphic sections. He suggests that by inspection of one or more relatively complete and continuous local sections, a sequence of events can be suggested and adopted as a working hypothesis. Using the working hypothesis a matrix can be constructed showing the number of times in the local stratigraphic sections under consideration one biostratigraphic event (i.e., the lowest or highest occurrence of a fossil group in a local section) occurs below another and the number of times the reverse is true (Fig. 4-3). Thus, in each cell of the matrix in Figure 4-3a the number (n) in the upper left corner is the number of sections in which the biostratigraphic event listed on the left vertical axis of the matrix occurs above the biostratigraphic event listed on the horizontal axis (bottom) of the matrix. The number (N) in the lower right corner of each cell is the number of sections in which the biostratigraphic events are present and separated (i.e., it can be determined that one occurs before the other).

By inspection of the initial matrix (Fig. 4-3a) based on an initial working hypothesis of the sequence of events (usually the observed sequence in a particular section with missing events interpolated from other sections), the sequence can be reevaluated, one pair of events at a time, in order to determine the most likely sequence of events. The most likely sequence of events will give a matrix in which all pairs of events show n/N values of $\frac{1}{2}$ or less in the lower right half of the matrix and n/N values of $\frac{1}{2}$ or greater in the upper left half of the matrix (Fig. 4-3b). Hay's suggested method is to scan the initial matrix for any n/N values in the lower right half of the matrix that are greater than $\frac{1}{2}$. Any event that shows a value greater than $\frac{1}{2}$ in relation to another event should be moved in the sequence so as to come below the other event. Thus a new working hypothesis of the sequence of events is constructed, and from this new sequence a matrix can again be constructed. If the revised sequence gives a matrix in which all n/N values in the lower right half are $\frac{1}{2}$ or less, the most likely sequence has been determined. If some values are still greater than $\frac{1}{2}$, the procedure can be repeated until the most likely sequence is arrived at.

Hay (1972, p. 263) suggests that once

the most likely sequence has been established, it is desirable to determine the probability that each pair of biostratigraphic events is separable and known in true sequence. This can best be done by assuming that the events are not sequential, but that the relations of all pairs are random; then the probability of this being untrue can be determined. If the sequence of a pair of biostratigraphic events is random, the probability of one biostratigraphic event preceding the other is $\frac{1}{2}$.

According to Hay (1972, p. 263), for any given pair of biostratigraphic events, if N is the number of observed sections in which the two events being compared are separable and n is the number of sections in which the events occur in the predicted order, then "the probability P that the observed arrangement is caused by random distribution is:

$$P = \sum_{r=n}^{N} (N!/(r!(N-r)!)) \, (1/2)^{N-1}$$

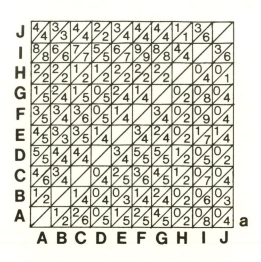

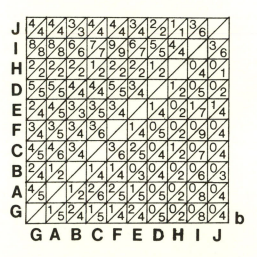

Figure 4-3. Matrices of the observed occurrences of biostratigraphic events (after W. W. Hay, 1972, pp. 262–263, figs. 3, 4): *(a)* initial matrix (the hypothesized sequence of events [A, B, C, etc.] is listed from lowest to highest on the vertical [top to bottom] and horizontal [left to right] sides); *(b)* revised matrix.

where *r* is a variable integer running from *n* to *N*." (W. W. Hay [1972, p. 263, footnote] notes that "using the above formula, probability values for *n/N* values of ½, ²⁄₄, ³⁄₆, ⁴⁄₈, etc. are greater than 1.0. This is a peculiarity inherent in the formula; for practical purposes all of these values can be reduced to 1.0 (indicating complete randomness)." The probability that the sequence is nonrandom is $1 - P$.

Using Hay's technique of probabilistic stratigraphy, the single most likely (in the sense of recurring in many local sections) composite sequence of a series of biostratigraphic events recorded in a number of local sections can be established. As L. E. Edwards (1982*a*, 1982*b*) has pointed out, however, probabilistic methods such as Hay's produce most-probable-range hypotheses that are shortened relative to the total observed stratigraphic ranges of the taxa under consideration. This is because "probabilistic methods have an averaging effect. If a taxon is long-ranging in some stratigraphic sections and short-ranging in others, its most probable range will be intermediate" (L. E. Edwards, 1982*a*, p. 147). Of course, the total stratigraphic range of a particular fossil taxon in a stratigraphic section or series of sections need not correspond to the "true" biostratigraphic range of the taxon.

Lack of preservation, incomplete sampling, and geographic and facies restrictions cause observed first occurrences to be displaced upwards, whereas these same factors cause last occurrences to be displaced downwards. Conversely, contamination and reworking cause observed first occurrences to be displaced downwards and last occurrences to be displaced upwards relative to the actual range of the species involved. (L. E. Edwards, 1982*a*, p. 149)

Furthermore, the true sequence of events may differ from one local stratigraphic section to another; for instance, species A may have reached, by migration, the area of deposition of local stratigraphic section I before species B, but species B may have reached, by migration, the area of deposition of local stratigraphic section II before species A. In such cases the first appearances of species A and B would occur in, and should occur in, reversed order in local stratigraphic sections I and II; from the point of view of probabilistic stratigraphy the observed arrangement is random ($P = 1.0$).

In sum, W. W. Hay's (1972) method is a tool that can be utilized in order to help determine the probability that the sequence of an observed pair of biostratigraphic events is due to a random distribution. This method has obvious utility for those who would utilize biostratigraphic data in postulating morphocline polarities or actual ancestor-descendant sequences for fossil organisms. It is unlikely that any investigator would place much weight on a sequence of biostratigraphic events if there was a high probability that the sequence was describable by a random distribution.

As alluded to in the beginning of this section, the relative and absolute ages of single fossil specimens may be very precisely and accurately datable. Within a single stratigraphic section, by the law of superposition, it may be known with certainty that the individual represented by a particular fossil lived prior to an individual represented by another particular fossil. Reworking and stratigraphic leaks of younger fossils into older strata (or vice versa) may be identifiable through sedimentological analysis. Likewise, independent means of chronostratigraphic correlation (e.g., paleomagnetic anomaly patterns, geochemical patterns of isotope variation, radiometric age determinations, lithostratigraphic correlation of ash layers [bentonites]: see Harland et al., 1982; Odin, 1982) between sections in different areas and basins or on different continents may indicate very precisely the temporal ordering of particular fossil specimens. However, the temporal ordering of individual specimens does not necessarily indicate the temporal ordering of the first and last appearances of the taxa they represent, especially if the taxa under consideration have overlapping ranges. One must also be very wary when dealing with a series of nonrepeated facies and the accompanying facies-restricted fossils in a stratigraphic section. A reconstructed sequence of presumed first and last appearances of certain taxa may actually record a series of appearances and disappearances of particular depositional environments and their corresponding rock records.

USE OF PHYLOGENIES IN
BIOSTRATIGRAPHIC CORRELATION

Biostratigraphy is the science of correlating rock units on the basis of fossil content. Here correlation (= chronocorrelation) refers to the determination of the temporal equivalence of the rock bodies in question: Two rock bodies that are correlated with one another are thought to have been deposited or formed roughly contemporaneously. The basic method, or assumption, of empirical biostratigraphy is that identical (or closely similar) organisms in different rocks (= biocorrelation) implies some kind of equivalency of the rocks in question (Eldredge and Gould, 1977). In most cases this equivalency is interpreted as a temporal equivalency, although it need not necessarily be temporal (for instance, it may be a facies or ecological equivalency; such can be misleading when incorrectly interpreted as indicating temporal equivalency). The identical organisms used to establish the equivalency of different rock bodies are usually the occurrence of the same species-level taxa in the different rock bodies. Thus, the basic working unit of information for the empirical biostratigrapher is the stratigraphic range of a distinct species-level taxon in a particular stratigraphic section; on the basis of such information for two or more sections the sections are correlated.

Among the classic discussions of biostratigraphy in English are those of Arkell (1933) and more recently Shaw (1964). These authors define and utilize the following terms (for recommendations on current, standardized, formal stratigraphic nomenclature, see International Subcommission on Stratigraphic Classification [ISSC], 1976; and North American Commission on Stratigraphic Nomenclature [NACSN], 1983). The local vertical stratigraphic range of a taxon in a local stratigraphic section (or closely related series of sections) is its teilzone; the time that the teilzone represents is the teilchron. The teilzone is empirically determinable. The biozone is the true or theoretically total stratigraphic range of a taxon; the time that the biozone represents is the total temporal duration of the taxon, from the first individual of the taxon to the last individual of the taxon, the biochron. The biozone can never be established with certainty (we can never be sure that we have the first and last individual) and thus is a theoretical concept that is unattainable in actual practice (Shaw, 1964, p. 103). What is attainable, however, is the sum of the correlated teilzones in correlatable local sections; this Shaw (1964) refers to as the *total stratigraphic range* (the interval between the lowest and the highest documented occurrences of a taxon is often referred to as the *taxon range zone* [ISSC, 1976, p. 53; NACSN, 1983, p. 862]).

It is often assumed that many species (or higher taxa) reach some point in their history when they attain a maximum abundance of individuals. This point in time is referred to as the hemera of the species; the rock body or unit deposited during, and recording, the hemera is referred to as the epibole (also referred to as an *abundance zone* [NACSN, 1983, p. 863] or an *acme zone* [ISSC, 1976, p. 59]). The rough equivalent of the epibole for a number of different species that occur together is a faunal or floral zone (cf. *assemblage zone, cenozone, Oppel zone, concurrent range zone;* ISSC, 1976, pp. 50–57; NACSN, 1983, pp. 862–863). The time represented by a faunal or floral zone is referred to as a secule or moment. A faunal or floral zone may variously refer to either the acme or the total duration of a particular fossil assemblage.

Shaw (1964, p. 83) proposes that biostratigraphy, or indeed any kind of stratigraphy that is attempting temporal (as opposed to lithologic or ecologic, for instance) correlation, is based on the two concepts that Shaw refers to as epochs and eras. An epoch is any distinctive or recognizable, relatively instantaneous (in geologic time) event. An era is a period of time bounded by two epochs, the event that marks the beginning of the era and the event that marks the end of the era. Thus a teilzone (or more precisely, the teilchron that is represented by the teilzone) can be thought of as an era that is bounded by two events or epochs. The first event is the first (earliest or stratigraphically lowest) appearance of the taxon under consideration in a local area or section, and the second event is the last (latest or stratigraphically highest) occurrence of the taxon under consideration in the local section or area. Alternatively, on a very coarse scale, the entire teilzone of a low-level taxon might be thought of as an event or epoch marking the boundary of an era. Likewise epiboles, faunal, and floral zones can be thought of and treated in analysis as epochs or eras. Usually, epiboles are treated as epochs and faunal and floral zones are treated as eras. For the most refined biostratigraphic correlations, as discussed in Shaw (1964), one usually correlates the epochs that bound the eras formed by teilzones of species-level taxa. As Eldredge and Gould (1977) point out, species-level taxa are normally used because they are small enough—that is, limited to short ranges stratigraphically—to allow good stratigraphic resolution, but they are large enough to circumvent problems of intraspecies variation that are not temporally (and thus biostratigraphically) significant, such as polymorphism and geographic and ecological variation.

In the discussion thus far of empirical biostratigraphy, the concept of phylogeny has not been utilized. Most practical biostratigraphers, when they are doing biostratigraphy, do not worry (and need not worry) about the phylogenetic relationships per se of the organisms utilized (however, it must be pointed out that many systems of biostratigraphic zonation are based on supposed or assumed evolutionary sequences of particular fossil groups: W. W. Hay, 1972; cf. *lineage zone* of ISSC, 1976, p. 60). As long as identical epochs and eras can be recognized in different rock bodies, these epochs and eras, and thus the rock bodies themselves, can be correlated. To recognize identical, contemporaneous, and correlatable epochs and eras the empirical biostratigrapher is dependent upon finding identical fossils in the different rock bodies to be correlated. As long as identical fossils are found, and given the assumption that identity means temporal equivalence, it does not even matter if the "fossils" have an organic origin. However, what of cases where identical fossils—that is, of the same species—are not found in the rock bodies to be correlated? Such may be the case if the rock bodies are from widely separated geographic areas, perhaps from two different continents. The approach that is usually followed in such cases is still to correlate on approximately the same basis, but to use maximal similarity instead of identity of forms. Often the concept of stage of evolution of individual taxa or of entire faunas is used to correlate different rock bodies. As Eldredge and Gould (1977, p. 33) note, "this approach is a complex mixture of empiricism and a priori assumption." The approach can be extremely useful and accurate and has been very important in mammalian biostratigraphy (see Savage, 1977; Tedford, 1970), but when carried to an extreme it can lead to a reliance on false data and a priori assumptions to the exclusion of empirical data. In its worst aspects, the concept of stages of evolution can be extended "until 'intermediacy' in one or more features will constitute *primae facie* evidence that a taxon comes from sediments intermediate in age in the stratigraphic sequence containing all taxa" (Eldredge and Gould, 1977, p. 34, italics in the original).

If identical forms are lacking in two rock bodies to be correlated, a more profitable approach might be to try to identify sister taxa in the different rock bodies and correlate them on that basis, as suggested by Eldredge and Gould (1977). If two groups are sister groups in the strict sense (that is, relative to all other organisms), then their times of origin will be identical because they both descended from a common species that underwent a speciation event giving rise to two daughter species, each of which formed the stem species for one of the sister groups under consideration. If we have two rock bodies that have no species in common but do have a number of related species between them, and if there are sister groups where one of the sister groups is represented in one rock body and the other sister group is represented by species in the other rock body, we might use the first appearances of these sister groups in the respective rock bodies as rough epochs useful in correlating the rock bodies. As was just pointed out, given two sister groups, one in each of two rock bodies, if each sister group was completely represented (or at least its base was represented) in its respective rock body, then the appearances (first occurrences) of the sister groups should be temporally equivalent. Of course, the actual first appearance of a sister group may not be preserved in the record (just as the first appearance of a species may not be recorded), but correlations of the first recorded appearances may serve to give a rough correlation where perhaps there was no correlation previously. When correlations based on first appearances of many different pairs of sister groups are compared, they may be found to be compatible, thus strengthening our confidence in the proposed correlation. Correlations based on sister taxa may also be compared with those based on more traditional stages of evolution and, of course, with correlations based on nonbiostratigraphic data.

Dealing only with morphological (*sensu lato*—that is, any usable intrinsic characters of an organism or organisms comprising a taxon) characters, we can never put a minimum age on an organism; we can never say how young any particular organism might be. Even a totally plesiomorphic organism may have survived for hundreds of millions of years and be extant today. However, one can put a maximum age on an organism or taxon. Any organism can be no older than the youngest, or latest acquired, synapomorphy that it bears, and no taxon can be older than the youngest synapomorphy that unites the taxon. If we can calibrate, relatively or absolutely, when certain synapomorphies arose in various groups, then rocks bearing those fossil organisms can be no older than the time of acquisition of the synapomorphies under consideration. Furthermore, if we can establish that certain organisms bearing certain other synapomorphies are not present in the rocks under consideration because the synapomorphies had not yet appeared (not because of an artifact or poor sampling or preservation, or because organisms bearing the synapomorphies had not yet migrated into the area where the particular rocks under consideration were being deposited), then the rocks must have been deposited before such synapomorphies appeared. But here it is always difficult to evaluate such a lack of evidence (negative evidence). It seems that this form of analysis can be utilized to give at least rough or coarse temporal correlations of rock units. Such an approach differs from some more traditional biostratigraphic approaches in relying exclusively, or at least as much as possible, on the analysis of the appearance of synapomorphies. An isolated occurrence of a single primitive and otherwise unknown form in an otherwise uncorrelated rock body does not necessarily tell us anything about the age of the rock, except that it must be equal in age or younger than rocks deposited during the time of origin of the lowest level previously known taxon to which the new specimen is referable. For example, if we collect a single nontherian and nonmonotreme mammal jaw from a rock, the rock cannot be older than the origin of the Mammalia. But just because all previously known nontherians (except for monotremes) were extinct by the beginning of the Oligocene, this does not mean that this new specimen has to be early Oligocene or older. It may turn out to be the first Miocene specimen of a nontherian and non-monotreme. Granted the chances of this being so might be small, and we might feel safe assuming that it is early Oligocene or older. However, we cannot be sure, and the above example is based on a real situation. Until recently the Multituberculata, a group of nontherian, nonmonotreme mammals, were believed to have gone extinct during the Eocene (Van Valen and Sloan, 1966); however, within the last few years well-documented reports of Oligocene multituberculates have been published (Krishtalka et al., 1982; Savage and D. E. Russell, 1983).

SHAW'S GRAPHIC EXPRESSION OF
BIOSTRATIGRAPHIC CORRELATION

Shaw (1964; see also F. X. Miller, 1977) proposed a simple but extremely useful method to express presumed time equivalence in two stratigraphic sections graphically. Shaw applied his method to correlations based on the stratigraphic ranges of identical species in different sections, where the first occurrences and last occurrences of species are the epochs that are regarded, until demonstrated otherwise, as time equivalents in different sections. As such, we need not know or be concerned with the phylogenetic relationships of the constituent species as long as we can distinctly recognize identical species in the sections to be correlated. However, Shaw's (1964) methodology is applicable in graphically expressing presumed time equivalence of any potential epoch markers, including those that might arise out of phylogenetic analysis. As was mentioned in the previous section, the earliest (lowest) appearances of members of sister groups in two different sections may potentially be treated as time-equivalent events (epochs), at least at a coarse level. By correlating many such independent presumed time-equivalent events using Shaw's methods, we may be able to arrive at an approximate correlation between the sections where there was no correlation previously. Furthermore, Shaw's methods may demonstrate that certain presumed time-equivalent events are incongruent with the overall correlation. These incongruences are then subject to reconsideration and perhaps rejection. Because of their potentially wide applicability, Shaw's basic techniques are briefly reviewed.

Given two stratigraphic sections that have some temporal overlap, if we can identify a point in one section (X) that is temporally equivalent to some point in another (Y), then we can plot such a correlation (equivalence) as a single point on a two-dimensional graph (with X and Y axes perpendicular to one another) at the intersection of the coordinates X and Y. The positions of points X and Y in their respective sections can be expressed in terms of any convenient units measured from a fixed point on the particular section in question, such as in meters from the base of the section or from a distinctive bedding surface. Due to the law of superposition, given simple sections that are undeformed, points higher in both sections will represent later events. In each section infinitely closely spaced points X_1, X_2, X_3, ... , and points Y_1, Y_2, Y_3 ... will be contiguous (or identical in the case of nondeposition) and when plotted together as pairs on the graph will form a continuous line. Shaw points out that such a line of correlation will exist for two sections with temporal overlap whether we can locate it or not. Once the line is located, the sections have been temporally correlated completely.

In an ideal case, if two identical sections are correlated with one another (or if the same section is correlated with itself) and the points are plotted on a two-dimensional graph with arithmetic scales, the graphic correlation will take the form of a straight line with a slope of 1. In general, if we can accurately locate equivalent epochs in the two sections to be correlated, and deposition was regular in each of the two sections, such a plot will take the form of a straight line. The slope of the line of correlation will be a function of the relative rates of accumulation of rock in each section. The more equivalent epochs that we can locate accurately, the better our correlation will be. The line drawn through the points of correlation that we can locate accurately will approximate the true, total correlation of the sections in question.

When the relative rate of accumulation between two sections changes, the slope of the line of correlation between the two sections will change. Abrupt changes in slope are referred to as *dog-legging* by Shaw (1964, p. 134). If dog-legging occurs between two sections, both can be correlated to a third section in order to try to determine in which section the greatest change of rate of sediment (rock) accumulation has occurred. For example, given sections A, B, and C, if when A and B are correlated with each other no dog-legging occurs, but if dog-legging occurs whenever C is correlated with either of the other sections, we may suspect that it is C that has suffered the change of rate. Of course, it may also be the case that both A and B suffered the same rate of change of sediment accumulation at the same time-equivalent point, while C is the section that was characterized by a constant rate of rock accumulation. Sedimentological studies may help resolve such problems, but a priori it may be suggested that the first alternative (that C is the section that suffered the change of rate) is more likely.

If there is a break in rock accumulation in one section relative to another section that is characterized by continuous accumulation, this fact will show in the graph of correlation as either a horizontal or vertical line. All of the time represented by the break within one section will be compressed into a single plane. Thus, even if not immediately evident in the field, breaks in sections should be readily apparent when the sections are correlated to other sections and the time represented by breaks in some of the sections is represented by a net rock accumulation in other sections.

When presumed time-equivalent points (epochs) are plotted against one another for two sections, a majority of the points may fall on a fairly smooth line or on a set of dog-legged lines. A few points, however, may not fall onto the plot for the rest of the array. We may decide to omit such points from further consideration as not truly time equivalent, perhaps on the basis of special knowledge of the particular points in question. Such points may represent unfilled ranges for one or the other of the sections and thus should be eliminated (L. E. Edwards, 1978; F. X. Miller, 1977; see above). It may only be the plotting of all of the presumed time-equivalent points, however, that point up the deficiencies of certain presumed epochs (Shaw, 1964, p. 159).

In his book Shaw elaborates in detail on his technique of graphic expression of correlation, including the quantification of the graphic pattern of correlation and tests of the significance of the correlation. Shaw's methods, along with L. E. Edwards's (1978) and W. W. Hay's (1972) methods for determining sequences of events in the stratigraphic record, have been discussed, refined, and elaborated upon by many subsequent investigators such as L. E. Edwards (1982*a*, 1982*b*, 1984), L. E. Edwards and Beaver (1978), Harper (1981), F. X. Miller (1977), and Southam, W. W. Hay, and Worsley (1975).

ECOSTRATIGRAPHY

A recent development in the methodology, approach, and philosophy of organismal-based stratigraphy is ecostratigraphy (for discussion and reviews on this subject, see Cisne and Rabe, 1978; Hoffman, 1980; Krassilov, 1974, 1978; Martinsson, 1973, 1978, 1980*a*, 1980*b*). Martinsson (1973, p. 442) has defined ecostratigraphy as ecosystem stratigraphy, or "the construction of time-planes with the greatest possible precision and frequency through environmentally defined stratigraphic units." In ecostratigraphy, stratigraphic units are defined on a synecological basis (Martinsson, 1978) and are interpreted as paleoecosystems (Krassilov, 1974). Ecostratigraphy depends on a reference set of time-planes in the stratigraphic record, each of which records a remarkable or significant geobiologic event (such as a persistent paleocommunity) in the stratigraphic distribution of the organic species preserved in the rocks (Hoffman, 1980). The ecostratigraphic time-scale is built up from these time-planes, each of which helps delineate a natural unit in geobiohistory.

In contrast to ecostratigraphy, in traditional chronostratigraphy the reference time-planes are purely arbitrary and abstract; they are marked by designated and internationally agreed on reference points (golden spikes) in type sections. Such time-planes bound divisions of the geologic time-scale (Hoffman, 1980). However, Hoffman (1980, p. 98) has pointed out, in actuality most higher-order chronostratigraphic units (such as erathems, systems, series, and many stages) do approximate ecostratigraphic units. This is because for the most part the boundary time-planes were originally picked at the points in the stratigraphic record where particularly significant biogeologic events (such as major extinctions) were observed to occur. The major discrepancy between eco-stratigraphically based units and more traditionally based units (for example, concurrent fossil ranges) has been at the chronozone level (i.e., below the level of the stage: see ISSC, 1976, p. 68; Hoffman, 1980).

The basic assumption of the ecostratigraphic approach is that there are "real" ecostratigraphic units to be found in the rocks. Either there is a process of community evolution that is reflected in the fossil record, or there is min-imally significant co-evolution among the various species in a particular ma-crohabitat, such that ecostratigraphic units will have an independent existence. If these assumptions are invalid, this would imply that the ecostratigraphic units are merely epiphenomena (without an independent and real existence in nature) that result from the random association (through the common phe-nomena of migration, evolution, extinction, and so on) of independent lineages of organisms (cf., traditional concurrent range zones of chronostratigraphy; Hoffman, 1980).

Hoffman (1979; 1980, p. 99) has argued that on the whole, ecological communities are merely epiphenomena "of the overlaps in distributional pat-terns of various organisms controlled primarily by the environmental frame-work." Consequently, if Hoffman's view of ecological communities is accepted, then community paleoecology is nothing more than an epiphenomenal sci-ence, and one of the primary assumptions of the ecostratigraphic approach to the fossil record is undermined. Hoffman (1980) has also attempted to demonstrate that in many cases co-evolution between organisms within a macrohabitat may be relatively insignificant. In particular, he compared planktonic organisms to shallow-marine benthonic organisms and found that while the former do not appear to show significant co-evolution, the latter do. Thus, ecostratigraphy may be applicable to describing rocks of the shallow-marine realm, but it is not applicable to the record of the pelagic realm. One must be cautious when applying ecostratigraphic methods and approaches to the stratigraphic record; although potentially very powerful and informative, the entire record is perhaps not adequate or appropriate for the use of these techniques.

CHAPTER 5

Historical Biogeography

Neontological biogeography deals with and ultimately attempts to explain the distribution of extant biological taxa on the face of the earth. Paleobiogeography covers the same subject matter, but includes a temporal component—taxa through time in various geographical areas (as documented by actual fossils and stratigraphic relationships; see Jablonski, Flessa, and Valentine [1985] for a recent review). Biogeography as a whole must be concerned with the spatial aspects, temporal aspects, and phylogenetic (evolutionary) relationships of organisms (Croizat's [1964] "Space, Time, Form": see Ball, 1975; Croizat, 1981; Nelson, 1978a). Zoogeography refers to the biogeography of animals, and phytogeography refers to the biogeography of plants. Fundamental to any biogeographic study is to first know what the fundamental (natural) biological taxa are whose distributions call for explanations. Here such taxa are regarded as monophyletic groups, as described in preceding sections.

There are two basic activities in or aspects of the study of biogeography: descriptive and interpretive biogeography (S. A. Cain, 1944; Wiley, 1981). Descriptive biogeography is the basic data gathering of biogeography. It includes the documentation of the ranges of specific organisms and the delimiting of biogeographic regions, provinces, and realms. Interpretive biogeography attempts a synthesis of the data of descriptive biogeography, ultimately formulating some combination of hypotheses, scenarios, and explanations to account for the data of descriptive biogeography. In some cases interpretive biogeography may make predictions as to what taxa, presently unknown, will be found where on the basis of the distributions of other taxa, or predict on the basis of present-day distributions of biotas what the geological and ecological history of an area must have been.

In regard to interpretive biogeography, there are two basic approaches to biogeographical explanations: ecological biogeography and historical biogeography. Ecological biogeography attempts to understand and explain the distribution of a taxon in terms of its ecological requirements and the present-day features of the environment. Historical biogeography, on the other hand, is concerned with explaining the distributions of taxa in terms of the history of the taxa and the history of the areas in which they live or have lived. Historical biogeography attempts a unification of the history of life and the history of the earth (Nelson and Platnick, 1981). The history of life refers to the phylogenetic relationships of the organisms and taxa involved as deduced from the attributes (characters) of the organisms themselves and interpreted in a historical context. The subject of the history of the earth is that of the field of historical geology. But hypotheses of the relationships of taxa can be interpreted as hypotheses of relationships of biotas and geographic areas; as such, data from historical geology (e.g., stratigraphy, paleomagnetism, geochemistry: Platnick and Nelson, 1978) can be reciprocally illuminating relative to distributional data of taxa. In this context Patterson (1983a) distinguishes two principal aims of biogeography: (1) to seek pattern among the distribution of organisms and to use the pattern in investigating earth history and (2) to explain biogeographic distributions of organisms through the application of theories of evolution and earth history.

For historical perspectives on the science of biogeography, the reader is referred to the work of Nelson (1978a, 1981, 1983b) and Nelson and Platnick (1980, 1981).

BASIC BIOGEOGRAPHIC TERMS
AND CONCEPTS

Following Wiley (1981), some of the basic terms and concepts used in biogeographic analysis are reviewed here.

The range of a species is the general geographical area where the species commonly occurs; for a migratory species, the entire migratory route may be included within its range. A locality is a particular area or place where a population of a species (a deme) lives. A typical species range will include a number of localities and the empty spaces between them. A track is an area or band that entirely circumscribes the total range or geographic distribution of a specific monophyletic taxon. At times tracks may be drawn as narrow bands or even lines on a map connecting the areas where members of a monophyletic taxon are found. A generalized track is a combination of two or more congruent tracks of independent monophyletic groups (see Croizat, 1964; Rosen, 1975).

The assemblage of organisms that naturally occurs in a particular geographic region is termed the biota; the animals constitute the fauna and the plants constitute the flora. As Wiley (1981, p. 281) points out, for biogeographical analysis these terms are usually used in conjunction with the concept of a natural geographic area that is distinguished from other natural geographic areas by a unique combination of organisms and physical attributes. A natural geographic area or biogeographical unit may also form an ecological unit from the point of view of the organisms. Some taxa will be found in several biotas, while other taxa will be found in a single biota—they are endemic to the biota in question. Furthermore, a taxon endemic to a certain biota may not be found throughout the entire area of the biota. An area of endemism is the specific area where a certain endemic taxon is found.

Several different concepts of dispersal are used in biogeography (Udvardy, 1969; Wiley, 1981). Organismic dispersal refers to the movement of organisms within the range of the species; it is of major concern to those interested in ecological biogeography. Species dispersal refers to the movement of organisms from the present species' ranges to habitats outside of the species' ranges. If the individuals involved can colonize the new areas, they may successfully extend the species range, or the individuals involved in such dispersal may become isolated from other members of the species and undergo allopatric speciation. Biotic dispersal refers to the spreading of the geographic area covered by an entire natural biota. Biotic dispersal is the result of concurrent species dispersal.

Traditionally, certain biogeographers have believed that there were (or are) certain areas on the earth where most new species are produced; such areas act as species factories (Wiley, 1981, p. 283). Species produced in these centers of origin then spread out to populate all other geographical areas. A variation on this theme is that any particular monophyletic group may have its own center of origin. The concept of centers of origin is associated with dispersal explanations of historical biogeography; this notion has been rejected by certain investigators (e.g., Croizat, Nelson, and Rosen, 1974).

METHODS IN BIOGEOGRAPHY

Patterson (1983a, p. 7) has pointed out that even descriptive biogeography involves theory, primarily theory concerning the systematics (relationships) of the organisms concerned: ''recognition of a disjunct distribution demands both an estimate of relationship between the disjunct populations, and an estimate of the taxonomic level at which they form a unique taxon." Likewise, the delimitation of areas of endemism depends upon taxonomic and systematic judgments.

Patterson (1983a; see also Patterson, 1981c) distinguishes four basic methods of biogeography: Wallace's method, equilibrium methods, phenetic methods, and cladistic methods. Equilibrium methods primarily involve ecological biogeographic explanations, while the other three approaches are primarily historical in nature.

Wallace (1876, 1880) is often considered the founder, or inventor, of modern biogeography (particularly zoogeography: George, 1964). Wallace's method of biogeography might be considered the classical evolutionary method. Wallace analyzed the degrees of relationships between biotas and the patterns of geographic distributions of organisms, using evolutionary theory and scenarios, by comparing lists of genera (primarily mammals in Wallace's case) known from various regions (an early statistical method) and by incorporating fossil data whenever possible. The works of such investigators as Darlington (1957, 1970), Matthew (1915), and Simpson (1940, 1953b, 1965, 1980) follow in the tradition established by Wallace (Patterson, 1983a).

As the prime example of equilibrium methods in biogeography, Patterson (1983a) puts forth MacArthur and Wilson (1967). Theirs is a combination of ecological biogeography and dispersalism based on the mathematical modeling of biogeographic patterns which in turn has its foundations in ecology, population genetics, chance dispersal, colonization, and extinction. Rosen (1978) and Patterson (1983a) suggest that equilibrium theory is to historical biogeography as population genetics is to phylogeny reconstruction—at this stage they are basically independent of one another.

Phenetic methods of biogeography involve statistical measures of biotic similarity between various areas or realms, such as coefficients of faunal resemblance (e.g., Simpson, 1980), generalized tracks of organisms between areas (e.g., Croizat, 1964; see discussion by Ball, 1975), or more simply, the comparing of long lists of organisms (taxa) from different areas (Keast, 1973, 1977). In other words, phenetic estimates of affinity or pattern are comparable to the phenetics of the numerical taxonomists among systematists. Such workers appear to rely more on quantity than quality. As Patterson (1983a, p. 10) states, "What such estimates of affinity mean, no one is quite sure. If all the taxa involved are natural (monophyletic in Hennig's sense) they may contribute qualitative information. . . . But how much noise is introduced by the quantitative method no one knows."

The cladistic vicariance method of biogeography is essentially that outlined by Nelson and Platnick (1981; Platnick and Nelson, 1978). Areas of endemism are related hierarchically in area cladograms; the geographic distribution of species and their phylogenetic relationships are the characters in theories of area or biota relationships (Rosen, 1984). This method is a direct outgrowth of cladistic analysis as a means of reconstructing phylogenetic relationships and is discussed in detail below.

EXPLANATIONS FOR THE DISTRIBUTIONS OF ORGANISMS IN HISTORICAL BIOGEOGRAPHY

There are two basic approaches to the explanation of organismal distributions in historical biogeography: dispersal explanations and vicariance explanations (Nelson, 1984; Nelson and Platnick, 1978, 1980, 1981; Nelson and Rosen, 1981; Patterson, 1983a; Platnick and Nelson, 1978; Vuilleumier, 1978). Dispersal hypotheses account for the geographic distribution of taxa in terms of centers of origin and the subsequent dispersal of the organisms into new areas (Darlington, 1957). As species disperse, part of an ancestral population may cross a barrier. Once crossed, the barrier may effectively separate the populations such that they undergo allopatric speciation. Platnick and Nelson (1978) have referred to dispersal biogeography as colonialistic biogeography (taxa originate in an area and colonize other areas) or vacuum biogeography (areas are assumed to have been originally devoid of the organisms that later dispersed there).

Wiley (1981, pp. 285–291) distinguishes two forms of biogeography that both seem to fit, more or less, under Platnick and Nelson's (1978) category of dispersal biogeography: evolutionary biogeography and phylogenetic biogeography. Evolutionary biogeography (analogous to evolutionary taxonomy) is based on the following assumptions or tenets: (1) There are certain limited areas on the earth (centers of origin) that tend to produce the majority of new species and thus higher taxa. (2) New species, as they are continually produced in the centers of origin, tend to spread out (disperse) geographically and displace the older and more primitive species toward the periphery of the range of the higher monophyletic taxon. (3) Consequently, the most recently derived members of a larger taxon will tend to be found in or near the center of origin, and there may be numerous species (perhaps living sympatrically) in and near the center of origin. (4) If a fossil record is known for the group under consideration, it is expected that the earliest (and thus most primitive) fossils attributed to the group will be found in the center of origin. If one can trace the history of the group through time using fossils, one should observe a steady expansion of members of the group outward from the center of origin. (5) Organisms will naturally tend to disperse as widely as they possibly can (given their innate adaptations and the environmental conditions they face); consequently, progressively more derived and better adapted forms will tend to displace ancestral, more primitive forms. Papers dealing in an evolutionary biogeography framework include Briggs (1974), Darlington (1957,

1970), Hubbs (1958), Matthew (1915), Mayr (1952), Simpson (1965), and Wallace (1876).

Phylogenetic biogeography has three basic tenets: (1) "Closely related species tend to replace each other in geographic space" (Wiley, 1981, p. 287). (2) If independent taxonomic groups show similar biogeographic patterns, they may share the same biogeographic history (they may have been affected by the same external events). (3) Allopatric speciation via peripheral isolation is usually assumed in phylogenetic biogeography. Phylogenetic biogeographers, unlike evolutionary biogeographers, assume or suggest that it is the primitive members of a monophyletic group that will be found near the center of origin of the group. Proceeding outward from the center of origin, more derived forms will be encountered (this is a chorological progression). Such should be the case if at the periphery of an ancestral species' range a small population was isolated and underwent allopatric speciation. The new species at the periphery would be derived relative to the ancestral species at the center. The new species may then expand or disperse in a geographic area adjacent to that of the ancestral species. On the periphery of the new species (where the species does not border with the original ancestral species) a small population may become isolated and undergo allopatric speciation. This newest species will be derived relative to both the immediately ancestral species from which it evolved and the original ancestral species and furthermore will be the farthest away geographically from the ancestral center of origin. Such a scenario may continue through numerous speciation events. Phylogenetic biogeography methods have been developed and utilized by Ball (1975), Brundin (1966, 1972), Cracraft (1973, 1974b, 1975, 1983a), and Hennig (1966a, 1966b).

The second approach to explanation in historical biogeography, the vicariance approach (Croizat, Nelson, and Rosen, 1974; Nelson, 1984; Nelson and Platnick, 1978, 1980, 1981; Platnick and Nelson, 1978; Rosen, 1974b, 1978, 1979; Wiley, 1981; and various articles in Nelson and Rosen, 1981, especially Patterson, 1981c), suggests that the ancestors of taxa occurred in the areas where their descendants are subsequently found. Related species (sister groups in the strict sense) represent isolated parts of a once widely distributed ancestral species whose population was subsequently divided by the appearance of a geographic or ecological barrier within the species range (that is, the ancestral species has undergone vicariance). With the separation of two or more populations in the ancestral range, the populations may undergo allopatric speciation. In this scenario, sympatry of closely related forms would be evidence of dispersal subsequent to the disjunction and speciation event.

Platnick and Nelson (1978; Nelson and Platnick, 1981) have pointed out that dispersal hypotheses really take two forms, one of which is better considered a vicariance model. Sometimes, in the context of a dispersalist explanation, it is suggested that an ancestral species enlarges its range through time (without crossing any barriers [note that the concept of barriers is a relative term—what may be a barrier to one species may not be a barrier to a second species]), but subsequently the enlarged range of the species is fragmented into two disjunct ranges such that the separated populations differentiate with time and undergo allopatric speciation. This is really a vicariance explanation because the dispersal takes place before the appearance of any barriers. Vicariance explanations suggest that an ancestral species will be distributed over a certain range, which is subsequently fragmented; such explanations do not rule out the possibility that the population of the ancestral species had to disperse over the ancestral range to obtain the initial distribution that is subsequently acted upon by vicariance events. In the classic dispersalist models, however, it is postulated that there is a preexisting barrier over which part of an ancestral species crosses (usually accidentally, by chance). Once separate, isolated populations are established on either side of the barrier, they can undergo allopatric speciation. In the vicariance model, the barriers that come to separate populations of an ancestral species always form within the ancestral species range. In summary, in dispersal explanations disjunctions in distributions of taxa are caused by organisms dispersing over preexisting barriers; in contrast, according to vicariance explanations any disjunctions in the distribution of taxa are caused by the appearance of barriers within the range of an ancestral species.

If dispersalist explanations are postulated, different (possibly ad hoc) explanations of dispersal may be hypothesized to explain the distribution of unrelated groups of organisms. Of course, a certain mechanism (perhaps a corridor such as a land bridge across an ocean body) may be postulated as having served as a common dispersal route for a group or groups of organisms over a certain period of time, but in general there is no a priori reason why there should be a strong congruence between the distributions of unrelated taxa. In contrast, the vicariance model predicts that, in general, ancestral taxa that occupy a general geographic area will respond to geological and ecological vicariance events that occur within the species range by undergoing allopatric speciation. If numerous ancestral species originally occupy a certain geographic area, they may all (or at least the majority) respond to the same vicariance events by speciating. Thus, the distributions and phylogenetic relationships of the members of any one monophyletic group in a certain area(s) would be expected to be congruent with the phylogenetic relationships of other groups in the same area(s); furthermore, the phylogenetic interrelationships of the members of a taxon should reflect the history of the ancestral biota in terms of the vicariance events that were responsible for disjunctions in ancestral populations.

Thus far in considering different explanations and models of historical biogeography we have considered the mechanism of speciation to be some form of allopatric speciation involving the disjunction (either through dispersal or vicariance) of an ancestral species' range with resulting isolation and divergent evolution of the separated populations. It is by recognizing degrees of phylogenetic relationship between various taxa relative to others that we can hope to reconstruct biogeographic history (whatever model, vicariance or dispersalist, we favor). Given the assumption of allopatric speciation (Platnick and Nelson, 1978), sympatry of closely related forms is evidence of some kind of dispersal event, while allopatric and parapatric distributions can potentially be explained in terms of vicariance events without the need for dispersal events (of course, allopatric distributions can also be the result of dispersal). If no speciation has occurred, then we cannot reconstruct biogeographic history using such information. If speciation does not occur via allopatric speciation (perhaps it may form sympatrically: see section on speciation patterns), such information will not help in this approach to reconstructing biogeographic history or may even be misleading.

In order to test biogeographical hypotheses against distributional data of organisms in the context of a Popperian falsifiability system, Platnick and Nelson (1978) have suggested that two assumptions must be made: (1) The histories of various taxonomic groups can be compared meaningfully, and we can hope to recognize disjunctions and sister taxa between closely related organisms, only if it is possible to recognize monophyletic groups and their interrelationships. We must assume that we can make reasonable hypotheses as to the phylogenetic relationships of the organisms whose geographic distributions we are studying and wish to explain. As Nelson and Platnick (1981, p. 43) aptly point out, we begin with cladograms of the distribution of character states of organisms, from these we derive phylograms (historically interpret the cladograms in terms of recency of common ancestry), and finally we base our studies of historical biogeography on the phylograms. Each level is further removed from our initial observations and must ultimately be subservient to the levels before it. (2) It must be assumed that there is some mechanism of evolution that results in the systematic distribution of related forms in related areas—that there is a reason to expect a correlation between phylogenetic relationships and biogeographic patterns. The assumed mechanism of evolution is allopatric speciation.

Given a certain distribution and hypothesis of relationships among a group of organisms, and assuming allopatric speciation, how can we decide between a dispersal or vicariance explanation? First, let us assume that the distributions of the species involved are nonsympatric; thus, no dispersal need be hypothesized initially. A first suggestion might be that the dispersalist versus the vicariance explanations could be tested by knowing the relative ages of the particular taxonomic disjunctions and the ages of the particular barriers involved (Platnick and Nelson, 1978). The vicariance explanation predicts that taxonomic disjunctions should be approximately the same age as the barriers;

it is falsified if the disjunction is shown to be older than the barrier. The dispersal explanation predicts that the barriers should be older than the taxonomic disjunctions; if the taxonomic disjunctions are older than or the same age as the barriers, the dispersal explanation is falsified. Note that both explanations are falsified if the taxonomic disjunction is older than the putative barrier that caused it; thus this test is not a very good discriminator on this account. The test hinges upon knowing fairly accurately whether the taxonomic disjunction and putative barrier were roughly contemporaneous or the barrier is older. However, calculating or estimating such ages is highly problematical and open to wide margins of error (Platnick and Nelson, 1978). The timing of the rise of species-level barriers cannot generally be estimated with the required accuracy by historical geologists. The maximum known age of a taxon (the minimum age) is the age of the oldest known fossil that can be attributed to that taxon or the age of the oldest known fossil attributable to the sister group of the taxon (sister group in the strict sense), whichever is oldest. Such maximum ages of taxa can always be changed by the discovery of an older fossil, and there is not necessarily any assurance that we have the oldest fossil in hand. If the taxonomic disjunction appears to be too young relative to the particular barrier, we can always postulate that older fossils will be found. Platnick and Nelson (1978, p. 4) also note that even if older fossils are found or a particular barrier proves to be younger than hypothesized, this does not falsify either a dispersal or a vicariance model per se. All that is actually falsified is one particular correlation between a barrier and a taxonomic disjunction. We can also postulate that there was an older, perhaps as yet unknown, barrier affecting the distribution and disjunction of the organisms under study. Thus, use of the dating of barriers and taxonomic disjunctions may often be insufficient to distinguish unambiguously between dispersalist and vicariance explanations.

Platnick and Nelson (1978) suggest that in theory it might be possible to decide between dispersal or vicariance explanations by the initial adoption and rigorous testing of one explanation. However, when they explore this line of inquiry, they find that the initial adoption of either explanation is not sufficient always to distinguish cases of vicariance from cases of dispersal. They conclude that the only kind of information that can be obtained unambiguously from distributional data of taxa combined with the phylogenetic relationships of the taxa is information on the relative recency of area interconnections. The same allopatric pattern might be produced by sequential biotic dispersal, by vicariance, or possibly by some combination of both; "distributional data seem sufficient to resolve a pattern of interconnections among areas that reflects their history, but not to specify the nature of those connections" (Platnick and Nelson, 1978, pp. 7–8). Similarly, given a cladogram of taxa that is interpreted phylogenetically, we can state what the interrelationships of the taxa are, but without supplementary information we cannot specify the exact nature of those interrelationships (e.g., that species A gave rise to species B versus A and B both being direct descendants from an unknown species X). Platnick and Nelson (1978) suggest, however, that once general patterns of area interconnections are found, independent sources of data, specifically from historical geology, may be able to be used to resolve the nature of interconnections between areas and biotas. I believe that it is important and beneficial to review Platnick and Nelson's reasons for arriving at these conclusions. The analysis that follows is based on Platnick and Nelson (1978; see also Nelson and Platnick, 1981).

Platnick and Nelson (1978) begin their analysis by assuming that there are three allopatric taxa, A, B, and C, which are found today only in areas a, b, and c, respectively, and whose interrelationships are of the form (A (B, C)) (Fig. 5-1a). A dispersal explanation could be as follows: The common ancestor of taxon ABC originally lived in area a and is today represented by taxon A (if A, B, and C are species-level taxa, A may be identical to the common ancestor or a direct, though not necessarily immediate, descendant of the common ancestor). Some members of A migrated to area b and there speciated to form a new taxon, which is today represented by B; subsequently

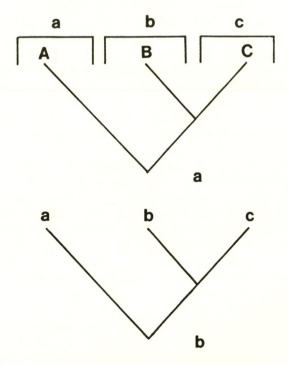

Figure 5-1. The phylogenetic relationships of three hypothetical allopatric taxa (A, B, C) found today in three areas (a, b, c): *(a)* cladogram of taxa; *(b)* general area cladogram.

some members of the taxon today represented by B migrated to area c and speciated to form a taxon today represented by C. How can we test such a dispersal explanation? As Platnick and Nelson (1978, p. 4) point out, to test any hypothesis we must be able to deduce some prediction from it with which additional data can agree (thereby corroborating the hypothesis) or disagree (thereby falsifying the hypothesis). It might be suggested that we could make predictions of the distributional pattern of this group, perhaps as compared to those of other groups occupying the same area, on the basis of the dispersal abilities of the organisms under consideration. In this scenario we are concerned solely with the dispersal abilities of potentially unknown ancestors; thus it is a tricky business at best to hypothesize what these dispersal abilities might have been. If the dispersal abilities were similar to those of other groups occupying the same area, we might expect to be able to compare the distributions of different groups and find that they are congruent—but even such comparisons may not help us decide unambiguously between dispersal versus vicariance explanations. If all groups had the same dispersal capabilities, they may have all dispersed together or they may have all vicariated together, but we may not be able to distinguish one scenario from the other. If we hypothesize that certain groups had a greater capacity for dispersal than other groups, we might hypothesize that the distributions of the groups with a greater capacity for dispersal are more likely to be explained in terms of dispersal. But this is not rigorous testing, just scenario building.

Perhaps a better line of inquiry would be to predict in what geographical areas fossils attributable to the various taxa should be found. Ideally, a fossil referable to the ancestral species of ABC (i.e., one that is completely plesiomorphous relative to A, B, and C but that shares the synapomorphies uniting A, B, and C) should be found in areas a and b (the ancestral species of BC is postulated to have migrated from area a to area b, and after having arrived in area b speciated, thus gaining the synapomorphies uniting B and C). Likewise, the ancestor of B and C should be found in areas b and c. Any fossil taxon that bears apomorphies of taxon A, B, or C (that is, is referable to taxon A, B, or C) should be found only in area a, b, or c, respectively. At least, such would be the case if only single, one-way dispersal of ancestors of taxa BC and C occurred. But if it is hypothesized that the present distribution of the organisms can be explained by dispersal of the organisms (thus, the ancestors were capable of dispersing), then any anomalous fossil occurrences (such as a fossil referable to taxon C found in area a) could be accounted for by postulating another dispersal event (in this case dispersal of a species from area c to area a after taxon C had formed). Using this line of reasoning, "it appears that any distribution pattern whatsoever can be explained by dispersal if we are willing to postulate a sufficient number of separate dispersal events" (Platnick and Nelson, 1978, p. 5). To avoid postulating numerous separate dispersal events, Platnick and Nelson suggest that a methodological rule be adopted that requires minimizing the number of parallel dispersal events (that is, in most cases, not allow more than one dispersal from any given area to another given area). As an example of this rule, suppose that a fossil taxon that was possibly ancestral to A, B, and C was found in areas a, b, and c. Our dispersal hypothesis states that such a taxon should be found only in areas a and b (the taxon originally occurred in area a and subsequently invaded area b, where it speciated) but not in area c (a species that originated in area b is supposed to have invaded area c). To account for the presence of such a taxon in all three areas, plus the distribution and relationships of the taxa A, B, and C, we must hypothesize that there were two independent invasions of area c, one by a potential ancestor of ABC and one by a potential ancestor of BC. However, adopting a methodological rule such as that suggested by Platnick and Nelson would not allow such independent, parallel dispersals, and we should perhaps seek a vicariance explanation. In general, the presence of a plesiomorphic cosmopolitan fossil taxon will serve to reject dispersal-type explanations. But Platnick and Nelson (1978, p. 5) point out that even if such a methodological rule may work in theory, in practice the relevant fossils may not be available in many cases. If the relevant fossils are not available, we have no way to rule out a dispersal explanation, and if we initially assume dispersal explanations, we may falsely conclude that dispersal explains the distributions of many groups whose distributions are actually solely the result of vicariance.

The other logical choice is to assume initially that an allopatric distribution is the result of vicariance. If such is the case, then the ancestral form of ABC primitively had a cosmopolitan distribution (occurred in areas a, b, and c). Area abc was divided by a vicariance event into two smaller areas, a and bc, and the isolated populations in these two areas underwent allopatric speciation. Subsequently, area bc was divided by a vicariance event into areas b and c, with the isolated populations in these areas undergoing speciation. If we consider the predicted distribution of fossil occurrences, a form that is plesiomorphous relative to all of the extant taxa (A, B, and C) would be expected to occur throughout the area abc. A form that bears the synapomorphies uniting B and C but lacks any apomorphies that are restricted to just B or just C should be found throughout the area bc, but not in area a (unless it dispersed there). Finally, any forms that bear the apomorphies restricted to taxa A, B, or C should be found in areas a, b, or c exclusively. If we have what appears to be a fairly complete fossil record for the group under study and we have all the right fossils in the right places and only in the right places, this would tend to corroborate our vicariance hypothesis. If we do not have any fossils, then this lack of fossils says nothing about our hypothesis. If we have only a few fossils, but they are in the right places, this might be seen as weakly corroborating the vicariance explanation. But having some fossils in the wrong places does not necessarily falsify the overall vicariance explanation. Even if vicariance events were primarily responsible for the taxonomic disjunctions observed, we can still postulate some movement (dispersal) of the organisms (or their fossils, see below concerning "Beached Viking Funeral Ships"). In this case also, fossil evidence will not necessarily distinguish unambiguously between vicariance and dispersal explanations.

Platnick and Nelson (1978, p. 6) suggest that we may be able to test a vicariance explanation by converting a cladogram of taxa (such as that of our example for A, B, and C) into a cladogram of areas (such as that for a, b, and c in our example). In this case the resulting cladogram of areas (Fig. 5-1*b*) is represented (a (b, c)) and suggests that areas b and c share a more recent common biota and series of geological and ecological events than either does with area a. If the cladogram of areas is true, then the component taxa of other monophyletic groups occupying the areas a, b, and c should in general show the same set of relationships. That is, in general for groups with distinct taxa in areas a, b, and c, the taxa found in areas b and c should be more closely related to each other than either is to those found in a. If the other groups occupying the area do not show the same pattern, this still does not necessarily mean that the pattern is not a general pattern (that is, reflects a true area cladogram, and the vicariance explanation holds). If the pattern seen in one group is a unique pattern, it can be suggested that the organisms of the other groups examined might not have responded, by speciating, to the vicariance events. Alternatively, the ad hoc hypothesis could be made that there had been parallel extinction, or possibly dispersal also, in the other groups examined such that this general pattern was obscured. If numerous different groups in the areas show various different patterns, this might suggest that they all underwent dispersal and speciation independently, thus refuting the general vicariance hypothesis. If numerous different groups in the areas a, b, and c show the same pattern, this might be seen as corroborating the vicariance explanation. But it is also possible that even if the patterns between independent groups are congruent, no vicariance was involved. Many different groups in an area may share the same allopatric patterns if they were all involved in the same form of unidirectional and sequential dispersal, as we initially hypothesized. Thus, if we initially suggest vicariance explanations for taxa with allopatric distributions, this may result in considering some cases that are actually the result of dispersal as the result of vicariance.

Platnick and Nelson (1978) have also posed the problem that in some instances one may find numerous groups that show an identical general allopatric pattern of distribution and relationships within a certain area abc, but that the groups do not all correspond in their higher-level relationships. For example (Fig. 5-2), numerous groups may show congruent patterns within the restricted area abc, but the higher-level affinities of some of the groups may be with taxa in areas to the north, while the higher-level affinities of some of the other groups may be with taxa in areas to the south. Such a case may be best explained by some combination of biotic dispersal and vicariance. Area abc may be a composite (composited by a vicariance event in reverse—formerly separated areas join together) of two smaller areas, each of which originally belonged to one of the larger (northern or southern) areas, in which dispersal (promoting an initial cosmopolitan biota in area abc) and subsequent vicariance, or possibly sequential dispersal, has taken place. Area abc may have originally been a part of either the northern or the southern area, and a vicariance event brought the two areas, or at least components of the two areas, together. For example, if abc were originally a part of the southern area, abc may have been shifted north and joined to the northern area, or a piece of the northern area may have shifted south to join to abc. Biotic dispersal could then take place, followed by allopatric speciation (by further vicariance or by sequential dispersal) of the biota in area abc. Or, area abc may represent a new area that is younger than either the northern or the southern area. For example, the northern and southern areas might represent two large landmasses separated by ocean, and area abc may represent an island chain that has emerged between them and been populated by elements of both the northern and southern biotas. Subsequent to the initial population of area abc, either a series of vicariance events or sequential dispersal may occur, giving rise to the congruent patterns of the groups residing in area abc.

Based on the above type of reasoning, Platnick and Nelson (1978) conclude

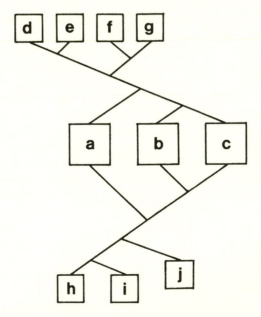

Figure 5-2. General area cladogram for areas a, b, and c, areas to the north of area abc (areas d–g), and areas to the south of area abc (areas h–j); see text for explanation.

that in general one cannot always, or even usually, distinguish unambiguously between instances of vicariance and instances of dispersal solely from distributional data of extant (or even extinct) taxa. Only in exceptional cases, with extremely complete supplementary data (such as a complete fossil record in all areas), can vicariance versus dispersal explanations be distinguished with reasonable certainty. What one can hope to arrive at, however, is area cladograms, or cladograms of the relative recency of connections between entire biotas. Such biotic cladograms, when well corroborated by the congruence of patterns in many independent groups, may be combined with geological and ecological evidence (if available) to refine or choose between vicariance and dispersal scenarios. With this in mind, Platnick and Nelson suggest that what is needed is a method of analysis to help determine whether distributional patterns in independent groups occupying the same general area correspond to one another, are congruent, and thus corroborate a hypothesized general pattern of areas and biotas. They then proceed to develop such a method of analysis, which is briefly reviewed here.

Platnick and Nelson (1978) begin by analyzing the simplest case, that of two-area cladograms. If we are given a pair of closely related taxa, A and B, each of which is endemic to a separate area, a and b, does this observation lead to any hypothesis as to the pattern of interconnections between areas a and b? We might suggest that areas a and b have close interconnections—are sister areas—and try testing this hypothesis by examining other pairs of taxa found in areas a and b. Other pairs of taxa that show exactly the same pattern—that is, of two closely related relatives, one endemic to area a and the other endemic to area b—will appear to corroborate the hypothesis. If we find a set of closely related taxa, one of which is endemic to area a and the other of which is found in area b but also in area x (i. e., in area bx), this pattern could still be viewed as compatible with the original pattern (although not exactly congruent) for our taxa A and B; we need only postulate that originally B was found throughout area bx but is now extinct in area x. Similar arguments could be made for taxa found in area a and somewhere else (area x), or taxa found in a and somewhere else and b and somewhere else. By hypothesizing the appropriate extinctions, any distribution could be compatible with a two-area pattern. In other words, any two areas we choose to examine will be connected (related) at some level; thus, the most basic unit of analysis must be at the level of three areas.

If we have three taxa, A, B and C, each of which is endemic to the area a, b, and c, respectively, and the relationships of the taxa are are (A (B, C)), then we can hypothesize that areas b and c were more recently connected to each other than either was to area a. Platnick and Nelson discuss possibilities for critically testing this hypothesis using other groups that occur in areas a, b, and c.

A group that is represented by a single taxon that occurs throughout the area does not test our hypothesis. Such a taxon would be compatible with the hypothesis because there is no reason why every group should necessarily speciate in response to a dispersal or vicariance event. As Platnick and Nelson (1978, p. 9) state, "a barrier effective for isolating flies may not also isolate birds." Furthermore, even if a single taxon of a group is restricted to area a, b, c, ab, ac, or bc, this still does not refute the hypothesis; the taxon may have previously occurred throughout the total area but is now extinct in some parts of the area.

By similar reasoning, any group that has only two taxa represented within the area abc cannot refute the hypothesis; such a group may simply have failed to respond to one or the other dispersal or vicariance events, or members of the group may now be extinct in one of the areas. If a group represented in the area under study has one taxon endemic to area a and another taxon endemic to area bc, it may have speciated in response to the event that separated areas a and bc, but it may have failed to speciate in response to the event that separated areas b and c. If a group has a taxon endemic to area ab and another endemic to area c, it may have failed to speciate in response to the first event but have speciated in response to the second event. If the areas were arranged sequentially (Fig. 5-3a) such that areas a and b shared a border and areas b and c shared a border, but areas a and c did not, then a group composed of two taxa in the area under study that had one taxon that occurred in areas a and c but not in b and another taxon that occurred

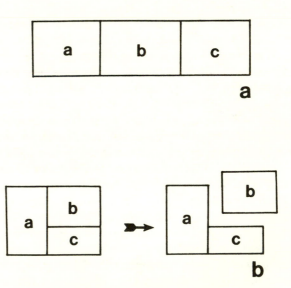

Figure 5-3. Two different arrangements of hypothetical areas a, b, and c: *(a)* areas lined up sequentially; *(b)* areas arranged such that each one is in contact with the other two.

only in area b would refute the hypothesis that areas b and c were more recently connected to each other than either was to area a (given that extinction has not occurred in either of the taxa presently under consideration). Our hypothesis would be refuted because such a group would suggest that area b was isolated relative to areas a and c, but as long as the areas are lined up sequentially (Fig.5-3a), any event that would isolate b from c would also isolate a from c. However, the areas need not be lined up sequentially; they may be arranged in such a manner that each one is in contact with both of the others (Fig. 5-3b). If this is the case, any group that does not respond by speciation to the first event (the event that separates a from bc) but does respond to the second event (the event that separates b and c) might end up with separate taxa in ac and b or in ab and c.

The event that isolates b from c, for instance, need not also isolate a from c for the particular group in question. As an example, imagine an initial uniform continent as area abc; after 50 million years a mountain chain arises in the middle of the continent separating area a on the west and area bc on the east. Some organisms will respond to this vicariance event and others will not. Let us imagine a terrestrial mammal that is well adapted to high altitudes and so does not respond to this first vicariance event. Now imagine that a rift valley develops, splitting area bc into a northern area (b) and a southern area (c), and that eventually area b detaches from the mainland and drifts off to the northeast and is completely surrounded by water. Some organisms that responded to the first vicariance event will also respond to this vicariance event; the phylogeny of such organisms will correspond to the general area cladogram of (a (b, c)). But our terrestrial mammal, which did not respond to the first vicariance event, may be unable to swim or cross any body of water; thus it may respond to the second vicariance event even though it did not respond to the first. The particular mammal taxon will show the relationships ((ac) b) (note that this is different from ((a, c) b)—the first indicates that there are two closely related taxa or areas, ac and b; the second indicates that there are three taxa or areas of which a and c are more closely related, or more recently connected, than either is to b), and ((ac) b) in this case is still compatible with the general area cladogram (a (b, c)). The moral of this story is that any group with only two taxa in an area cannot serve to refute a hypothesis of area relationships involving three areas.

It turns out that the only groups that can possibly test our initial hypothesis are groups that have three or more taxa in the area abc. Furthermore, such groups must have separate taxa that are endemic to each of the smaller areas, a, b, and c; otherwise we will be reduced to one- or two-area statements for our test group. Given such a group, there are only three relevant (and resolved) cladograms: The taxa in areas b and c may be more closely related to each other than either is to a, the taxa in areas a and b may be more closely related to each other, or the taxa in areas a and c may be more closely related to each other. If we find mumerous groups where the first cladogram holds, then our initial hypothesis (that areas b and c were more recently connected than either one was to a) is corroborated. (Note that as there are only three possibilities, one-third of the cladograms may conform to the hypothesis by chance alone [Nelson and Platnick, 1981].) If our hypothesis is not corroborated, we may feel forced to reject it. It should also be noted that, as Platnick and Nelson (1978, p. 10, footnote) point out, unresolved (in this case trichotomous) cladograms for test groups do not test our hypothesis unless all test groups are characterized by unresolved cladograms. In such a case we may suspect that some event caused an area to be broken up into several smaller areas simultaneously, but such a scenario should be tested using data from historical geology. It may also be that we are merely unable to find the relevant, perhaps subtle, synapomorphies—that is, unable to resolve the cladograms—perhaps because several events took place in rapid succession.

Three-taxon and three-area statements and hypotheses form the basis for historical biogeographic analysis along the lines outlined above, just as they do for phylogenetic analysis using cladistic methodology. Biogeographic analysis involving more than three areas and/or three taxa within a monophyletic group can be broken down into component three-taxon and three-area statements, which are then subject to analysis and testing (this topic is discussed in more detail by Platnick and Nelson, 1978, and Nelson and Platnick, 1981).

METHODS OF BIOTIC DISPERSAL AND VICARIANCE

In the latter part of the nineteenth century and the beginning of the twentieth century, many concepts were developed relating to how organisms and biotas (those restricted to land in particular, but the same principles apply to water and ocean-dwelling organisms) got where they are today. Much of this early work (see Matthew, 1915; McKenna, 1973) was based on a stabilist model of the earth with the continents and oceans fixed in their present positions. Such a scenario often made it necessary to postulate the existence of land bridges and corridors. With the general acceptance of plate tectonics (Hallam 1983), land bridges or chance raftings need not always be invoked to explain the transoceanic distributions of terrestrial taxa; instead, the ocean may have arisen in the middle of their ancestral range. A few possible mechanisms of dispersal and vicariance, which may be invoked as parts of historical biogeographic scenarios, are briefly mentioned here.

Simpson (e.g., 1940, 1953b, 1965; reviewed by McKenna, 1973) outlined three major types of biotic dispersal routes: corridors, filters, and sweepstakes. Simpson (1953*b*, p. 21) stated that "a corridor is a route along which the spread of many or most animals [the concept can also be applied to plants] of one region to another is probable." An example of a corridor for terrestrial organisms would be a large landmass, such as a continental body or wide land bridge connecting two geographic areas. It was expected by Simpson that prompt, two-way dispersal of organisms would take place along the corridor. Corridors might form and be destroyed (open and close) in geological time through vertical crustal movements. A chain of mountains, forming at first islands and then a continuous bridge of land, might rise between two landmasses that had previously been separated by water. Such a resulting land bridge, if substantial enough and at the correct latitude, could form a corridor for interchange between most members of the previously isolated biotas. On the other hand, the same mountain chain arising in the middle of a large body of water might serve to separate a once unified oceanic realm into two biotic provinces. Likewise, a mountain chain arising in the middle of a continental landmass would serve to split the biota occupying the continent.

At the opposite end of the spectrum from corridor dispersal, Simpson (1953*b*) distinguished sweepstakes, or waif, dispersal. Here the concept is that even if there are extremely formidable barriers to dispersal for some particular form of organism, in many cases it is not entirely impossible that an individual of the species could cross the barrier. What is at any one time extremely improbable may become realistically possible or even probable given long periods of time. The classic example is that of a gravid female of some wholly terrestrial animal being rafted across a large ocean body, surviving the trip, and founding a new population in a previously uninhabited (by the taxon under consideration) area. Such sweepstakes dispersal is random, unpredictable, and due to chance.

Between the extremes of corridors and sweepstakes, Simpson (e.g., 1953*b*) recognized filters. Filters are bridges or connections between distinct biotas that are relatively continuous but do not allow the free interchange of all elements of both biotas. A series of lesser barriers may be encountered in dispersal along the connection; such connections selectively filter out certain organisms from dispersing. A classic filter is an isthmian link between terrestrial biotas; the land bridge is not complete, and only those organisms that can cross moderate-sized bodies of water will be readily dispersed. Or if the connection between two low-latitudinal biotas passes through high latitudes, certain organisms may be able to withstand the passage through the high latitudes while others will not.

In many cases, as dispersal barriers and connections form and are destroyed, they will vary from corridors (before a barrier is formed) to filters (as a barrier is being formed) to sweepstakes routes (once the barrier is in place) and back again.

As McKenna (1973, 1984) has noted, under the new plate tectonics model all of Simpson's types of dispersal routes may still be operative, but they are supplemented by other types of dispersal such as what McKenna terms "Noah's arks" and "beached Viking funeral ships." In a plate tectonics framework, organisms do not move about over the surface of the earth only as long as they are alive and then die to be buried and possibly preserved as fossils in one unmoving place; they are also passengers (both as living and fossil forms) on plates that move across the face of the earth. Great modifications to the distributions of organisms can be caused by repeated separations and recombinations of tectonic plates and the landmasses riding upon them. Biotas may also migrate across a plate, tracking a consistent climatic regime, as the plate crosses latitudinal and climatic belts. For instance, as a plate moves north, the fauna may migrate south at the same rate (of course, the same trend can be caused by increasing northern cold, as during the Ice Ages: Schoch, 1984b, p. 55). One-way dispersal of a balanced biota can be possible under a plate tectonics regime. For instance, we may envision a landmass containing a single, unified biota. One part of this landmass may separate from the main body and drift to, and collide with, a second landmass. A representative sample of the biota of the first landmass may thus be transported by the "Noah's ark" to the second landmass in a one-way dispersal. Likewise, fossils may be transported from one landmass to another and thus be found to occur on the second landmass even though in life the organisms never occupied the second landmass (McKenna's concept of "beached Viking funeral ships"). It is also possible that a landmass can be subducted; as it is subducted, the biota occupying the surface will continually migrate away from the area of subduction. Thus, while the extant members of the group survive, the original area of origin and early fossil forms of the group will be destroyed (McKenna, 1984). One can imagine numerous variations on such themes; repeated collisions and separations of tectonic plates can make for very complicated distributions of recent and fossil organisms.

It should be pointed out that when vicariance biogeographers are concerned with vicariance events, many such large-scale events may be the result of plate tectonic processes literally splitting ancestral distributions—for example, when the pieces of a continent move away from one another (Hallam, 1983). But vicariance events may be any physical or ecological events that are of a large enough magnitude to split an ancestral population; what might cause a disjunction in a taxon is dependent upon the physical attributes of the taxon. A change in the course of a stream or river, the flooding of a low-lying area of land, a change in climatic conditions, the rise of a mountain range, or the opening of an ocean basin may or may not cause disjunctions among certain organisms. McKenna (1973, p. 303) has astutely pointed out that there may not always be an exact correlation between a taxonomic disjunction and its ultimate, or primary, cause. For instance, mid-continental rifting, which ultimately results in the formation of an ocean basin, may begin in a preexisting epicontinental sea; thus isolation of taxa on the resulting landmasses may be more ancient than the formation of the true ocean separating them. On the other hand, the beginning of rifting and separation of two landmasses may precede the last actual terrestrial biotic link between them.

CHAPTER **6**

Systematics, Taxonomy, and Classification

The process of classifying is a necessary component of any scientific activity, indeed of human endeavor in general. Classification is the process of ordering sensations or perceptions and imparting meaning to the world around us. Classification involves grouping entities or phenomena on the basis of some (any—arbitrary or not) kind of relationship; usually the resulting groups are given names (Wiley, 1981). The words of our language are a form of classification. As Simpson (1961, p. 5) has aptly said,

The whole aim of theoretical science is to carry to the highest possible and conscious degree the perceptual reduction of chaos that began so lowly and (in all probability) unconscious a way with the origin of life. In specific instances it can well be questioned whether the order so achieved is an objective characteristic of the phenomena or is an artifact constructed by the scientist. . . . Nevertheless, the most basic postulate of science is that nature itself is orderly. In taxonomy as in other sciences the aim is that the ordering of science shall approximate or in some estimable way reflect the order of nature. All theoretical science is ordering and if . . . systematics is equated with ordering, then systematics is synonymous with theoretical science.

Simpson (1961, p. 3) has distinguished two basic ways of classifying or associating objects or phenomena. *Association by contiguity* involves a functional, structural, or historical relationship between the entities associated to-

255

gether, although the things grouped together might be quite dissimilar. For instance, the various organs of a single individual, the presidents of the United States, or the organisms of a certain ecosystem are associated by contiguity. *Association by similarity* is the grouping of things together because they possess one or more common characteristics; for instance, we associate all red fruit together, all chairs together, or all stars together. In many cases, of course, we may group or classify entities together on the basis of a combination of contiguity and similarity, and in many instances we might deduce that certain things share an association of contiguity by the observed presence or possibility of an association by similarity.

Wiley (1981, pp. 194–197) has distinguished three major types of classifications: (1) natural classes, (2) historical groups and individuals, and (3) convenience classes. Classifications based on natural classes are composed of natural classes (groups) of individuals (or instead of individuals, other natural classes ultimately composed of individuals) that fit the class definitions. A class is a spatiotemporally unrestricted group of entities or phenomena that fit the particular class definition. In Wiley's (1981, p. 194) terminology in this context,

A natural class [as opposed to an artificial or convenience class] is one that contains individual entities that fit the definition and whose origins and behavior are governed by natural law. That is, their origins and actions can be predicted by what we term "natural law" or hypotheses concerning natural processes. The similarities displayed by members of a particular class are the products of the workings of particular laws, the entities are not necessarily similar because they share a similar history.

Examples of classifications of natural classes are the periodic table of elements and systems of mineralogy. Wiley (1981) makes the point that classifications of natural classes are inherently nonhierarchical—that is, they are not inherently structured in a groups-within-groups arrangement, although they may be formed into a groups-within-groups arrangement as a matter of convenience.

Classifications of individuals and historical groups have their basis in history; entities are classified together according to their historical connections and relationships (often inferred). For example, 20 years after their graduation we might classify together (for certain purposes) the persons who were seniors in college during a particular year at a certain institution, or we classify parents, their children, and their grandchildren together because they are all historically related (in this case genealogically). Wiley (1981) points out that classifications of individuals and historical groups are inherently hierarchical; they form groups within groups. Of course, the process behind the pattern need not be known, and the pattern need not even be considered historical in the strict sense. As Wiley (1981, p. 196) notes, the same hierarchical organization of organisms can be arrived at by believing that it was produced by the logical mind of God (nonevolutionary views, for instance as perceived by Linnaeus and Agassiz) or that it was produced by ancestry and descent among historically connected organisms and groups (evolutionary views). Of course, the point is that we first recognize the pattern independently of any hypothesis of process behind the pattern—or else we run the risk of producing what Wiley (1981, p. 197) terms an *ad hoc classification* (see below).

Wiley (1981, p. 197) distinguishes two types of classifications of convenience classes. Those that do not purport to have any particular basis in terms of natural process or history he calls *convenience classifications*. Convenience classifications are useful in allowing us to communicate, store, and retrieve data. Most convenience classifications are based on Simpson's association by similarity; the majority of improper nouns in our language are convenience classifications. Ad hoc classifications, the second type of classifications of convenience classes, "purport to reflect either process or history and fail to do so" (Wiley, 1981, p. 197). As such, ad hoc classifications may be extremely deceptive and possibly difficult to distinguish from true classifications of natural classes or from true classifications of individuals and historical groups.

It is often suggested (e.g., Simpson, 1961) that an important criterion for biological classifications is that they be practical and convenient. I reject such a notion and rather argue (see below) that in order for biological classification to be scientifically meaningful, it must reflect real phenomena of nature. It may be unfortunate if certain investigators consider such a natural classification inconvenient, but we depart from the realm of science if we modify our classifications merely for the sake of convenience (cf. Brundin, 1966, p. 20).

BIOLOGICAL SYSTEMATICS AND CLASSIFICATION

As Nelson (1970), Nelson and Platnick (1981), and Wiley (1981), among others, have pointed out, there are two basic areas of biological research: general and comparative biology. General biology focuses on particular biological processes, such as how muscles contract, how the brains of mammals function, and how genes are coded and passed from one generation to the next. In contrast, comparative biology is concerned with the study and explanation of the diversity of organisms observed on earth. Comparative biology is inherently concerned with comparing different organisms or different species. In the most general sense, systematics refers to the ordering, and in some cases explanation, of all phenomena (Hennig, 1966a) and in this sense is synonymous with theoretical science (Simpson, 1961). Biological systematics (from here on referred to simply as systematics) is usually considered either synonymous with comparative biology or as a subcategory of comparative biology. Thus Mayr (1969, p. 2) states that "systematics is the science of the diversity of organisms," and Simpson (1961, p. 7) defines systematics as "the scientific study of the kinds and diversity of organisms and of any and all relationships among them." Rosen (1974a, 1974b) considers systematics to be biogeography plus evolution.

As part of the study of the diversity of organisms—as part of systematics—one usually describes and orders—that is, classifies (either formally or informally)—the organisms being studied. Taxonomy is the theoretical study of the ordering of organisms; it includes the theory and practice of describing organisms and their diversity, ordering the diversity on the basis of some relationship between the entities ordered, and the conversion of the ordering into a classification. Simpson's (1961, p. 11) classic definition is: "Taxonomy is the theoretical study of classification, including its bases, principles, procedures, and rules." Biological classification in general is simply the process of ordering organisms into groups on the basis of some principle or relationship thought to occur among the entities so ordered. A biological classification is usually a graphic system or means of conveying a particular ordering or arrangement, a particular way of classifying, a set of organisms. Usually, a biological classification takes the form of a set of words (names) that denote hierarchical groupings of organisms. Biological nomenclature is the application of names (formal or informal) to the groups recognized in a given biological classification (Simpson, 1961, p. 9). Most workers (e.g., Wiley, 1981, p. 6) regard a taxon as any grouping of organisms that is potentially nameable (what a taxon is or is not may depend on the particular classificatory scheme being used). By the last definition, a taxon is usually regarded as a class composed of actual organisms. Alternatively, a taxon can be defined as a concept (Løvtrup, 1977a, p. 20): "A taxon is a concept defined by a set of properties distinguishing a particular class of animal [or organism]. All individuals, past, present or future, possessing the whole set of properties are usually said to be members of the taxon."

The most common formal convention—indeed, it is almost universal—for classifying organisms is the Linnaean hierarchical system. In the Linnaean system organisms are grouped into taxa, and these taxa in turn are grouped into progressively more inclusive (higher-level) taxa. Each taxon is given a formal name (formed according to specific nomenclatural conventions) and also assigned to a certain named category. These categories specify the level of inclusion, or the ranking, of a particular taxon. Some of the common categories used today for zoological nomenclature (these differ in some cases for nonanimals [plants, bacteria]) in order of subordination (most inclusive to least inclusive) are:

Empire
  Kingdom
    Division
      Phylum

(Division for plants)
Class
  Legion
  (= "Division" of some authors)
    Cohort
      Order
        Family
          Tribe
            Genus
              Species.

Other terms occasionally used by various authors for miscellaneous supra-ordinal categories include branch, grade, and series. More categories can be placed within these main categories by adding prefixes such as magna-, super-, grand-, mir-, sub-, infra-, parv- (see McKenna, 1975; Wiley, 1981). Thus, between Cohort and Order there could be:

Cohort
  Subcohort
    Infracohort
      Parvcohort
        Magnaorder
          Superorder
            Grandorder
              Mirorder
                Order.

Of course, new categorical names can be coined to denote additional levels of ranking.

To illustrate the Linnaean system, extant humans are classified as the genus and species *Homo sapiens,* and certain fossil men are classified as *Homo erectus* (that is, in the same genus but in a different species from ours). Still more distant relatives of ours are classified as *Australopithecus africanus* (and there are several other species of *Australopithecus*). *Homo* and *Australopithecus* are generally regarded as both belonging to the family Hominidae. The common chimpanzee, *Pan troglodytes,* the common gorilla, *Gorilla gorilla,* and the orang-utan, *Pongo pygmaeus,* are often included in the family Pongidae. Together the families Hominidae and Pongidae, along with numerous other families of monkeys, lemurs, tarsiers, and so on, make up the order Primates. Primates is included along with numerous other orders, such as the Carnivora, Artiodactyla, Perissodactyla, Edentata, and Rodentia, in the class Mammalia.

In the Linnaean system the species category need not always be subordinate to the genus category; this is because it is required that every species have a binomial name composed of a generic name and a specific name. Thus one cannot have a species without a genus; however, sometimes in practice a species is named and only questionably assigned to a certain genus. Still, a species must always be assigned (even tentatively) to some genus, while a new genus and species may be described but need not necessarily be assigned to a complete series of higher categories (this depends on the applicable code; see below); for instance, in the case of botanical nomenclature it may be required that a species be assigned to a family-level taxon). Thus, a new species of mammal might be described that is recognizable as a mammal but does not fit the concept of any of the known families of mammals. The species may be described as a new genus and species and assigned to the class Mammalia without being assigned to any lower category.

The Linnaean hierarchy is just one of many possible conventions for setting up formally expressed classifications of organisms. At least theoretically, any desired convention could be used consistently to classify organisms, but if some form of the Linnaean system is not used, the number of other investigators who will easily understand, accept, and use a given classification is limited. The Linnaean system has the great advantage that it has been adopted almost universally as the basic framework for classifying virtually all organisms (note that in particulars, adopted conventions [such as the categorical ranks] differ for different types of organisms [e.g., plants versus animals], but the spirit and many particulars are the same). The Linnaean system, at least in a modified or annotated form, is a useful, practical means of classifying organisms; the essentials of nomenclature are well codified and governed by several international commissions (separate commissions govern Linnaean nomenclature of animals, plants, and bacteria, as discussed below). Because it is universally understood and accepted, the Linnaean system should serve as the basis for any modern biological classification and nomenclature. It is imperative that the various international codes and rulings be followed and upheld by systematic biologists, otherwise the virtues of Linnaean based systems—universal acceptance, comprehension, and stability—will be lost.

Acceptance of the Linnaean system in general principle, however, by no means strait-jackets the investigator. For example, the *International Code of Zoological Nomenclature* applies only to family-level taxa and below (i.e., below the level of approximately the superfamily); all ordinal-level and higher taxa are not governed by the code. At the ordinal level an investigator is free to utilize the named categories of the Linnaean system, as outlined above, or to dispense with the named categories and rank taxa using some other convention. In a modified Linnaean system ordinal-level taxa (using conventional and identical Linnaean latinized names) may be ranked by simple indentation or by using some form of numerical prefixes that take the place of categorical names (e.g., Hennig, 1966a; Løvtrup, 1977a; Schoch, 1984b). Thus, as part of a classification of the vertebrates, Schoch (1984b, pp. 9–10) included the taxa:

16. Mammalia
  17. Eotheria
    18. Triconodonta
    18. Docodonta
  17. Apotheria

The same classification could have been expressed more conventionally by using categorical names in place of the numbers (however, for the classification outlined by Schoch [1984b], this would have required 25 supraordinal categories), perhaps as such:

Class Mammalia
  Subclass Eotheria
    Infraclass Tricondonta
    Infraclass Docodonta
  Subclass Apotheria

Or the classification could be expressed by pure indentation:

Mammalia
  Eotheria
    Triconodonta
    Docodonta
  Apotheria.

  But more important, for the systematist who adopts the Linnaean system (either in strict or modified form), the Linnaean system is only a convention for naming and expressing subordination of taxa once the systematist has decided what organisms to group together into taxa at various levels. The Linnaean system does not legislate to the systematist how organisms should be grouped into taxa or what taxa should be formally recognized. It is in deciding how to group organisms into formal taxa that most substantial controversies between systematists arise. As was noted earlier, adherents of the three major schools—evolutionary systematics, phylogenetics (cladistics), and phenetics—have very different views about how organisms should be grouped into taxa and what taxa should be formally recognized and named.

  To review, in general cladists recognize exclusively monophyletic taxa (*sensu* Hennig) in their classifications, and sister groups are given equal ranks. Evolutionary systematists routinely recognize paraphyletic taxa, and sister groups need not be, and often are not, given equal ranks. Pheneticists group organisms into taxa on the basis of measures of overall similarity without particular regard to the phylogenetic significance of the groupings. A fundamental question here is whether biological classifications are objective and real systems that (ideally) reflect nature, or whether they are inherently arbitrary and subjective at some point: Need they incorporate some art? Wiley (1981, p. 197), a phylogeneticist, has expressed his opinion that scientific classifications, such as biological classifications, must "be justified in terms of natural process or pattern"; they should not be arbitrary, artificial, or mere convenience classifications. In contrast, Simpson (1961) has emphasized that in his opinion there is an artistic component to biological classification. Given the same group of organisms and the same information about them, two different investigators may classify them in different ways. "Taxonomy is a science, but its application to classification involves a great deal of human contrivance and ingenuity, in short art. In this art there is leeway for personal taste, even foibles, but there are also canons that help to make some classifications better, more meaningful, more useful than others" (Simpson, 1961, p. 107).

Much of the controversy over how to classify organisms is actually concerned with the question of why we classify organisms; what is the use or purpose of biological classification? Many workers from different schools profess to be seeking natural classifications, but what they consider to be natural classifications can be very different.

Phylogenetic systematists generally hold that a classification should convey more or less completely and precisely any information that is known about the phylogeny (*sensu stricto*) of the group of organisms being classified. A classification follows directly from a hypothesis of phylogenetic relationships, and the series of phylogenetic relationships can be reconstructed completely from the classification—that is, there is an isomorphic relationship between the phylogeny and the classification (Eldredge and Cracraft, 1980, pp. 193–194).

Evolutionary systematists (e.g., Mayr, 1969; Simpson, 1961) believe that a classification should be based, to some degree, on the (hypothesized) evolutionary (as opposed to phylogenetic) history of the organisms being classified, but that a classification should also serve as an efficient and convenient storage and retrieval system of information about the organisms concerned. For Bock (1977) a natural classification must be based on, and be in agreement with, the whole of evolutionary theory and must also reflect the entire evolutionary history of the group of organisms classified, not just the cladistic branching sequence (phylogeny). Evolutionary systematists in general wish to include something more, or something other, than strict phylogeny in their classifications. The argument is often made that one does not need a classification to express a hypothesis of phylogeny; it is perhaps better expressed in the form of a phylogram or narrative scenario (of course, such an argument misses the fundamental point that it is not that a classification must represent phylogeny completely per se, but that only phylogenetic classifications will be composed of "real," "natural," or "nonarbitrary" taxa). What evolutionary systematists generally wish to include, other than phylogeny, in their classifications is some kind of evaluation of the evolutionary history of the group in question. Components such as degree of morphological divergence, adaptive radiations, and grade levels of organization are seen as important. Thus, evolutionary systematists tend to recognize, in certain instances, paraphyletic groups and also will give different ranks to acknowledged sister taxa. Evolutionary systematists believe that their classifications are the most useful, have the highest empirical content and the greatest capability for making predictions and broad generalizations, and are the best possible classifications to be used as a basis for general comparative studies of organisms.

Pheneticists (e.g., Sneath and Sokal, 1973) advocate classifications in which members of a group (taxon) share many correlated attributes (as many as possible, without distinguishing between primitive and derived characters). Such classifications are purported to have a high content of information (or the information implied by the classification is said to be high) and thus are "natural." According to Sneath and Sokal (1973, p. 25) "a 'natural' taxonomy is a general arrangement [based on the above criterion of correlated attributes] intended for general use by all scientists." These authors argue that classifications based on some special purpose (such as expressing phylogeny or evolution) are special or "arbitrary" classifications of restricted use (in general, each is good only for its one special purpose).

PHYLOGENETIC VERSUS EVOLUTIONARY CLASSIFICATIONS

If we wish to express some form of phylogenetic or genealogical information in our classifications—that is, if we wish to base our classifications to some extent on phylogeny—then we have two basic choices of classifications that explicitly incorporate such information: phylogenetic (cladistic) classifications and evolutionary systematic (sometimes referred to as eclectic or syncretistic) classifications. Phenetic (numerical taxonomic) classifications are not explicitly concerned with phylogeny, genealogy, or evolution, except as a side product that might result from such classifications; however, it seems that when actual classifications are made using phenetic philosophy they are not completely divorced from considerations of phylogeny and evolution.

As has been stated earlier (and see discussion by Eldredge and Cracraft, 1980), evolutionary systematists (e.g., Mayr, 1969; Simpson, 1961) believe that a diversity of evolutionary information (such as information concerning rates and degrees of morphological and adaptive evolution), not just phylogeny *sensu stricto,* should be used in constructing a classification, and consequently such information is supposedly stored in the resulting classification. However, as Eldredge and Cracraft (1980, p. 194, footnote) have stressed, the only information that can be incorporated into, and retrieved from, the Linnaean hierarchy is nested set membership—which taxa are members of which higher taxa. In other words, only hierarchical patterns can be retrieved from a Linnaean classification (Bonde, 1977). Consequently, such a system can be used to show only one type of information (the only possibility for showing more than one type of information is if the two or more types of information were completely and perfectly correlated with one another, as nested sets of synapomorphy and phylogeny may well be). The information incorporated into a classification, and thus retrievable from the classification, might be a general index of overall similarity, information regarding biogeographical or ecological associations, or phylogeny (more precisely, nested sets of synapomorphy). But one cannot hope to combine more than one type of information in a classification simultaneously and then recover all of the information completely. Yet this is exactly what evolutionary systematists attempt to do; however, they fail in having no rules or methods for retrieving from the classification (in isolation; i.e., without supplementary materials or discussion) the information that went into the construction of the classification.

Evolutionary systematists often stress the importance of the capability of classifications to make predictions and testable generalizations concerning the distribution of characters among taxa (see Eldredge and Cracraft, 1980, for specific examples); they then further claim that evolutionary classifications are superior to other classifications in this aspect. But it is unclear what kind of predictions the evolutionary systematists believe that evolutionary classifications are superior at making. If one is concerned with gross, overall similarities, in terms of numbers of similarities this criterion leads logically to phenetic or numerical taxonomic classifications. If instead one is interested in predicting the occurrence of special, evolutionary similarities or nested sets of similarities among taxa, this criterion leads to adoption of a cladistic program and classification. In this context Farris (1977b, 1979, 1980, 1982) has demonstrated that parsimonious cladistic classifications are maximally informative (therefore of maximum predictive value) as compared to both evolutionary and phenetic classifications. Farris (1977b) demonstrated that even utilizing Gilmour's (1940, 1961) concept of a natural classification as generally adopted by pheneticists, phenetic clustering by overall similarity does not always yield an optimal Gilmour-natural classification, whereas clustering by special similarity (i.e., by synapomorphy to form cladograms) apparently does yield optimal Gilmour-natural classifications. To quote Farris (1983; reprinted in Sober, 1984b, p. 684):

I have elsewhere (Farris, 1979, 1980, 1982) already analyzed the descriptive power of hierarchical schemes. I showed that most parsimonious classifications [i.e., parsimonious cladograms] are descriptively most informative in that they allow character data to be summarized as efficiently as possible. That conclusion has aroused some opposition, as syncretistic taxonomists had been inclined to suppose that grouping according to (possibly weighted) raw similarity gave rise to hierarchies of greatest descriptive power. . . . As I have observed before (particularly Farris, 1980), the presence of a feature in the diagnosis of a taxon corresponds to the evolutionary interpretation that the feature arose in the stem species of that taxon. There is thus a direct equivalence between the descriptive utility of a phylogenetic taxon and the genealogical explanation of the common possession of features by members of that group.

The cophenetic correlation coeffient (Farris, 1977*b*) compares the correlations between taxa of a classification and the raw similarities or distance matrices upon which a classification may be constructed; thus, the cophenetic correlation coefficient can be viewed as a measure of the information content of a classification (Phillips, 1984). Using this approach, Farris (1979), Mickevich (1978), and Mickevich and M. S. Johnson (1976) have all found that cladistic-based classifications have higher cophenetic correlation coefficients (i.e., retain more information) than do classifications based on other criteria (such as overall similarity). Furthermore, in many cases a sound cladistic classification may be more stable than a phenetic-based classification as more characters are added to an analysis (Phillips, 1984).

Cladistic classification will predict any phylogenetically based similarities that evolutionary classification will, but cladistic classification will predict the similarities at the correct hierarchical level. Evolutionary classifications contain grade groups united by symplesiomorphy; the same symplesiomorphies are taken into account in a cladistic (phylogenetic) classification, but at the (correct) level where they are synapomorphies. Unlike a phenetic or evolutionary classification, a cladistic classification predicts all possible character-state associations among the taxa classified (Phillips, 1984). Phylogenetic and evolutionary classifications both differ from pure phenetic classifications in explicitly sorting out and rejecting the use of similarity based on homoplasy.

Phylogenetic classifications predict that groups share certain similarities at certain levels, but they do not predict or indicate how much the same groups may have diverged in subsequent differences (apomorphies). Evolutionary classifications attempt to indicate such evolutionary divergence by the unequal ranking of sister groups and by the separating out of subgroups from monophyletic groups, raising the subgroups to a high rank and leaving the remainder of the group as a paraphyletic group. I have argued elsewhere (see section on monophyly, polyphyly, and paraphyly) for the importance of monophyletic taxa, an importance that evolutionary classifications deny by freely utilizing paraphyletic taxa.

It should be pointed out here that logically in the Linnaean hierarchy, monophyletic sister groups must be given equal ranks (Hennig, 1966*a*). If we have a monophyletic taxon A composed of two monophyletic taxa, B and C, then we cannot classify them as such:

Order A
  Family B
  Suborder C

If we are to recognize C as a suborder and we wish to call B a family that does not belong to C (both together comprise A), we must erect a redundant second suborder containing only the family B as such:

Order A
  Suborder X
  Family B
  Suborder C

But if such is the case, then family B equals suborder X, and we have ended with a system where the two sister groups are equally ranked. What sometimes seems to happen in the case of evolutionary classifications is that, given the evolutionary systematist wishes to recognize B as a family and C as a suborder, but not B as a suborder in and of itself, some other organisms will be associated with B in order to flesh out the suborder X. These organisms may be primitive or derived relative to B, but most likely they will be close to B in terms of some kind of overall similarity; they may be forms transferred from C to be included with B in X. Such rearrangement of organisms will usually lead to the erection of nonmonophyletic taxa; if a set of primitive forms was transferred from C to X, C and B may remain monophyletic, but X will be a nonmonophyletic assemblage.

RANKING OF TAXA

There has been much discussion in the literature on proposed criteria for assigning absolute ranks to taxa, but no consensus has been reached (Bonde, 1977). In general, investigators working on particular groups of organisms follow the traditions of ranking established over the years within the particular group. Hennig (1966a, pp. 154 and 187) suggested that the rank of a taxon could be assigned according to the absolute age of origin of the taxon; taxa that originated earlier would be assigned higher ranks, and those that originated later would be assigned lower ranks. For example, taxa that originated during the time period from the Upper Carboniferous through the Upper Permian might be ranked as orders, and taxa that originated during the time period from the Triassic through the Upper Cretaceous might be ranked as families (Hennig, 1966a, p. 187). This criterion has not been adopted generally, however. In many instances it is not possible to date the origination of a group sufficiently precisely to apply the method, and furthermore it would radically change the rankings within many classifications (not that such changes would be detrimental, but they do tend to meet resistance by specialists of the particular groups concerned). For example, Hennig (1966a, p. 187) suggested that the commonly accepted orders of mammals should be downgraded to tribes.

Griffiths (1974) suggested assigning ranks of taxa on the basis of morphological divergence (following the tradition of the evolutionary school). Farris (1976) has suggested a system where sister taxa need not have the same ranks and where absolute ranks of taxa are determined by the life-span of the taxa (temporal duration from origination to extinction or to the present). The major shortcoming of both of these suggestions is that they do not treat historically equivalent entities (sister groups) equally; taxa of the same rank may not be meaningfully comparable, whereas sister groups may be assigned different ranks (see chap. 7). Løvtrup (1973, 1975, 1977a, 1977b; cf. Hennig, 1966a, 1966b) has dealt with the problem of rankings by abandoning the Linnaean system of named categorical ranks. Beginning at the apex of a phylogeny (which is assigned the lowest numerical rank), Løvtrup raises the rank by one at each dichotomy. The net result is that the terminal taxa (species) may have varying ranks. A similar system of numerical rankings of taxa was used by Schoch (1984b, pp. 7–10) in his classification of the vertebrates.

DIFFICULTIES IN THE CLASSIFICATION
OF FOSSILS

Given a phylogeny of extent monophyletic taxa, such as that shown in Figure 6-1, and assuming that we wish to classify the taxa phylogenetically utilizing the Linnaean hierarchical system, the classification must take the form:

Category 1, taxon I
 Terminal taxon A
Category 1, taxon II
 Category 2, taxon III
  Terminal taxon B
 Category 2, taxon IV
  Category 3, taxon V
   Category 4, taxon VI
    Terminal taxon C
   Category 4, taxon VII
    Terminal taxon D
    Terminal taxon E
  Category 3, taxon VIII
   Terminal taxon F
   Terminal taxon G.

Since A lies on a totally separate branch relative to the rest of the group, it must be ranked (category 1, taxon I) equally relative to the rest of the group. This poses no problems if we are dealing exclusively with extant taxa; A must just have a separate ancestry relative to taxa B through G. Or to put it another way, B through G share a common ancestor that is not shared with A; B through G are a monophyletic group and A is a monophyletic group, and the groups B through G (taxon II, above) and A (taxon I) are sister taxa.

Let us further assume that terminal taxa A through G are species-level taxa, and that none of the terminal taxa are known to be characterized by any autapomorphies. Does this mean that A cannot be excluded from an ancestral position (i.e., that taxon I is nonmonophyletic) relative to B through G? Again, if we are dealing solely with extant taxa, we know that A is con-temporaneous with B through G, and so the currently observed A is not ancestral to B through G. Of course, contemporary A may be the unchanged descendant of a species A that did give rise both to extant A and to an ancestor of B through G, and in this sense A may be the ancestor of B through G. Still, when dealing with contemporaneous species very few investigators are concerned with this subtle point. Obviously, extant A did not give rise to extant B through G, and in this sense A is monophyletic.

Now let us assume that we are dealing with species that are not contem-poraneous, and indeed A is older than B through G. It may now be considered a distinct possibility that A is the ancestor of B through G. Assuming for a moment that A is indeed the ancestor of B through G, classifying A in the separate taxon I of equal rank with taxon II would mean that taxon I is non-monophyletic (specifically paraphyletic). How do we deal with this sort of problem? As was discussed in chapter 3, it may be impossible to recognize ancestors objectively, but we still need a way to deal with wholly plesiomorphic taxa.

One way to deal with the problem is to deny or reject the concept of ancestors and treat all terminal taxa as monophyletic, nonancestral taxa (that is, treat them as if they were all contemporaneous) and then to proceed with classification from that stance. The approach has justification in the suggestion that ancestor-descendant relationships are not testable. As Patterson and Rosen (1977, p. 154) have aptly pointed out, a taxon that shares synapomorphies with all the members of a certain group but shares no apomorphies with any fewer members of the group (as for A relative to the group composed of A

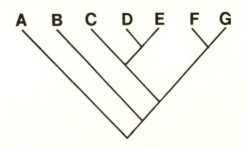

Figure 6-1. Hypothetical phylogeny of extant monophyletic taxa A through G.

through G in the above example) is equally closely related to all members of the group in question. However, cladistic analysis cannot determine or distinguish between the case where the taxon in question shared a common ancestor with the other members of the group and the case where it was itself the common ancestor of the other members of the group. To assume one choice over the other is an ad hoc assumption—that is, axiomatic. It might be argued that classifying such a taxon in a separate taxon of equal rank is choosing the first possibility over the second. But such a classification does not choose between the two possibilities; it merely expresses the agreed-upon cladistic relationships. However, from the perspective of the Recent the possibility is allowed that such a taxon (taxon I in the hypothetical example above) might not be strictly monophyletic. In such a case, if an investigator interprets all of the taxa as strictly monophyletic, he or she would, in essence, be choosing the first alternative over the second (i.e., that the taxon in question was not ancestral to its sister taxon). It must also be pointed out that even if a taxon, a single species, was truly ancestral to its sister group, such a species was (from its point of view) monophyletic and natural while it was living before it gave rise to anything else (cf. the proposal that fossils be classified separately from extant organisms, discussed below; but see also chap. 3 for a discussion of the problems with this argument). Partly in this sense it can be argued that species are inherently monophyletic—they are monophyletic during their life-times—or that the concept of monophyly does not apply to species-level taxa.

A number of conventions have been suggested for dealing with such ambiguous situations, where we are dealing with relatively primitive fossil forms, some of which may be potential ancestors, and which are not easily accommodated within a classification of extant forms.

Crowson (1970; discussed extensively by Patterson and Rosen, 1977; see also Bigelow, 1961, and Griffiths, 1974) has suggested that fossil organisms should be classified in different systems from those of extant organisms. Geologic time could be arbitrarily divided into segments, perhaps corresponding to the geologic periods (Neogene, Paleogene, Cretaceous, Jurassic, Triassic, and so on), and all of the organisms from any one period could be classified in their own system (classification). The same or similar names could be used, as much as possible, for persistent taxa that last from one period and classificatory system to another, but the categorical rank of a persistent lineage might be upgraded one categorical rank for each successive period. Thus a group that persists from the Cretaceous to the Recent and is ranked as an order today might be ranked as a tribe during the Cretaceous (in the Cretaceous classificatory system). The gist of this scheme is that the organisms of any period are classified as they would be if we lived during that period, without the knowledge of the future evolution of organisms (specifically, without the knowledge of what the Cretaceous forms, for instance, might or might not evolve into or whether lineages will even persist—yet for each period we cannot completely divorce ourselves from what we know of the life of other periods, as is evident in the suggestion of the successive upgrading in ranking of persistent groups).

The primary advantage of such a system is that what may be grade taxa or paraphyletic taxa (nonmonophyletic and therefore unclassifiable from the point of view of a single classification for all organisms) may be monophyletic from an earlier point of view. For example, the standard concept of the Reptilia is a paraphyletic group when recent and fossil organisms are included because some members of Reptilia gave rise to birds and some gave rise to mammals. But in a classification for just Permian organisms (before birds and mammals evolved), the same Reptilia might be monophyletic. Conversely, in a classification for just Recent organisms, we can classify extant birds, mammals, and reptiles without worrying about their ancestors (of course, certain reptiles may still be closer cladistically to certain nonreptiles than to other reptiles). In sum, having different classificatory systems for different time periods is a way to get around the necessity of classifying ancestors and descendants in the same classification.

A major problem with such a system is that it entails a number of different, complicated, complex classificatory systems that are not directly comparable. Patterson and Rosen (1977) have demonstrated that the practical complexity would be enormous. Further, within any one system, covering some period in time (perhaps on the order of tens of millions of years), problems of ancestors and descendants can still arise, although admittedly to a lesser degree than with a classification encompassing all organisms simultaneously. Furthermore, such a system will break up into arbitrary segments natural groups that do persist through time. In an extreme case, members of a single species-level lineage that crosses a period boundary may be classified differently in two different systems. From a practical point of view, such a system would eliminate much use of fossil organisms in biostratigraphic correlation. The geologic age of a fossil organism must be known before it can be classified in the appropriate system.

It has been suggested that all organisms can be classified in a single classification, but that such a classification should be based initially (and arbitrarily) on extant forms, and that extinct forms could then be fit into the classification using a set of conventions that apply only (or primarily) to fossil organisms (Patterson and Rosen, 1977). Nelson (1972a, 1972b, 1973) has noted that in the Linnaean hierarchy there are theoretically two ways that phylogenetic relationships might be expressed: through subordination of taxa (as is universally done and makes the Linnaean system a hierarchy) and through the ordering or sequencing of taxa (equally ranked subtaxa of a more-inclusive taxon). Classifications based on common ancestry (degrees of relatedness—phylogeny *sensu stricto*) but not on ancestors and descendants can be expressed completely using subordination alone (see above, and Patterson and Rosen, 1977). However, if we are dealing with a series of fossils, each of which alone might be the sister group of an inclusively larger set of organisms (as A and B) in our hypothetical example above, this scheme can lead to the proliferation of a large number of high-level taxa, each of which contains only one species or one species plus another, slightly lower, high-level taxon. In order to avoid this proliferation of categories and names, a series of low-level taxa can be sequenced such that, within a higher taxon, each low-level taxon in the sequence is the sister group of all of the remaining taxa. Such a scheme basically suggests first establishing a classification of Recent organisms on the basis of subordination alone and then fitting fossil organisms into the classification using a combination of subordination and sequencing in such a way as to try to minimize any changes in the initial classification based on recent forms. Of course, in such a scheme where equally ranked taxa are sequenced, one must indicate that the particular taxa in question form a sequence, rather than an unresolved polychotomy in the classification. The sequenced taxa should themselves each be monophyletic, except when we come to a sequence of species-level taxa. Then, excluding cases with apomorphies, we still run into the problem of whether any particular taxon is truly monophyletic, or potentially ancestral to its sister group.

As a refinement on Nelson's (1972*a*, 1972*b*, 1973) concept of sequencing fossil taxa, Patterson and Rosen (1977, p. 160, italics in the original) have proposed

that fossil groups or species, sequenced in a classification according to the convention that each such group is the (plesiomorph) sister-group of all those, living and fossil, that succeed it, should be called "*plesions.*" Plesions may be inserted anywhere (at any level) in a classification, without altering the rank or name of any other group. They may bear a categorical [i.e., Linnaean] name representing any conventional rank, from genus and species upward. . . , these ranks being those already existing in the literature, used only for reference and to avoid ambiguity. In effect, our proposal is that it should no longer be necessary to rank fossils formally, except within extinct monophyletic groups, where self-contained phylogenetic classifications may be built up.

It would seem somewhat arbitrary that the baseline for classification should be extant organisms. This sets up a fundamental distinction between extant organisms and organisms that happen to be extinct. Yet extant and extinct organisms do not differ (as far as anyone has ever demonstrated) from each other as biological organisms (see Van Valen, 1978). Recent organisms seem to be chosen primarily because they are what most biologists deal with, they have been classified longer than extinct organisms, and they are supposedly better known than extinct organisms. As stated in the quote above, the concept of plesions is intended by Patterson and Rosen (1977) to apply only to extinct species and monophyletic groups. All extant organisms are first to be classified using subordination. The equal ranking and listing of more than two subtaxa within a taxon would indicate an unresolved multichotomy (perhaps unresolved due to a lack of information, or unresolved because an ancestral species gave rise to more than two daughter species—there is no way ultimately to distinguish between the two cases; in order to do so, absolute knowledge of ancestors and descendants would have to be invoked). However, using a sequencing system for certain fossil taxa means that it is impossible to indicate unresolved multichotomies without introducing further notation. If, using the plesion system, there were an unresolved multichotomy, then the plesions at this level would be interchangeable (Patterson and Rosen, 1977, p. 161). One of the interchangeable taxa, usually the most complete or best studied, is included within the actual classification as a plesion, while the other(s) is relegated to the category of *incertae sedis* under the lowest named hierarchical category to which it can be allocated (this may be within a plesion that is composed of several species). For example, given the phylogeny shown in Figure 6-2, where the capital letters are extant species and the lowercase letters are extinct fossil species, a possible classification of only the recent forms would be:

Order I
  Family II
    Species A
  Family III
    Species E
    Species F

    Introducing the fossil taxa utilizing the plesion concept, our classification might be:

Order I
I *incertae sedis,* species b'
  Family II
    Species A
  plesion species b

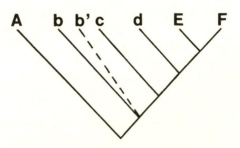

Figure 6-2. Hypothetical phylogeny of extant taxa A, E, and F and extinct taxa b, b', c, and d.

plesion species c
plesion species d
Family III
  Species E
  Species F

Or b and b' may be interchanged. The problem is that the above classification could equally well apply to the cladogram shown in Figure 6-3 where the multichotomy between a couple of extinct species and an extant group is at a different node. Patterson and Rosen (1977, p. 161) admit that the advantage of a cladogram over such a classification (using plesions as they propose) is that uncertainties in relationships, where taxa can be interchanged, are more precisely shown on the cladogram than in the classification. Indeed, if sequencing is used in a classification, the tacit assumption seems to be that a phylogeny can always ultimately be resolved into totally dichotomous relationships. Using a system of subordination only, multichotomous branches can be expressed precisely. Thus for Figure 6-2 a possible classification, treating all organisms (recent and fossil) on an equal basis, would be:

Order I
  Suborder II
    Species A
  Suborder III
    Infraorder IV
      Species b
    Infraorder V
      Species b'
    Infraorder VI
      Parvorder VII
        Species c
      Parvorder VIII

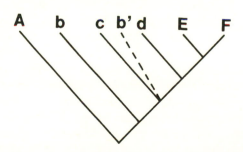

Figure 6-3. An alternative hypothetical phylogeny for the taxa shown in Figure 6-2; see text for explanation.

    Family IX
      Species d
    Family X
      Species E
      Species F

Such a classification may be considered burdensome with its numerous named and ranked taxa (and of course, there may be empty taxa, such as the genus and family required for each species, that have not been listed here), but it expresses precisely the relationships as known (or hypothesized to the best of our ability) of the taxa classified. Of course, using a system of sequencing, it could be indicated by using some convention, perhaps with braces, that several taxa are interchangeable or form a multichotomy. For example:

    plesion species b } interchangeable
    plesion species b' } interchangeable

could be substituted in the above sequenced classification corresponding to Figure 6-2 for plesion species b. Patterson and Rosen (1977) do make the important point that in many cases uncertainty as to the phylogenetic relationships of fossil forms in particular (which also may be true of some recent forms) exists because these are very imperfectly known and one should perhaps not try to classify them; rather, it is best to leave them in the category *incertae sedis*. The *incertae sedis* category is perfectly applicable in a classification based completely on subordination.

Plesions may or may not be monophyletic. Just like any other terminal taxa, if a plesion consists of two or more species, then these species must be united by a synapomorphy—must share an ancestor in common that is not shared with any other taxon—but if a plesion consists of a single species (perhaps represented by a single specimen) that does not exhibit any autapomorphies, it may potentially be an ancestor of its sister group, or it may not be (see comments above concerning this problem for terminal taxa at the species level). Patterson and Rosen (1977, quoted above) state that a plesion will be the primitive (plesiomorph) sister group of a group composed of, or based on, extant organisms (the actual sister group of the plesion may be comprised of extant and extinct forms). However, such a statement is meaningless for a plesion that is composed of more than one species or for a plesion composed of one species that bears autapomorphies. Of two monophyletic sister taxa (each of which is composed of more than one species united by synapomorphies or is a single species bearing an autapomorphy), one of which is extant and the other of which is extinct, the one that is extinct might be labeled a plesion just because it happens to be extinct. It is not labeled a plesion for any other reason (such as because it is a plesiomorph—unless a taxon is considered to be plesiomorphous merely by virtue of having had the bad luck to go extinct, but then we risk falling into adaptive/evolutionary scenarios) than that it happens to be extinct. If a plesion is a single species that is totally primitive relative to its sister group, then it may be characterized as the primitive sister group (plesiomorph), but these are not the only plesions. Effectively, the use of plesions merely excludes fossils from the necessity of being classified and ranked on the same level as extant organisms. The philosophical concept of plesions appears to be akin to the concept of having totally separate classifications for fossils and for extant organisms.

WILEY'S CONVENTIONS FOR ANNOTATED LINNAEAN CLASSIFICATIONS

Wiley (1979b; 1981, pp. 205–232) has explicitly formulated a series of ten conventions that he suggests be utilized in classifying organisms phylogenetically. These conventions cover many of the problems inherent in the phylogenetic classification of fossils in particular and are worth reviewing in detail here.

1. Wiley suggests that the Linnaean system and hierarchy be utilized as a base line, along with the following conventions, to classify organisms. As was discussed above, the Linnaean system has been adopted internationally; it appears to be too well ingrained to not utilize it in some form. Moreover, the Linnaean system works, and that is the ultimate reason it has lasted as long as it has and has been adopted as widely as it has.

2. Wiley suggests that any new classification should, as much as possible, be a modification of a commonly accepted existing classification. Redundant or empty categories (for example, a family composed of a single subfamily composed of a single tribe composed of a single genus with a single species) should not be introduced unnecessarily. Some redundant categories are necessary or useful conventions; Wiley (1981, p. 204) identifies the following five categories as either demanded by various codes or as generally useful to nonspecialists even if redundant: genus, family, order, class, and phylum. Thus, the zoological code requires that every species be assigned to a genus, and the concept of an order may be useful to nonspecialists (as when comparing orders of organisms) even if it is the sole order of a phylum and contains but a single species. Wiley (1981, p. 205) suggests that "natural taxa of essential importance to the group classified will be retained at their traditional ranks whenever possible, consistent with phylogenetic relationships of the taxonomy of the group as a whole." It is a little unclear what Wiley means by "of essential importance," obviously a subjective value judgment; presumably this would be well-defined monophyletic groups of extant organisms.

Thus the base line for any phylogenetic classification of a group that includes at least some extant members would be the traditional classifications of those extant members.

3. Wiley adopts the concept of sequencing of taxa (the sequencing convention) introduced by Nelson (1972*a*, 1973). Taxa "may be placed at the same categorical rank (or designated 'plesion' and indented [that is, lined up with other taxa of a given categorical rank]) and sequenced in phylogenetic order of origin" (Wiley, 1981, p. 209). When a list of taxa of the same categorical rank or a set of plesions is encountered in a classification constructed according to Wiley's conventions, it is understood that the first taxon in the list is the sister group of all of the remaining taxa of equal rank. In any individual case, it is up to the particular systematist to decide whether or not he or she feels that it is more advantageous to utilize the sequencing convention or to develop a hierarchical classification based on subordination. A sequenced classification may be much simpler, involving fewer categorical ranks and taxonomic names, and may superficially resemble a traditional, nonphylogenetic classification, but it has the distinct disadvantage of leaving many natural (monophyletic) groups unnamed. Furthermore, when the use of sequencing or subordination is left to the discretion of the particular systematist, given an agreed-upon phylogeny, there may be several equivalent, equally valid classifications that can be derived from the phylogeny.

4. As noted previously, if a sequencing convention is adopted, then either multichotomies (trichotomous or polytomous interrelationships) must be eliminated or a further convention to distinguish multichotomies from series of equally ranked and sequenced taxa must be adopted. In my previous discussion I suggested that the components of multichotomies might be labeled interchangeable. Wiley (1979*b*, 1981) has made essentially the same proposal, using the Latin term *sedis mutabilis* (of changeable position). At any point in a classification where there is an unresolved multichotomy involving categorically ranked taxa or plesions, all of the groups involved can be listed together and the term *sedis mutabilis* placed after each taxonomic name. Wiley (1981, p. 210) suggests that the taxa so involved should be given equivalent ranks, but traditionally ranked taxa and plesions can also be involved together in a multichotomy without having to apply equivalent traditional ranks to the plesions (the plesion being a variable rank that can be considered equivalent to any traditional categorical rank).

5. Wiley suggests that the traditional term *incertae sedis* can be used to place any monophyletic group of uncertain relationships within the hierarchy at some point where it is believed to belong. This may be a best guess or it may be that a relatively low level group shares synapomorphies with a relatively high level group, but the investigator cannot fit the low-level group in question into a cogent scheme of relationships with other low-level groups of the higher-level group. Thus, if Tribe A apparently belongs (on the basis of synapo-

morphies) with Tribes B, C, and D in Family X, and the relationships of the other tribes (e.g., (B, (C, D))) are well worked out relative to each other but it is not known where A fits into this scheme, then it may be classified as follows:

Family X
  X *incertae sedis:* Tribe A
    Subfamily Y
      Tribe B
    Subfamily Z
      Tribe C
      Tribe D

Plesions may also be inserted into a classification *incertae sedis* in the same manner.

6. Wiley notes that it may often be the case that there are paraphyletic or polyphyletic groups or assemblages, or groups of unknown status, that have traditionally been formally or informally named or classified. A systematist may not be able to resolve fully the relationships of all of the individual members of the group (perhaps the material or descriptions are inadequate) but may have a vague idea as to the relationships of some of the forms and approximately where they should be placed in the classification. Wiley suggests that such groups not be formally ranked or further named and that the traditional name for the group be placed in quotation marks and inserted in the classification at the appropriate hierarchical level in the status of *incertae sedis.*

7. Wiley discusses the problem of classifying fossil groups and decides that they should be classified with extant organisms, but in a different way. Wiley (1981, p. 205) redefines a plesion as "a name of variable rank accorded a fossil species or a monophyletic group of fossil species when classified with one or more Recent species or groups of species" and further states (Wiley, 1981, p. 219) that "the status of 'plesion' will be accorded to all monophyletic fossil taxa." Taking the last phrase literally, all classifications of fossil organisms would consist of nothing but either sequences of plesions or plesions within plesions, a suggestion that must be rejected as equivalent to not ranking taxa at all and eliminating any concept of subordination, except perhaps by some other convention, such as varying degrees of indentation in the written classification. However, in his examples Wiley (1981) appears to utilize the concept of plesion in the same sense and way as do Patterson and Rosen (1977; see discussion above).

8. Wiley (1981, p. 222) correctly points out that even if in most cases we can never absolutely distinguish or recognize a true ancestor, if we ever could (or if we should observe one species giving rise to others), there should be a way to classify a true ancestor. The taxonomic boundaries (the synapomorphies) that distinguish a true ancestor (a stem species: Hennig, 1966a) will be the boundaries that include all the descendant species, the higher taxon to which it gave rise; the stem species must be included in the higher taxon to which it gave rise. But the stem species will not be referable to any subgroups of the larger taxon (since it gave rise to all and is equally related to all). In this sense, the stem species will be equivalent to the higher taxon itself. Thus Wiley (1981, p. 223) suggests that a stem species of a taxon above the level of the genus should be placed in a monotypic genus and placed within the classification in parentheses beside the taxon containing its descendants. A stem species that gave rise to the members of a genus-level taxon should be placed in the genus-level taxon and placed within the classification in parentheses beside the genus-level taxon. Wiley suggests that if a species persists unchanged after giving rise to a certain taxon and then gives rise to another taxon, the same species may be placed in parentheses beside two different taxa within the same classification. For example, given the phylogenetic tree in Figure 6-4, where species A initially gave rise to species B (which then gave rise to numerous other species) but A remained unchanged, and later A gave rise to species C and D (each of which gave rise to numerous species), Wiley (1981, p. 225) would classify these organisms in the following manner:

Taxon X (A)
 Taxon Y (B)
  Taxon Q
 Taxon Z (A)
  Taxon R (C)
  Taxon S (D)

However, with the current state of methodologies of phylogenetic analysis, unless we could actually observe the evolution of the organisms, the above sequence of events could not be reconstructed. If species A remained unchanged (remained the same species and there were no other species between A and B, C, D), there would be no way to know that B branched off prior to C and D, that C and D branched off simultaneously, and that A went extinct at that time. There would be no basis for saying that C and D (and their clades) are more closely related to each other than either is to B (indeed, if they both descended simultaneously from an unchanged A and share no synapomorphies among themselves, they are no more closely related to each other than either is to B or A). The taxa A, B, C, D (and their descendants) would share only the synapomorphies of A in common, resulting in an unresolved multichotomy, A would appear to be relatively plesiomorphic relative to B, C, D, and thus we might wish to hypothesize that A is the ancestor of B, C, D. A classification of these taxa might be:

Taxon X
    Taxon including only A, *sedis mutabilis*
    Taxon Q (including B and close relatives), *sedis mutabilis*
    Taxon R (C and close relatives), *sedis mutabilis*
    Taxon S (D and close relatives), *sedis mutabilis*

or hypothesizing A as the stem species of taxon X:

Taxon X (A)
    Taxon Q, *sedis mutabilis*
    Taxon R, *sedis mutabilis*
    Taxon S, *sedis mutabilis*

I see no justification for considering one and the same stem species to have given rise to more than one other taxon without invoking a multichotomy.

    9. Wiley proposes a way to deal with taxa that are of hybrid origin. A hybrid taxon will be placed in the classification with one (or both) of the parental taxa, the names of the parental species that crossed will be placed in parentheses separated by an "×" (representing "hybridization between"), and the parental names in parentheses will be placed beside the hybrid taxon's name. Thus if taxon X is the result of hybridization between species A and B, it would be placed within a classification as such: Taxon X (A × B). This

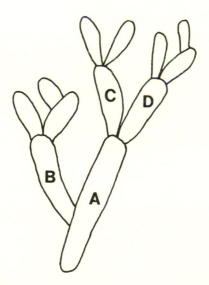

Figure 6-4. Hypothetical phylogenetic tree for species A through D and their unlabeled descendants.

convention can be used for species-level taxa that are the result of hybridization or for higher-level taxa that are thought to be the result of hybridization. A similar convention, but using a "+" sign instead of "×," can be used to indicate that a certain taxon resulted from symbiosis between two other taxa (the taxa that joined in a symbiotic relationship to form a new taxon may not be known [probably will not be known] at the species level, and thus supraspecific names may be used).

10. Finally, Wiley states that information other than genealogical information (which is already fully reflected in a phylogenetic classification), such as the biogeographic or temporal (stratigraphic) distribution of organisms, may be included within parentheses within a classification at the appropriate places. When dealing with sequenced taxa in a classification, where all of the natural taxa are not formally named, such information can be introduced between taxa within the sequence and would apply to all of the taxa within the sequence that follow the introduction of the information. Thus, in the following classification of species A, B, and C, based on the sequencing convention, where A occurs in North America, B occurs in South America, and C occurs in Asia, such information could be included as follows:

(N. Amer., S. Amer., and Asia)
Species A (N. Amer.)
(S. Amer. and Asia)
Species B (S. Amer.)
Species C (Asia)

Including much extraneous information in a phylogenetic classification soon becomes very cumbersome, and such extraneous information is not a part of the classification per se. Here it also can be noted that in many classifications involving both extant and extinct taxa, the extinct taxa are often flagged by a dagger (†) placed just in front of the name of the taxon.

RULES AND CODES OF NOMENCLATURE

The application of formal Linnaean names to certain groups (taxa) of organisms is governed by internationally recognized codes of nomenclature. These codes do not restrict the use of individual taxonomic judgment on the part of a particular systematist in deciding upon the composition of taxonomic groups to be formally recognized. But once it is decided what groups are to be formally recognized in a Linnaean classification, the codes of nomenclature provide rules for arriving at unique and universally accepted formal names to be applied to the taxonomic groupings recognized. Any investigator who becomes involved with the application of formal Linnaean nomenclature to taxa must consult the relevant codes; this section presents the briefest of introductions to this complex and highly legalistic sphere.

Today internationally recognized Linnaean nomenclature is governed by three independent and complementary codes (Nelson and Platnick, 1981; Wiley, 1981): *The International Code of Zoological Nomenclature* (1964, International Trust for Zoological Nomenclature, London; at this writing a new edition of the *Zoological Code* is due to be published at any time [*see* Melville, 1984]), *The International Code of Botanical Nomenclature* (1983, Bohn, Scheltema, and Holkema, Utrecht), and the *International Code of Nomenclature for Bacteria* (1969, 1971, *International Journal of Systematic Bacteriology* 16(4):459–490 and 21(1):111–118).

A recent discussion of the general principles upon which these codes are based and the rules that they legislate is provided by Wiley (1981, pp. 383–400). General principles governing the use of, and forming the nucleus of, all of the codes are as follows. All biological taxa are given Latin or latinized names. The name of a species (the basic category of the Linnaean hierarchy) is minimally a binomial composed of a generic name and a specific name. In some cases subgeneric names may be inserted between the generic and the specific name, and for a sub- or infraspecific taxon a trinomial (a sub- or infraspecific name placed after the specific name) may be utilized. Supraspecific taxa are given uninomial Latin or latinized names. Any particular taxon assigned to a certain hierarchical category, if governed by an applicable code, can have only one correct name (not all categories for all organisms are governed by a code, however). In general, unless there has been a ruling by an international commission to the contrary, the principle of priority is followed; the earliest proposed and correctly published ("available" [zoology] or "legitimate" [botany and bacteriology]) name is the valid or correct name for a taxon. What constitutes publication of a taxonomic name is specified by each code. Junior synonyms (younger names for the same taxon) are not generally used unless conserved at the expense of a senior synonym that is suppressed. For every group of organisms, the principle of priority extends back to a specified taxonomic work (designated in the relevant codes) whose publication is considered the starting point of Linnaean nomenclature for that group (Wiley, 1981). In zoology, most nomenclature begins with the tenth edition of C. Linnaeus's *Systema Naturae,* which is arbitrarily assigned the publication date of 1 January 1758. Fossil plant names are dated from G. K. Sternberg's *Flora der Vorwelt,* which is assigned the date of 21 December 1820. The nomenclature of various groups of Recent plants is dated from a number of different works, but many of the groups (including bacteria) mark their beginning with the publication of the first edition of C. Linnaeus's *Species Plantarum,* 1 May 1753.

Within the group of organisms governed by a particular code, two different taxa may not have the same name. However, as the codes are independent of one another, it is possible, for example, that the same name may be used for a taxon of plants and a taxon of animals. Certain hierarchical categories of taxa are established with reference to types. As Stoll et al. (1964, p. 59) state: "The 'type' affords the standard of reference that determines the application of a scientific name. Nucleus of a taxon and foundation of its name, the type is objective and does not change, whereas the limits of the taxon are subjective and liable to change." Every species-level taxonomic name is based on, or tied to, a particular type specimen or series of type specimens. Type specimens serve as objective reference points; in making taxonomic and nomenclatural decisions, an investigator must ultimately refer back to the original type specimen(s). Likewise, a genus-level name is based on a type species, and a family-level name is based on a type genus. Of course, ultimately these higher-level names refer back to a particular type specimen(s).

The code that governs the work of the majority of paleontologists is the *International Code of Zoological Nomenclature.* This code applies one universal system of nomenclature to extant and extinct taxa of animals (no distinction is made between living and fossil forms). The *Zoological Code* is concerned only with names of the family, genus, and species groups (i.e., from approximately the level of superfamily to the level of subspecies) and does not govern the formation and application of names above the level of the superfamily (ordinal-group and higher names). There are currently no internationally agreed upon rules governing the application of zoological names at the ordinal level and above. Rules and recommendations for the formation of names of taxa at certain levels governed by the *Zoological Code* include specified endings; for example, family names are formed by the addition of the suffix -idae to the stem of the name of the type-genus, and subfamily names are formed by the addition of the suffix -inae to the stem of the type-genus.

The work of systematic and nomenclaturally oriented paleobotanists is governed by the *Botanical Code.* This code does distinguish, to a certain extent, between fossil and living organisms. Paleobotanists often recognize form genera—fossil remains that are unassignable to a specific plant family—and organ genera—genera erected for specific fossil parts of plants, such as seeds, roots, or trunks. By using the concept of organ genera, the trunk and roots of a single fossil species of tree may be assigned to different genera and species. At present, form genera are recognized by the *Botanical Code,* whereas organ genera no longer are (Wing and Hickey, 1984). The *Botanical Code* governs the formation of names from the phylum or division level to subspecific levels, and particular endings are specified for many of the taxonomic names. For example, families of plants usually end with the suffix -aceae, and subfamilies end with -oideae.

Trace fossils (ichnofossils—traces of the life activities [movement, behavior, physiology, digestion, reproduction, and so on] of organisms left in sediments [e.g., tracks or burrows] or on hard surfaces [such as borings]) are sometimes classified and named using a distinct (independent of zoological or botanical nomenclature) and informal Linnaean-like binomial nomenclature composed of ichnogenera and ichnospecies (see Sarjeant and Kennedy, 1973). However, many paleoichnologists working with animal traces follow the rules of the *International Code of Zoological Nomenclature* (Basan, 1979). At present there is no formal, distinct, and generally adopted code of trace fossil nomenclature nor an international commission or trust for trace fossil nomenclature (Sarjeant, 1983). Of course, as far as phylogenetically based systematics and nomenclature is concerned, a trace fossil should be classified with the organism that is responsible for the trace. The very concept of trace fossils as a separate and distinct group of entities, including the traces of life activities of both plants and animals, is not natural in a phylogenetic sense.

In many cases it may be that trace fossils as well as many organic remains that the paleontologist may encounter (for example, sponge spicules, otoliths, conodont elements) are specifically indeterminate within the context of formal Linnaean nomenclature. Some paleontologists (e.g., Sarjeant and Kennedy, 1973; Vialov, 1972) advocate the use of an artificial classification (parataxonomy) composed of form genera and form species that mimics Linnaean nomenclature. Such elements as isolated sponge spicules, which are morphologically identical, are grouped together and given a single Linnaean-like binomial name even if it is fully realized that the grouped entities originate from distinct biological species. Such a system is merely a classification of convenience classes, but using Linnaean-like names can give such a system an aura of naturalness, legitimacy, and validity and may contribute to endless confusion between parataxa and natural taxa. It appears that a better way to handle such entities is to erect a pure convenience classification that does not mimic Linnaean nomenclature (e.g., sponge spicule, type 2a; otolith, type 3CB: see, for example, Tway and Zideck, 1982, 1983, who present a non-Linnaean utilitarian classification system for Late Pennsylvanian ichthyoliths [microscopic fish skeletal remains]).

In dealing with fragmentary fossil material, especially in conjunction with the older literature, one often runs across the case where a species has been erected on the basis of a type specimen that is now considered to be specifically indeterminate (in some cases, the type may presently be specifically indeterminate because part or all of it has been lost since the original description of the taxon). In such cases, "a name not certainly applicable to any known taxon" (Stoll et al., 1964, p. 151) is considered a *nomen dubium* (Simpson, 1945, uses the term *nomen vanum* for the same concept). A name that has been published without the required characterizations to fulfill the requirements of the applicable code, thus not making the name available or legitimate, is considered a *nomen nudum*. A *nomen oblitum,* a forgotten name, is a senior synonym of a name that has not been used in the primary biological literature for longer than a designated amount of time (specified by the applicable code) and thus loses its validity. *Nomina dubia, nomina nuda,* and *nomina oblita* are names that are commonly ignored in most instances (i.e., other than in comprehensive and monographic systematic revisions and nomenclatural histories of particular groups).

CHAPTER 7

The Importance of Phylogeny Reconstruction to Evolutionary Studies

The formulation of corroborated hypotheses of phylogenetic relationships between organisms is a fundamental goal of biological and paleontological research and needs no further justification (Eldredge and Cracraft, 1980, p. 241). Some would even argue that the central question of paleontology is: "What has been the history of life?" (Cracraft, 1981a, p. 456). Yet there is further justification for undertaking rigorous phylogenetic analyses. In themselves, phylogenies explain the distributions of characters only in terms of inheritance from common ancestors. Phylogenies per se do not explain the whys or hows of the origins of characters and traits. However, corroborated hypotheses of phylogenetic relationships (evolutionary patterns) can be used to investigate evolutionary processes and biogeographic and paleogeographic patterns and can possibly be used in practical biostratigraphy to correlate fossil-bearing rocks that may contain fossils of related groups, but share no identical species.

Above all else, phylogeny reconstruction searches out order among the diversity of organisms; this is one of the basic subjects of comparative biology (Janvier, 1984; Nelson and Platnick, 1981; Wiley, 1981). Such order as is discovered and elucidated can be conveniently expressed in phylogenetic classifications. Phylogenetic analysis distinguishes between real, natural (monophyletic) groups that are worth investigating and groups of organisms that are neither real nor natural, but merely taxonomic artifacts imposed upon the natural diversity by the investigator. The only thing that is real and non-arbitrary (nonsubjective) in a biological classification is the set of phylogenetic relationships reflecting the history of the organisms classified (Janvier, 1984). It is natural groups and relationships that provide the raw data for large-scale diversity studies and provide the evolutionary patterns against which to test predictions deduced from theories and hypotheses of evolutionary and bio-geographic studies.

In studying taxic diversity through time (e.g., Sepkoski, 1978; Van Valen, 1984), especially when using refined, quantified, mathematical characterizations, it is important that the taxa used be real units in nature—monophyletic groups in a strict sense. If the units are not natural entities, the validity and results of such studies are problematic (cf. Cracraft, 1981a). It is not enough to note that when using a large systematic data base some of the taxa represent the work of "lumpers" and some of the taxa represent the work of "splitters," and then to hope that the extremes balance out. The question is: Are the taxa used monophyletic (natural)? What does it mean to study the changing diversity within groups, or among groups, if the groups are paraphyletic—mere artifacts of a certain classificatory scheme based on a certain investigator's biases? Probably very little—or such studies may be positively misleading.

Another related problem is that of the equivalence or nonequivalence of equally ranked taxa, even if the taxa themselves are strictly monophyletic (Cracraft, 1981a; Novacek and Norell, 1982). As an example, an investigator may recognize four strictly monophyletic families (A, B, C, D). If the families originated simultaneously from a multichotomous speciation event (Fig. 7-1a), we are perhaps justified in considering the families equivalent and comparing them to one another. However, it may be that cladistically the families are related as follows: (A (B (C, D))) (Fig. 7-1b). In such a case each family is not equivalent to each of the others. Rather, A is the sister group of, and therefore historically equivalent to, the taxon (B, (C, D)), and likewise B is the sister group of, and equivalent to, (C, D). It is not enough to use only monophyletic groups; the classification must also be phylogenetic (sister taxa are given equal ranks). Unfortunately, all too often the ranking of taxa is based on historical tradition and the predilections of a particular investigator. Taxa assigned equal ranks (either within or between larger monophyletic groups) may not be historically equivalent; even if they are (i.e., sister taxa), the ranks assigned to them may be quite arbitrary. Also, there need not be any correlation between species diversity within a higher taxon and the rank of the higher taxon.

Phylogenetic patterns are subject to further analysis; from the study of a single monophyletic group an investigator may inductively arrive at general hypotheses of change, which can then be further tested against phylogenetic patterns for other, independent monophyletic groups. It is important that such corroborated hypotheses of relationships be arrived at with a minimum of assumptions and without incorporating putative data and hypotheses that the same phylogenetic patterns will later be used to test, in order to avoid circularity of reasoning. Thus, if we wish to test biostratigraphic and biogeographic hypotheses using phylogenetic patterns, we must initially reconstruct phylogeny without relying on stratigraphic and biogeographic data—that is, we must base our phylogenetic reconstructions solely on the intrinsic morphologies of the organisms involved. For example, if we want to test the hypothesis that potential ancestors always occur earlier (stratigraphically lower) than potential descendants, we will not base our putative transformation series or determinations of polarities of morphoclines on stratigraphic data. Likewise, if we wish to test various hypotheses of dispersal or vicariance to explain biogeographic patterns, we will not initially use biogeographic assumptions in order to determine the phylogenetic relationships of the organisms involved. The basic data of vicariance biogeography is independently corroborated phylogenetic hypotheses for various monophyletic groups of organisms. If we wish to test the hypothesis that the members of a particular monophyletic group exhibit an evolutionary trend in a certain character, then we must construct our phylogeny for the organisms under consideration on the basis of other characters, independent of the one for which a trend is hypothesized. To give one more example, if we wish to explore and test hypothesized evolutionary

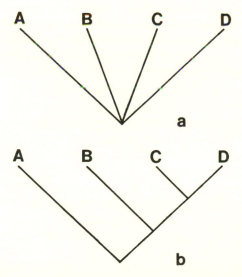

Figure 7-1. Two of the many possible interpretations of the phylogenetic relationships of "families" A, B, C, and D.

processes (evolutionary theory) using phylogenetic patterns, we cannot assume the hypothesized evolutionary process that we wish to test when reconstructing phylogeny.

Eldredge and Cracraft (1980, p. 242) have noted that for too long, and in too many cases, the study of evolutionary theory "has been plagued by an assumption-ridden inductive process." Typically, an investigator may observe the sequences of fossils in local stratigraphic sections, from this arrive at a suggested phylogenetic (evolutionary) story for the group under consideration, next hypothesize a narrative explanation for the evolution of the group (usually based on prior assumptions concerning natural selection and adaptation, and the extrapolation of genetical and microevolutionary processes occurring within populations in ecological time to the macroevolution of species in geological time), and finally find further cases that are made to fit the scenario (based on prior assumptions and notions of process). If another case is not viewed as readily compatible with the first case, then a combination of ad hocisms and another scenario can be formulated to explain the second case. In opposition to the methodology outlined above is the approach espoused here (and also by Eldredge and Cracraft, 1980; see also Schoch and Meredith, 1984). We must formulate well-corroborated hypotheses of evolutionary relationships (historical patterns) that are independent of the evolutionary assumptions and hypothesized evolutionary mechanisms we want to test. Any scientific hypothesis of an evolutionary mechanism must make predictions; these predictions deduced from the hypotheses of mechanisms can then be tested against the independently formulated historical patterns. Unfortunately, as Rosen (1984, p. 92) notes, most current theories of the processes and mechanisms of evolution are not sufficiently deterministic for this purpose; "few, if any, phylogenies are incompatible with the proposed mechanisms." Evolutionary theory in its current state serves primarily as an attempt to explain historical (phylogenetic) events, but not to predict them (Scriven, 1959).

PHYLOGENY AND THE ANALYSIS OF EVOLUTIONARY MORPHOLOGY

Phylogeny has a potentially important role to play in the analysis of evolutionary morphology (i.e., the reasons for the forms of particular organisms, essentially equals "functional morphology" according to Fisher, 1985). As Lauder (1981; see also Lauder, 1982) has pointed out, three basic factors are often invoked to explain organic form and structure: functional (adaptational considerations: e.g., Rudwick, 1964), fabricational (adaptively neutral by-products of morphogenesis or developmental processes: Seilacher, 1970, 1973, 1979; see also Rachootin and Thomson, 1981, p. 183), and historical (an organism is constrained in its morphology due to its inheritance) factors.

Of these, historical aspects have received the least attention (Lauder, 1981). We may wish to correlate organismal structures with specific environmental adaptations and use as evidence for such a correlation the fact that within a certain monophyletic group those organisms with similar structures are found in similar environments. However, we must have a well-corroborated hypothesis of relationships for the taxa in question before drawing such conclusions. Similarities in structure and occupied habitat of two organisms may be the result of common ancestry rather than adaptations of the organisms in question per se (the similarities in structure might not even reflect adaptations to the environments presently inhabited: Lauder, 1981, p. 431). On the other hand, if the organisms under consideration are only distantly related yet share common structures and environments, perhaps the similarities in structures call for some kind of additional (nonhistorical, such as functional/adaptational) explanation. As Lauder (1981, pp. 431–432) aptly states,

A phylogenetic hypothesis allows the reconstruction of the historical sequence of structural change through time and thus serves as a null hypothesis from which significant deviations may be detected and indicates the appropriate level of generality at which structure-environment correlations must be explained.

From a phylogenetic hypothesis we can reconstruct the relative order of acquisition of certain derived features within a monophyletic group. If some kind of functional or adaptational explanation can be hypothesized for these features that corresponds to some environmental change, then we may attempt to predict from a phylogeny the relative sequence of environmental changes that a group underwent. This prediction may be potentially testable by the independent reconstruction of a sequence of environments in the rock record.

Lauder (1981, p. 433, italics in the original) proposes that

the goals of a historical approach to structural transformation are (1) to generate testable historical hypotheses about the way in which form changes and (2) to discover general regularities in the transformation of organic design. This research program focuses on the limits of change imposed by structural, functional, and epigenetic interactions within the organism [cf. Rachootin and Thomson, 1981] and on a search for *generalized historical pathways (tracks) of structural change.* . . . Historical analysis in morphology becomes the search for general patterns of structural transformation which may apply to a wide variety of organisms.

Such a research program must of course be based on rigorous, well-corroborated phylogenetic hypotheses.

Lauder (1981) contends that emergent, general features of design in a stem species (at the base of a monophyletic group) may influence the subsequent evolution of the group in a lawlike way. Such hypotheses can potentially be tested by studying and comparing the historical pathways of change (the sequence of change) and the morphological diversity in independent lineages (independent monophyletic groups) that are judged to be characterized by similar emergent, general features. (Lauder [1981, p. 436] is careful to point out that these are not unique and specific attributes of lineages whose effects can be tested, but only generalized features.)

Lauder (1981) presents several examples to illustrate these types of hypotheses. Decoupling of the elements of a primitively constrained functional/ mechanical system may allow for increased morphological diversity in derived forms. Similarly, metamerism and the duplication of parts in an organism may allow for increased morphological diversification in a lineage. Complexity of organization may influence patterns of morphological change—perhaps "more complex systems can have more potentially stable intermediate states and thus more options for change in design" (Lauder, 1981, p. 439).

MACROEVOLUTION AND PHYLOGENY

The concept and science of evolution is often divided into two realms or areas of study and theory: microevolution and macroevolution. Microevolution is usually considered to be evolution below the species level, such as within populations, subspecies, and species. Macroevolution is usually viewed as evolution above the species level, such as the evolution of one species into another and the origin of higher taxonomic groups, such as genera, families, orders, phyla, and kingdoms. (Some paleontologists and biologists use these terms differently. Clarkson [1979], for instance, considers microevolution to be the "origins of new species from existing ones largely through changing interactions between the animal and its environment" [p. 31], macroevolution to be "broadly synonymous with adaptive radiation: the diversification of an initial stock into smaller stocks, each becoming adapted to its own environment" [p. 32], and megaevolution as "the origin of new systems of animal organisation" [p. 34].) Various investigators hold quite different opinions about the connection and relationship between microevolution and macroevolution. Classically, as Eldredge and Cracraft (1980) have recently discussed, many workers have indulged in across-the-board extrapolations from microevolutionary phenomena to macroevolutionary phenomena. In this sense, macroevolution is viewed as nothing more than a large amount of microevolutionary change, or microevolutionary change accumulated over geologic time (cf. Bock, 1979).

Microevolution has been well studied and continues to be studied, in the laboratory, in natural populations that can be monitored by investigators, and through mathematical modeling. At present the basics of microevolutionary change appear to be understood; microevolution deals directly with changes in gene frequency and content in organisms making up populations over time and usually involves adaptation via the mechanism of natural selection. As Eldredge and Cracraft (1980) note, the study of microevolution requires continuous within-population, generation-by-generation data that is not generally available to the paleontologist. The fossil record is not complete enough to pursue true microevolutionary studies, and we cannot always be sure what the selective regimes were in any particular case. Those who view macroevolution as simply a large amount of microevolution tend to extrapolate microevolutionary processes that occur in ecological time to explain macroevolutionary patterns that occupy geologic time. Thus we can attempt to explain the evolution of a group of organisms solely through the standard processes of microevolution, especially natural selection resulting in increased fitness and adaptation of individuals to their environments.

Alternatively, it has been forcefully argued (see Eldredge and Cracraft, 1980, for a review) that it is not valid to extrapolate microevolutionary, within-population phenomena to the macroevolutionary level. Macroevolution, as far as this argument is concerned, involves not only the progressive adaptation (or at least, change even if nonadaptive) of a population, but the process of speciation—the splitting of lineages. Microevolutionary theory of within-population phenomena leading to increased adaptations, via natural selection in particular, does not appear to cause speciation per se (although once speciation has begun, natural selection may reinforce it, especially if hybrids are of reduced viability). Speciation does not appear to be fundamentally a process of adaptation (although once it has occurred, it may have adaptive value). Rather, the splitting of lineages—speciation—may be the result of some larger forces, external to the organisms concerned, at work (for example, vicariance events). There are thus two distinct phenomenological levels to evolution: microevolution involving changes within a population or set of populations (species) and macroevolution involving the splitting of lineages (speciation and the origin of higher taxa; all higher taxa are the result of splitting events at a species level). Eldredge and Cracraft (1980) have equated this phenomenological dichotomy with their view that species are discrete individuals unlike either populations below the species level or higher taxa above the species level (which these authors regard as real but historical entities rather than as individuals per se). Accordingly, Eldredge and Cracraft (1980, p. 277) define macroevolution as "change in species composition within a monophyletic group in space and time."

As I view it, the fundamental point is that macroevolution involves lineage splits; it does not matter whether one views species as of the same status as any other monophyletic groups or as special individuals. According to my views, macroevolution is qualitatively, not just quantitatively, different from microevolution. Microevolution (via natural selection) is usually confined to levels below the species level, and macroevolution usually involves levels at and above the species level (namely, splitting of species, speciation). Macroevolutionary (splitting) processes may originate externally (extrinsically) relative to a species (e.g., vicariance events), or they may originate within a species (at the species level); they may not always result in successful speciation. It is perhaps conceivable that there may be some way to have speciation in sympatry, truly initiated by some form of microevolutionary process that leads to macroevolutionary change (splitting: see chap. 2); therefore, splitting and macroevolution in such a case may be the direct result of what would be considered microevolution. However, a macroevolutionary splitting event may be imposed upon a species despite microevolutionary processes "opposing" such a splitting event (e.g., stabilizing selection that tends to promote uniformity among the individuals belonging to the species). In general, macroevolutionary processes may be equated with vicariance events and the like—events that are outside the intrinsic bounds of the species as a biological entity. In contrast, microevolutionary processes are partly, or wholly, intrinsically based; these include adaptation via natural selection, social selection, genetic drift, and the like. It must also be realized and remembered that any phylogeny, any genealogy or evolutionary history, may have a random component. Indeed, as I perceive much of macroevolution, the process would be random as far as the organisms involved are concerned although deterministic at other levels (that is, the speciation events are directed by other, nonbiological processes such as plate tectonics and accompanying environmental changes; earth and life evolve together).

To reiterate somewhat, various theories of macroevolution, as expounded by various workers, have usually taken one of two basic approaches. (1) Traditionally there has been a direct extrapolation of microevolutionary mechanisms and processes to account for macroevolutionary patterns (what Eldredge and Cracraft, 1980, refer to as the transformational approach). (2) Some authors have distinguished, or decoupled, microevolution from macroevolution and argued that it is not valid to extrapolate from one level of phenomena to the other. How does phylogeny reconstruction bear on theories of macroevolution? Any analysis of phylogeny results in a pattern of macroevolutionary relationships between taxa and ultimately reflects speciation events—that is, the material and probable mechanism of macroevolution. Any theory of macroevolution should make predictions about the resulting relationships of species, their characteristics, and their distributions in space and time. Thus, the historical patterns that result from phylogenetic analysis provide the tests of macroevolutionary theory. Such patterns do not test microevolutionary theory directly (as noted above, this requires generation-by-generation data), and they also do not test hypotheses concerning mechanisms of speciation (for that, one needs well-corroborated trees involving ancestors and descendants; these trees may be derivable, however, given certain assumptions or additional information, from phylogenetic hypotheses). Following is a brief review of some theories of macroevolution, most of which are no mere extrapolations of microevolution.

THEORIES OF MACROEVOLUTION

Numerous possible mechanisms and scenarios have been proposed to explain macroevolutionary change, both in the sense of speciation and evolution above the species level, and in the sense of obvious, radical changes or jumps in morphology or adaptation between individuals not separated by more than a relatively few number of generations. Besides the textbook models of vicariance and allopatric speciation, different hypothetical mechanisms that have been proposed include: genetic assimilation; heterochronic changes; developmental "preadaptive potentialities" latent in an organism and demonstrable experimentally; "hopeful monsters" where intermediate stages may be difficult or impossible to imagine (a bone is either one piece or two, there being no intermediates: see, for example, Frazzetta, 1970, 1975); maintenance of nonadaptive phenotypes at low frequencies in populations of species which, given the right environmental conditions, may suddenly be selected for; genetic revolutions in small, peripheral, isolated populations; speciation by flush-crash-founder cycles under reduced selective pressure; speciation and morphological changes brought about by hybridization (which seems to be especially common in plants and may also occur in animals) or endosymbiosis; speciation by polyploidy (also common for plants and demonstrated for some lower vertebrates); and epigenetic scenarios (Brooks, 1983; Goldschmidt, 1940; Ho and Saunders, 1979, 1984; Pollard, 1984; Rachootin and Thomson, 1980, 1981; Ridley, 1984; Sober, 1984b; Stanley, 1979; Telford, 1984; Van Valen, 1974). These possible mechanisms have been incorporated into numerous general theories of macroevolution. Recently Vrba (1982a, 1982b, 1983, 1984a; Vrba and Eldredge, 1984; see also Cracraft, 1982) has classified and discussed some of the more common theories of macroevolution. Here it is worth summarizing, and adding to, Vrba's analysis.

Vrba (1984a) attempts to make a fundamental distinction between those theories of macroevolution that are based on, or stress, organism-level processes and those theories that are based on species-level processes. In fact, however, various macroevolutionary theories range from one extreme to another. At one end of the spectrum, certain workers in the synthetic tradition consider the sorting of phenotypes at the organismal level to be the primary driving force of evolution at all scales. In contrast, sorting of phenotypes at the organismal level may be viewed as relatively inconsequential as far as macroevolution is concerned; macroevolutionary patterns may be attributed to the random, or nonrandom (e.g., species-selection), differential species survival within and among clades. Brief descriptions of and comments on ten broad approaches to a theory of macroevolution follow; seven of these theories (all but those of nonequilibrium evolution, devolution, and stressed environments) have been previously reviewed by Vrba. Due to space limitations, it is not possible to do justice to macroevolutionary theory here; for more detailed descriptions of these various theories, hypotheses, and models, the reader is referred to the papers cited in the text. As Vrba (1984a, p. 126) notes, proponents of any of these models acknowledge that more than one process or factor "may be acting simultaneously to produce a given macroevolutionary pattern."

   1. *Intrinsically directed organismal variation.* Certain investigators (e.g., Riedl, 1977, 1978; Zuckerkandl, 1983) have suggested that the genetic and epigenetic systems of organisms might intrinsically direct

the introduction of phenotypic variation at the organism level. . . . A variety of mutational combinations will lead to the same result in gene expression. The variety of morphological phenotypes derivable from change in cellular interactions is also severely restricted. Consequently the probability that similar phenotypic characters should appear repeatedly within one species or in closely or even distantly related species is high. (Vrba and Eldredge, 1984, pp. 162–163—cf. Roth's [1984, discussed in chap. 3] problem of distinguishing homology at the species level.)

It is suggested that this independent arising of certain phenotypic characters from an intrinsic genetic and epigenetic base might give rise to parallel macroevolutionary trends (cf. traditional parallelism, homoplasy). Riedl (1978) and Zuckerkandl (1983) are particularly concerned with gradual morphologic change within lineages.

   2. *Nonequilibrium evolution.* Considering the evolutionary process to be a consequence of particular physical properties of living systems, Wiley and Brooks (1982, 1983; Brooks, 1983; Brooks and Wiley, 1984) have suggested that evolution may be a nonequilibrium entropic phenomenon that follows the second law of thermodynamics. "Nonequilibrium evolution considers descent with modification to be an intrinsically generated, spontaneous process that will take place in the *absence* of ecological interactions" (O'Grady, 1984, p. 573; italics in the original). According to nonequilibrium evolutionary theory, the hierarchical ordering of organismal morphology was created or intrinsically generated by "a basic instability of the organisms involved" (O'Grady, 1984, p. 574). Order is not (or need not be) imposed externally (by a process of extrinsic effects such as natural selection), but is an emergent property of the organisms. Organisms may not evolve structures in order to perform certain functions, but may be able to perform certain functions because they possess certain morphologies. Phylogenetic history and the resulting structures may be prior to function and not vice versa (Brooks, 1983; O'Grady, 1984). In part, the susceptibility of a species to speciation may be determined by the degree of morphogenetic complexity of the species (Cracraft, 1982; Wiley and Brooks, 1982).

   3. *Synthetic hypothesis.* The traditional synthetic hypothesis (e.g., Simpson, 1953a; Bock, 1979), essentially that macroevolution is nothing more than an accumulation of microevolution, has been discussed to some extent above. Phenotypic variation at the organismal level is the result of mutations and recombinations of genes; sorting (via natural selection or random processes) among organismal phenotypes is the cause of evolution at both lower and higher levels. The primary emphasis of the synthetic theory is on intrapopulational natural selection and adaptation.

It should be mentioned here that Brooks and Wiley (1985; see also Nelson, 1985) have suggested that three basic beliefs or notions commonly held by adherents of the synthetic theory of evolution have been generally falsified by systematic (specifically phylogenetic systematic) analysis. These three notions are: (1) the discreteness and naturalness or reality of grades as taxa (cf. chap. 3); (2) the belief that the morphology of organisms often correlates better with functional considerations than with heritage or phylogenetic (historical) considerations (cf. Lauder, 1981, discussed above); (3) systematic relationships among organisms are random with respect to geography (cf. Croizat, 1964, and chap. 5).

4. *Devolution.* Meredith (1982) and Schoch and Meredith (1984; Meredith and Schoch, 1984) have proposed a theory of macroevolution based on the concept that there are two primary reasons that species go extinct (excluding events random to the organisms, such as sudden environmental catastrophes): incompetent species that are not sufficiently adapted to their environment (they may fail to keep abreast of constant environmental change) may suffer traditional Darwinian scarcity-extinction; and competent but unrestrained species may exhaust the resources of their niches, undergoing a boom-and-crash cycle that may lead to their own extinction (termed success-extinction). According to this theory, it is suggested that natural selection acts only, or at least primarily, on individual organisms, increasing their reproductive performance despite any effect this may have on the long-term survival of the population and species. Individuals of a successful species can become extremely well adapted and abundant and then suddenly exhaust their resources and bring about the extinction of their own species and subsidiary species.

This theory of macroevolution can be summarized as follows. Evolution in general, which is brought about (or directed by) Darwinian natural selection, may well consist of two separate processes, at two distinct hierarchical levels, which balance each other. At the level of the organism, microevolutionary selection acts on the level of individuals, and ultimately genes, to maintain and increase the reproductive success, and adaptiveness, of particular genotypes and culminates in the survival of the fittest individuals. The tendency for individuals to maximize their impact on the niche (in terms of increasing numbers and/or increasing use of resources; through, for example, increasing the size of individuals) will directly affect organismatic evolution at the level of the species. As has long been realized (see Schoch and Meredith, 1984, for further discussion of this point), if individuals are not at least marginally fit, they and their species will suffer Darwinian scarcity-extinction. But it can be argued that if individuals are successful in the fitness game, they will continue to multiply and impact their niche (natural selection on the level of the individual lacks foresight) until they exhaust all the resources upon which they are dependent. The population may consequently suffer a partial or complete collapse. Ultimately, a complete collapse may cause the extinction of the species. Thus, processes at the level of individual organisms, acting toward the perpetuation of the individual and its progeny, can ultimately cause the demise of the species. It is this process of the extinction of species composed of ultimately adapted individuals that Meredith and Schoch (Meredith, 1982 Schoch and Meredith, 1984) call macroevolutionary deselection.

Thus, there are two aspects to natural selection: (1) Microevolutionary selection selects for fitness among individual organisms and eliminates the relatively unfit; (2) macroevolutionary deselection acts to eliminate species composed of individuals with a runaway fitness (over-adapted individuals). These two aspects balance one another, and the net result is that species that persist through geologic time are those that are composed of mediocre individual organisms. This has been termed the special theory of devolution (Meredith 1982): *special* because it is a corollary of, and does not contradict, Darwinian evolutionary theory, and *devolution,* recognizing the transference of a niche or ecospace from one species to another as the species with the fittest (overly fit) individuals are eliminated.

5. *Directed speciation.* Certain investigators (e.g., Alberch, 1980; Oster and Alberch, 1982; Rachootin and Thomson, 1981) have elaborated on the concept of canalization and suggested that unexpressed epigenes may be present in organisms. Phenotypically, this would suggest morphological equilibrium for long periods of time interspersed with a large amount of change (phenotypic revolutions) when change does occur. Moreover, when change does occur it may be directed into certain morphological paths, possibly giving rise to an observed pattern of intrinsically directed speciation.

6. *Effect hypothesis.* The effect hypothesis of macroevolution was introduced by Vrba (1980) and subsequently has been elaborated on by he (1982b, 1983).

The basic statement (Vrba, 1980, 1983) is this: Differences between lineages (of sexually reproducing organisms) in characters of organism[s] and their genomes may incidentally determin a pattern of among-species evolution in a monophyletic group. Such characters can only b spoken of as collective species characters. Processes at the among-organism level (directly b upward causation) cause speciation, characteristic speciation rate (S), and species extinction rat (E) in a lineage. Thus, any characteristically different probabilities of S and E in different subclade of a monophyletic group will lead to a trend toward higher intrinsic rate of species increase (R where $R = S - E$, as in Stanley 1979). If a pattern of punctuated equilibria predominates i such a phylogeny [see chap. 3], then a trend in directional phenotypic evolution will result a well, which is not adaptive at the species level. (If such a trend is toward an increase in specie survival, then it represents adaptation at the level of organisms. But if the trend is powered b increasing S, then it is not adaptive at any level.) The effect hypothesis embodies a controversial new idea: that life's diversity patterns, and through them some long-term directional tendencies, may result as incidental, nonadaptive consequences. (Vrba and Eldredge, 1984, p. 164; repeated with minor changes in Vrba, 1984*a*, p. 128).

Prior to Vrba's (1980) introduction of the effect hypothesis, differential survival of species was regarded as due primarily to higher extinction rates of less well adapted species (e.g., Stanley, 1979). Serious consideration was not given to the possibility that higher speciation rates per se (as compared to extinction rates) might produce macroevolutionary patterns and trends (Vrba, 1983). Vrba (1983) suggests that the effect hypothesis differs from species-selection (see below) hypotheses (e.g., Eldredge and Cracraft, 1980; Stanley, 1975, 1979) in that in species-selection species are selected (survive) as a result of their superior aptations (= adaptations plus characters currently useful to the organism or species that did not arise under the pressure of natural selection for their current use: see Gould and Vrba, 1982). In contrast, according to the effect hypothesis, "Darwinian selection of organisms and other lower level processes simply determine differential R [the net rate of increase of species in a monophyletic group] and patterns among species, called effect macroevolution" (Vrba, 1983, p. 388).

7. *Random hypothesis.* Using computer simulations, Raup and others (1973) were able to reproduce realistic-looking macroevolutionary patterns by allowing frequencies of speciation and extinction events to vary randomly. The point seems to be that in some cases for some particular macroevolutionary patterns there may be nothing to explain (see Schoch and Meredith, 1984). One must eliminate the possibility that certain macroevolutionary patterns are the result of stochastic processes before hypothesizing more elaborate explanations.

8. *Species selection.* The most common and popular theories of autonomous macroevolution (i.e., distinct from microevolution) are those that hypothesize or propose some form of species selection. That is, there may be actual interspecific competition between separate species, with the result that some species survive and some go extinct (Eldredge and Cracraft, 1980, p. 274; see also Vrba, 1984*b*). Or, certain species may be selected over other species (survive relative to other species) due to other selective factors (perhaps of the physical environment). Stanley (1975) and Gould and Eldredge (1977) treat species selection as analogous to natural selection among individual organisms. Speciation is often viewed as essentially random relative to the clade; species are comparable to mutations among individual organisms, and it is species that provide the variation on which species selection acts. Superiorly adapted, or more fit, species survive, while inferiorly adapted species go extinct. If the forces of species selection act in a constant direction over a long period of time (over the duration of many speciation events), a trend among species in a clade may result. Gould (1982) has also suggested that differential fre-

quencies of speciation against a background of constant extinction may constitute a form of species selection (cf. Vrba's effect hypothesis discussed above).

9. *Extrinsic control of macroevolution.* Cracraft (1982, 1983a) in particular has suggested that macroevolutionary patterns, specifically differences in speciation and extinction rates between clades, may be the result of the differing climatic and geologic histories that the clades have undergone. Cracraft (1982, p. 363) summarizes his hypothesis as follows: "Speciation rate is regulated by environmental complexity, and extinction rate by environmental favorableness." This suggestion is similar to my own suggestion (arrived at independently and further discussed in the next section) that vicariance events may be the primary cause of many of the macroevolutionary patterns we observe. Or, as Croizat (1964) has suggested, the earth and its life have evolved together. Also important to mention here is Valentine's (1973) concept that changes in the worldwide diversities of species (particularly in the marine realm) are related to plate tectonics and changing paleogeography. The gist of the theory is that, in general, uniform, equable conditions should lead to low species diversity, whereas nonuniform climatic and paleogeographic conditions should lead to greater provinciality and diversity of biotas (cf. Kurtén, 1967).

10. *Stressed environments.* Recently Jablonski et al. (1983; see also Jablonski, Flessa, and Valentine, 1985; Lewin, 1983; Moore, 1983; Zinsmeister and Feldmann, 1984a, 1984b; cf. Hickey et al., 1983, and Flynn, MacFadden, and McKenna, 1984) have suggested that species-poor, stressed environments may be the primary area or source of major (macroevolutionary or megaevolutionary) evolutionary novelties. Specifically, Jablonski et al. (1983) demonstrated that among Phanerozoic marine continental-shelf communities, ecological innovations first appeared among the environmentally unstable onshore communities and then spread to the offshore communities. This phenomenon occurred even though more stable offshore communities are generally characterized by greater species diversity and by higher speciation and extinction rates (P. W. Bretsky and Lorenz, 1970). From this observation, one can generalize that major evolutionary novelties (those that lead to distinct, new adaptations and organismal organizations—not just variations on an established theme) in general may arise in species-poor, environmentally stressed environments (Lewin, 1983). This idea goes against the conventional wisdom that areas of rich species diversity (such as the tropics, with little environmental stress) harbor a large amount of evolutionary potential and should be the source of major evolutionary innovations. It also suggests that perhaps not all speciation events are equal; some speciation events are more likely to give rise to major evolutionary innovations than others.

Jablonski et al. suggest two alternative mechanisms to account for the patterns they observe. First, the greater resistance of nearshore species and clades to extinction may permit certain novelties to exist long enough to diversify into major evolutionary innovations and may increase "the total number of speciation events within a clade over its lifetime" (Jablonski et al., 1983, p. 1124: note that here it would seem that these authors include within the term *clade* paraphyletic groups that could each exist for a certain amount of time, giving rise to species and then itself going extinct while its descendants persist). Second,

Although speciation rates are lower onshore, the temporal and spatial heterogeneity of nearshore environments may be conducive to the production of evolutionary novelties or new ecological associations; new community types could expand across the shelf in the wake of attritional extinction of offshore taxa. . . . The incessant local extinctions and recolonization in frequently disturbed nearshore habitats may promote the origin of major new community types through repeated sorting and recombination of new and established species. Alternatively, the evolutionary novelties themselves may arise preferentially nearshore because new isolates in those habitats are commonly small and drawn from panmictic populations, and are thus more likely to undergo genetic revolutions or transiliences that could produce rapid shifts in morphology or physiology than the more frequent speciation events in offshore environments. (Jablonski et al., 1983, p. 1124 and footnote, p. 1125)

Related to the topic of macroevolution, but far beyond the scope of this book, is the question of the causes of the extinctions and revolutions observed in the fossil record. Numerous ideas invoking both deterministic and stochastic, terrestrial and extraterrestrial explanations, have been put forth concerning this topic. Periodic mass extinctions may help to control, drive, regulate, or influence macroevolutionary patterns through differential survival and extinction of various types of organisms during such crises (Gould, 1985; Gould and Calloway, 1980; Jablonski, Flessa, and Valentine, 1985). For recent reviews and a sample of the diversity of current thinking on this subject, the reader should consult Ager (1976), Benton (1985), Hoffman (1985), Kitchell and Pena (1984), Torbett and Smoluchowski (1984), and chapters by various authors in Bendall (1983), Berggren and Van Couvering (1984), Hallam (1977), and Silver and Schultz (1982).

Eldredge and Cracraft's Theory of Macroevolution

In their book, Eldredge and Cracraft (1980) consider species to be individuals (see discussion in chap. 2) and advocate the concept of species selection. In this way these authors decouple macroevolution from microevolution and the concept of adaptation via natural selection. They then attempt to conceptually reunite micro- and macroevolution (as expressed through species diversity) into one cogent theory of evolution.

Eldredge and Cracraft (1980, p. 302) argue that "traits of organisms are either directly related to the occupation and exploitation of the population's niche, or, at the very least, neutral (irrelevant) with respect to the occupation of the niche." I believe that such an assumption is a fundamental error. Most species-level or higher traits of any particular organism are due to inheritance from an ancestor, and that ancestor may well have occupied a different niche than the descendant species. Low-level traits that are detrimental or irrelevant may be modified by microevolutionary processes, but many fundamental traits (in particular those that are used to reconstruct phylogeny) are there because they were present in an ancestor. It is questionable if the traits, when they arose de nova, always arose as adaptations. Many traits may have been fixed randomly or coincidentally in small, isolated allopatric ancestral populations that happened to survive. In many cases there may have been two or more ways to do something equally efficiently, two or more characters that were equally adaptive, and it is simply by chance that one continued to evolve and survived to be passed down to ancestral species. In some cases organisms and species may continue to exist despite their inheritance of slightly inadaptive (handicapping) characters from an ancestral form (Scriven, 1959). As another fundamental assumption of their theory of macroevolution, Eldredge and Cracraft (1980, p. 304) state the admitted tautology that "the number of species present is equal to the number of niches realized (occupied and exploited in a given area)."

Eldredge and Cracraft (1980) point out that given a certain phylogenetic hypothesis for a monophyletic group of organisms, one can state, for certain features or traits, which species are relatively apomorphic or plesiomorphic. One can then attempt to correlate the relative apomorphy with adaptive specialization and the relative sizes of niches. Organisms that have a large tolerance to various ecological conditions, are generalists, and are relatively widespread are often referred to as eurytopes. Organisms that are relatively narrowly adapted, are specialists, and have a restricted distribution are referred to as stenotopes. Of course, these various traits do not always correlate with one another. Thus, a species may be composed of generalists and yet have a restricted distribution (be endemic to a certain area), and likewise an extreme specialist may have a cosmopolitan distribution. But it has been suggested that there is a significant correlation between relative niche breadth and species distribution, both stratigraphically and geographically (Eldredge and Cracraft, 1980, p. 307).

Relative to generalists, narrowly adapted, specialized organisms are predicted to have restricted geographic distributions (they may be dependent on certain precise ecological conditions that occur only in a certain area) and are also predicted to persist for shorter periods of time in the stratigraphic record. Specialized organisms are predicted to be more easily affected by environmental disruption and thus more prone to extinction than generalists that can survive in any of a variety of environments. Vrba (1983) has suggested that specialist organisms may be prone to higher rates of speciation and extinction relative to generalists, and these differing speciation and extinction rates may produce differing patterns of macroevolution. Eldredge (1979b) has suggested that speciation rates may be related to niche strategy within a monophyletic group. This

hypothesis states that, within a given group eurytopic species react differently to interspecific competition than do stenotopes; eurytopes tend to react to interspecific competition by mutual exclusion, whereas stenotopes more commonly react to such competition by further subdivision of resource space (niches are further narrowed). Hence, within a monophyletic group, if there is a discernible spectrum of stenotopy and eurytopy, there are frequently many stenotopic species and fewer eurytopic species. The hypothesis is suggested by actual patterns of distribution in nature: within monophyletic groups relatively eurytopic species tend to be allopatric with respect to one another (or "vicariant") and far-flung, whereas congeneric stenotopes are more likely to occur sympatrically. (Eldredge and Cracraft, 1980, p. 309)

Note that this hypothesis comes close to suggesting that speciation is in and of itself a process of adaptation.

One can combine a corroborated hypothesis of relationships for a group of organisms with data on the stratigraphic and geographic distributions of the organisms in order to elucidate questions concerning the validity of the above proposed correlations. However, from information on the distribution of the organisms and the cladistic relationships of the organisms, one cannot directly determine whether the organisms were generalists or specialists (one of the factors that we may wish to compare to the other factors). Such determinations may incorporate functional morphologic and paleoecological studies, including comparisons to extant analogs whose habits and habitats can be observed directly. From an analysis of the phylogenetic relationships of a group of organisms, one can, however, determine the degree of apomorphy for any particular trait in any particular species relative to the homologous trait in other members of the clade. This information may be equated in some cases with specialization or generalization for a particular trait.

Many theories of macroevolution concentrate on adaptations of organisms to their environments. As Eldredge and Cracraft (1980, p. 311) state, the basic assumption is "that diversity and adaptation are a function of modes of ecological niche occupation and specialization." Unfortunately, when they derive and test their theories of macroevolution, many investigators deal with only one monophyletic group at a time; often workers do not look at patterns that are potentially common to many clades in a given area over a given period of time. On the basis of vicariance views of speciation, I suggest that in many cases there may be congruences between the phylogenetic relationships and distributions of members of independent monophyletic groups occupying a common set of areas. The diversity of life may not be a function of ecology as much as it is a function of vicariance events (cf. Croizat, 1964; Nelson and Platnick, 1981). Vicariance events may be the prime driving force behind much of macroevolution. This hypothesis is eminently testable by comparing the phylogenetic relationships of endemic members of independent monophyletic groups in a given set of areas, as described in chapter 5. As Rosen (1974a, p. 289), in a review of Croizat (1964), wrote: "In short, the history, or phylogeny, of the oceans and land masses generally coincides with the history, or phylogeny, of the world biotas, and to a large extent with the phylogenies of individual groups of organisms."

Concluding Remarks

In this book I have attempted to introduce the reader to various fundamental issues and problems in the reconstruction of phylogeny using fossil organisms. One of the first points made was that I do not consider phylogeny synonymous with phylogenetic tree or evolutionary history. Phylogeny refers to nothing more than the relative degrees of relatedness (in a genealogical or phylogenetic sense; i.e., in terms of recency of common ancestry) of organisms or taxa. In contrast, evolutionary history is usually thought of as including phylogeny (as defined above) plus various other components such as ancestor-descendant relationships; degrees of morphological, physiological, behavioral, or ecological divergence; adaptations; and tempos, patterns, and processes of evolution. I suggest that a hypothesized phylogenetic tree, evolutionary history, or scenario should be based on a rigorous hypothesis of phylogeny. However, it is possible (even if not valid, justifiable, or desirable) to go directly from a pattern of fossils in the rock record to the formulation of a narrative evolutionary history or scenario without an analysis of the phylogenetic relationships among the organisms under consideration. Indeed, too often in the past this seems to have been the case among paleontologists interested in evolution, leading to what Savage (1977, p. 430) has referred to as "paleontological phylogenies" (the interpretation of stratigraphically ordered fossils as forming direct documentation of the course of evolution).

Biological systematics in the broad sense is concerned with the evolution of organisms (i.e., their changing morphologies or forms) in space (biogeography) and through time (Croizat, 1964). Today, among systematically oriented biologists and paleontologists, there are three major schools of researchers, some of whom are interested, to greater or lesser degrees, in phylogeny reconstruction: the phenetic (numerical taxonomic) school, the classical evolutionary (synthetic) school, and the phylogenetic (cladistic) school. I have attempted to briefly outline and discuss the tenets, goals, and methodologies utilized by representative workers associated with each of these schools, as far as phylogeny reconstruction is concerned. Personally, I find that my sympathies lie within the cladistic camp. The position that only evolutionarily advanced features (shared and derived character states, synapomorphies, homologies) are useful in assessing phylogenetic relationships is advocated. It is suggested that cladistic parsimony (ad hoc hypotheses of homoplasy being minimized) be maintained. The rationale for the use of a cladistic program is supported by discussions of such issues as species concepts, homology, determination of character polarity, parsimony, homoplasy (nonhomology), ancestor-descendant relationships, and monophyly.

Topics supplementary to phylogeny per se, but of importance to paleontologists, are reviewed. The fossil record not only provides organisms that are otherwise unknown to the neontologist, but also places the fossil entities in a temporal and paleogeographic context. Although the fossil record may not always be as good as some investigators would wish (one must be wary of reading evolutionary histories directly from the rocks), stratigraphic and temporal data should not be ignored. In many cases it is possible to analyze the adequacy of the fossil record in qualitative or quantitative terms. One may be able to reconstruct the historical sequence of taxa with a fair degree of certainty. Phylogenetic hypotheses may also be of use to the stratigraphic paleontologist in attempts to correlate between widely separated fossil-bearing strata.

In the latter half of this book historical biogeography, systematics, and classification (with a special emphasis on the classification of fossils) are discussed. These topics are intimately related to—indeed dependent upon—the reconstruction of phylogeny. Finally, the importance of phylogeny reconstruction to macroevolutionary studies is reviewed.

Although great advances have been made in the analysis of phylogeny during the last 35 years, the subject is still in its infancy. Rigorous phylogenetic analyses, although increasingly common, are still comparatively rare. And unfortunately, it is all too easy to apply algorithms or computer programs to matrices of biological data in order to arrive at putative phylogenies without considering the fundamental issues involved. The importance of rigorously formulated phylogenetic hypotheses and the recognition of only natural (monophyletic) taxa to macroevolutionary studies is just being realized. Much work remains to be done. It is hoped that this book has helped to review for the reader what has been accomplished in this area thus far, thereby pointing up what avenues still need to be explored. If it stimulates the reader to consider seriously, to think a little harder about the theoretical and philosophical foundations of phylogeny reconstruction in paleontology, then it has served its purpose.

Bibliography

Afifi, A. A., and S. P. Azen, 1979, *Statistical Analysis: A Computer Oriented Approach*, Academic Press, New York.

Agassiz, J. L. R., 1833–1844, *Recherches sur les Poissons Fossiles*, 5 vols. and suppl., Neuchatel.

Ager, D. V., 1973, *The Nature of the Stratigraphic Record*, Macmillan, London.

Ager, D. V., 1976, The nature of the fossil record, *Proc. Geol. Assoc.* **87**:131–159.

Alberch, P., 1980, Ontogenesis and morphological diversification, *Amer. Zool.* **20**:653–667.

Alberch, P., S. J. Gould, G. F. Oster, and D. B. Wake, 1979, Size and shape in ontogeny and phylogeny, *Paleobiology* **5**:296–317.

Arkell, W. J., 1933, *The Jurassic System of Great Britain*, Clarendon Press, Oxford.

Arnold, E. N., 1981, Estimating phylogenies at low taxonomic levels, *Zeit. Zool. Syst. Evolut.-forsch.* **19**:1–35.

Arnold, R. A., and T. Duncan, 1984, An introduction to computer-assisted cladistic analysis, in *Cladistics: Perspectives on the Reconstruction of Evolutionary History*, T. Duncan and T. F. Stuessy, eds., Columbia University Press, New York, pp. 295–298.

Ashlock, P. D., 1971, Monophyly and associated terms, *Syst. Zool.* **20**:63–69.

Ashlock, P. D., 1984, Monophyly: Its meaning and importance, in *Cladistics: Perspectives on the Reconstruction of Evolutionary History*, T. Duncan and T. F. Stuessy, eds., Columbia University Press, New York, pp. 39–46.

Ball, I. R., 1975, Nature and formulation of biogeographical hypotheses, *Syst. Zool.* **24**:407–430.

Basan, P. B., 1979, Trace fossil nomenclature: The developing picture, *Palaeogeography, Palaeoclimatology, Palaeoecology* **28**:143–146.

Batten, R. L., H. B. Rollins, and S. J. Gould, 1967, Comments on "The adaptive significance of gastropod torsion," *Evolution* **21**:405–406.

Baum, B. R., 1984, Application of compatibility and parsimony methods at the infraspecific, specific, and generic levels in Poaceae, in *Cladistics: Perspectives on the Reconstruction of Evolutionary History,* T. Duncan and T. F. Stuessy, eds., Columbia University Press, New York, pp. 192–220.

Beatty, J., 1982, Classes and cladists, *Syst. Zool.* **31:**25–34.

Behrensmeyer, A. K., 1982, Time sampling in the vertebrate fossil record, *North American Paleontol. Conv., 3rd Proc.* **1:**41–45.

Behrensmeyer, A. K., 1984, Taphonomy and the fossil record, *Amer. Sci.* **72:**558–566.

Behrensmeyer, A. K., and S. M. Kidwell, 1985, Taphonomy's contributions to paleobiology, *Paleobiology* **11:**105–119.

Bendall, D. S., ed., 1983, *Evolution from Molecules to Men,* Cambridge University Press, Cambridge.

Benson, R. H., 1977, Evolution of *Oblitacythereis* from *Paleocosta* (Ostracoda: Trachyleberididae) during the Cenozoic in the Mediterranean and Atlantic, *Smithsonian Contrib. Paleobiology* **33:**1–47.

Benson, R. H., 1981, Form, function, and architecture of ostracode shells, *Ann. Rev. Earth Planet. Sci.* **9:**59–80.

Benson, R. H., 1982a, Deformation, Da Vinci's concept of form, and the analysis of events in evolutionary history, in *Palaeontology, Essential of Historical Geology,* E. M. Gallitelli, ed., Istituto di Paleontologia, Università di Modena, Modena, Italy, pp. 241–277.

Benson, R. H., 1982b, Comparative transformation of shape in a rapidly evolving series of structural morphotypes of the ostracod *Bradleya,* in *Fossil and Recent Ostracods,* R. H. Bate, E. Robinson, and L. M. Sheppard, eds., Ellis Horwood Ltd., Chichester, England, pp. 147–163.

Benson, R. H., 1983, Biomechanical stability and sudden change in the evolution of the deep-sea ostracode *Poseidonamicus, Paleobiology* **9:**398–413.

Benton, M. J., 1985, Interpretations of mass extinction, *Nature* **314:**496–497.

Berggren, W. A., and J. A. Van Couvering, eds., 1984, *Catastrophes and Earth History: The New Uniformitarianism,* Princeton University Press, Princeton, N.J.

Berlin, B., D. E. Breedlove, and P. H. Raven, 1966, Folk taxonomies and biological classification, *Science* **154:**273–275.

Biegert, J., 1963, The evaluation of characteristics of the skull, hands, and feet for primate taxonomy, in *Classification and Human Evolution,* S. L. Washburn, ed., Aldine, Chicago, pp. 190–203.

Bigelow, R. S., 1961, Higher categories and phylogeny, *Syst. Zool.* **10:**86–91.

Blackwelder, R. E., 1967, *Taxonomy, A Text and Reference Book,* Wiley, New York.

Blatt, H., G. Middleton, and R. Murray, 1972, *Origin of Sedimentary Rocks,* Prentice-Hall, Englewood Cliffs, N.J.

Bock, W. J., 1974, Philosophical foundations of classical evolutionary classification, *Syst. Zool.* **11:**375–392.

Bock, W. J., 1977, Foundations and methods of evolutionary classification, in *Major Patterns in Vertebrate Evolution,* M. K. Hecht, P. C. Goody, and B. M. Hecht, eds., Plenum Press, New York , pp. 851–895.

Bock, W. J., 1979, The synthetic explanation of macroevolutionary change—A reductionist approach, *Bull. Carnegie Mus. Nat. Hist.* **13:**20–69.

Bock, W. J., 1981, Functional-adaptive analysis in evolutionary classification, *Amer. Zool.* **21:**5–20.

Bonde, N., 1975, Origin of "higher groups"; Viewpoints of phylogenetic systematics, *Problèmes actuels de paleontologie—evolution des vertebres,* Colloques Internationaux du Centre National de la Recherche Scientifique, no. 218, Dijon, pp. 293–324.

Bonde, N., 1977, Cladistic classification as applied to vertebrates, in *Major Patterns in Vertebrate Evolution,* M. K. Hecht, P. C. Goody, and B. M. Hecht, eds., Plenum Press, New York, pp. 741–804.

Bonner, J. T., 1974, *On Development, the Biology of Form*, Harvard University Press, Cambridge, Mass.

Bookstein, F. L., 1977, The study of shape and transformation after D'Arcy Thompson, *Mathematical Biosciences* **34:**177–219.

Borissiak, A., 1945, The chalicothere as a biological type, *Amer. Jour. Sci.* **243:**667–679.

Boucot, A. J., 1975, *Evolution and Extinction Rate Controls*, Developments in Palaeontology and Stratigraphy series, vol. 1, Elsevier, Amsterdam.

Boucot, A. J., 1983, Does evolution take place in an ecological vacuum? *Jour. Paleontol.* **57:**1–30.

Brady, R. H., 1983, Parsimony, hierarchy, and biological implications, in *Advances in Cladistics*, N. I. Platnick and V. A. Funk, eds., Columbia University Press, New York, pp. 49–60.

Bretsky, P. W., and D. M. Lorenz, 1970, Adaptive response to environmental stability: A unifying concept in paleoecology, *North American Paleontol. Conv., 1st Proc.* **1:**522–550.

Bretsky, S. S., 1976, Evolution and classification of the Lucinidae (Mollusca; Bivalvia), *Palaeontographica Americana* **8:**219–337.

Bretsky, S. S., 1979, Recognition of ancestor-descendant relationships in invertebrate paleontology, in *Phylogenetic Analysis and Paleontology*, J. Cracraft and N. Eldredge, eds., Columbia University Press, New York, pp. 113–163.

Briggs, J. C., 1974, Operation of zoogeographic barriers, *Syst. Zool.* **23:**248–256.

Broadhead, T. W., and J. A. Waters, 1984, Heterochrony and the problem of missing links in evolutionary sequences, in *The Evolution-Creation Controversy: Perspectives on Religion, Philosophy, Science and Education*, K. R. Walker, ed., special publication no. 1, The Paleontological Society, University of Tennessee, Knoxville, pp. 84–96.

Brooks, D. R., 1979a, Testing hypotheses of evolutionary relationships among parasites: The digeneans of crocodilians, *Amer. Zool.* **19:**1225–1238.

Brooks, D. R., 1979b, Testing the context and extent of host-parasite coevolution, *Syst. Zool.* **28:**299–307.

Brooks, D. R., 1980, Allopatric speciation and non-interactive parasite community structure, *Syst. Zool.* **29:**192–203.

Brooks, D. R., 1981a, Raw similarity measures of shared parasites: An empirical tool for determining host phylogenetic relationships?, *Syst. Zool.* **30:**203–207.

Brooks, D. R., 1981b, Hennig's parasitological method: A proposed solution, *Syst. Zool.* **30:**229–249.

Brooks, D. R., 1983, What's going on in evolution? A brief guide to some new ideas in evolutionary theory, *Canadian Jour. Zool.* **61:**2637–2645.

Brooks, D. R., 1984, Quantitative parsimony, in *Cladistics: Perspectives on the Reconstruction of Evolutionary History*, T. Duncan and T. F. Stuessy, eds., Columbia University Press, New York, pp. 119–132.

Brooks, D. R., and D. R. Glen, 1982, Pinworms and primates: A case study in coevolution, *Proc. Helminthol. Soc. Wash.* **49:**76–85.

Brooks, D. R., and E. O. Wiley, 1984, Evolution as an entropic phenomenon, in *Evolutionary Theory: Paths into the Future*, J. W. Pollard, ed., Wiley, Chichester, England, pp. 141–171.

Brooks, D. R., and E. O. Wiley, 1985, Theories and methods in different approaches to phylogenetic systematics, *Cladistics* **1:**1–11.

Brower, J. C., 1973, Crinoids from the Girardeau Limestone (Ordovician), *Palaeontographica Americana* **7:**263–499.

Brown, G. B., 1950, *Science: Its Method and Its Philosophy*, Allen and Unwin, London.

Brundin, L., 1966, Transantarctic relationships and their significance, as evidenced by chironomid midges with a monograph of the subfamilies Podonominae and Aphroteniinae and the Austral Heptagyiae, *Kungl. Svenska Vetenskapsakademiens Handlingar, Fjarde Serien* **11:**1–472.

Brundin, L., 1972, Phylogenetics and biogeography, *Syst. Zool.* **21:**69–79.

Brundin, L., 1976a, A Neocomian chironomid and Podonominae-Aphroteniinae (Diptera) in the light of phylogenetics and biogeography, *Zool. Scripta* **5:**139–160.

Brundin, L., 1976b, Parallelism and its phylogenetic significance, *Zool. Scripta* **5**:186.

Bush, G. L., 1975, Modes of animal speciation, *Ann. Rev. Ecol. Syst.* **6**:339–364.

Buss, L. W., 1983, Evolution, development, and the units of selection, *Proc. Nat. Acad. Sci. USA* **80**:1387–1391.

Cain, A. J., 1958, Logic and memory in Linnaeus's system of taxonomy, *Proc. Linn. Soc., London* **169**:144–163.

Cain, S. A., 1944, *Foundations of Plant Geography,* Harper & Row, New York.

Camin, J. H., and R. R. Sokal, 1965, A method for deducing branching sequences in phylogeny, *Evolution* **19**:311–326.

Camp, W. H., and C. L. Gilly, 1943, The structure and origin of species, *Brittonia* **4**:323–385.

Carleton, M. D., and R. E. Eshelman, 1979, A synopsis of fossil grasshopper mice, genus *Onychomys,* and their relationships to recent species, *Univ. Mich. Pap. Paleontol.* **21**:1–63.

Carroll, R. L., 1982, Early evolution of reptiles, *Ann. Rev. Ecol. Syst.* **13**:87–109.

Carroll, R. L., 1984, Response to Wyss and de Queiroz, *Jour. Vert. Paleontol.* **4**:609–612.

Cartmill, M., 1981, Hypothesis testing and phylogenetic reconstruction, *Zeit. Zool. Syst. Evolut.-forsch.* **19**:73–96.

Cartmill, M., 1982, Assessing tarsier affinities: Is anatomical description phylogenetically neutral?, *Geobios, Mem. Spec.* **6**:279–287.

Cavalli-Sforza, L. L., and A. W. F. Edwards, 1967, Phylogenetic analysis: Models and estimation procedures, *Evolution* **32**:550–570.

Chapman, R. E., P. M. Galton, J. J. Sepkoski, Jr., and W. P. Wall, 1981, A morphometric study of the cranium of the pachycephalosaurid *Stegoceras, Jour. Paleontol.* **55**:608–618.

Charig, A. J., 1982, Systematics in biology: A fundamental comparison of some schools of thought, *Syst. Assoc. Spec. Vol.* **21**:363–440.

Cheeney, R. F., 1983, *Statistical Methods in Geology,* Allen and Unwin, London.

Cisne, J. L., G. O. Chandlee, B. D. Rabe, and J. A. Cohen, 1982, Clinal variation, episodic evolution, and possible parapatric speciation: The trilobite *Flexicalymene senria* along an Ordovician depth gradient, *Lethaia* **15**:325–341.

Cisne, J. L., and B. D. Rabe, 1978, Coenocorrelation: Gradient analysis of fossil communities and its applications in stratigraphy, *Lethaia* **11**:341–364.

Clarkson, E. N. K., 1979, *Invertebrate Paleontology and Evolution,* Allen and Unwin, London.

Cohen, I. B., 1960, *The Birth of a New Physics,* Anchor Books, Garden City, N.Y.

Colless, D. H., 1967a, An examination of certain concepts in phenetic taxonomy, *Syst. Zool.* **16**:6–27.

Colless, D. H., 1967b, The phylogenetic fallacy, *Syst. Zool.* **16**:289–295.

Copleston, F. C., 1972, *A History of Medieval Philosophy,* Methuen, London.

Cracraft, J., 1973, Continental drift, paleoclimatology, and the evolution and biogeography of birds, *Jour. Zool. (Zool. Soc. London)* **169**:455–545.

Cracraft, J., 1974a, Phylogenetic models and classification, *Syst. Zool.* **23**:71–90.

Cracraft, J., 1974b, Continental drift and vertebrate distribution, *Ann. Rev. Ecol. Syst.* **5**:215–261.

Cracraft, J., 1975, Historical biogeography and earth history: Perspectives for a future synthesis, *Ann. Missouri Bot. Garden* **62**:227–250.

Cracraft, J., 1981a, Pattern and process in paleobiology: The role of cladistic analysis in systematic paleontology, *Paleobiology* **7**:456–468.

Cracraft, J., 1981b, The use of functional and adaptive criteria in phylogenetic systematics, *Amer. Zool.* **21**:21–36.

Cracraft, J., 1982, A nonequilibrium theory for the rate-control of speciation and extinction and the origin of macroevolutionary patterns, *Syst. Zool.* **31**:348–365.

Cracraft, J., 1983a, Cladistic analysis and vicariance biogeography, *Amer. Sci.* **71**:273–281.

Cracraft, J., 1983b, Species concepts and speciation analysis, *Current Ornithology* **1**:159–187.

Cracraft, J., and N. Eldredge, eds., 1979, *Phylogenetic Analysis and Paleontology*, Columbia University Press, New York.

Craig, G. Y., and A. Hallam, 1963, Size-frequency and growth-ring analyses of *Mytilus edulis* and *Cardium edule,* and their palaeoecological significance, *Palaeontology* **6:**731–750.

Crisci, J. V., and T. F. Stuessy, 1980, Determining primitive character states for phylogenetic reconstruction, *Syst. Bot.* **5:**112–135.

Croizat, L., 1964 [dated 1962], *Space, Time, Form: The Biological Synthesis,* published by the author, Caracas.

Croizat, L., 1981, Biogeography: Past, present, and future, in *Vicariance Biogeography: A Critique,* G. Nelson and D. E. Rosen, eds., Columbia University Press, New York, pp. 501–537.

Croizat, L., G. Nelson, and D. E. Rosen, 1974, Centers of origin and related concepts, *Syst. Zool.* **23:**265–287.

Crowson, R. A., 1970, *Classification and Biology,* Atherton Press, New York.

Crowson, R. A., 1982, Computers versus imagination in the reconstruction of phylogeny, *Syst. Assoc. Spec. Vol.* **21:**245–255.

Cubitt, J. M., and R. A. Reyment, eds., 1982, *Quantitative Stratigraphic Correlation,* Wiley, Chichester, England.

Darlington, P. J., Jr., 1957, *Zoogeography: The Geographical Distribution of Animals,* Wiley, New York.

Darlington, P. J., Jr., 1970, A practical criticism of Hennig-Brundin "phylogenetic systematics" and Antarctic biogeography, *Syst. Zool.* **19:**1–18.

Darwin, C., 1859, *On the Origin of Species by Means of Natural Selection, or the Preservation of Favored Races in the Struggle for Life,* John Murray, London.

Darwin, C., 1871a, *On the Origin of Species by Means of Natural Selection, or the Preservation of Favored Races in the Struggle for Life,* 5th ed., D. Appleton, New York.

Darwin, C., 1871b, *The Descent of Man and Selection in Relation to Sex,* John Murray, London.

Darwin, F., 1887, *Life and Letters of Charles Darwin,* John Murray, London.

Dayoff, M. O., 1969, Computer analysis of protein evolution, *Amer. Sci.* **221:**86–95.

De Beer, G., 1958, *Embryos and Ancestors,* Oxford University Press, London.

Deevey, E. S., Jr., 1947, Life tables for natural populations of animals, *Quart. Rev. Biol.* **22:**283–314.

De Jong, R., 1980, Some tools for evolutionary and phylogenetic studies, *Zeit. Zool. Syst. Evolut.-forsch.* **18:**1–23.

Dene, H., M. Goodman, M. C. McKenna, and A. E. Romero-Herrera, 1982, *Ochotoma princeps* (pika) myoglobin: An appraisal of lagomorph phylogeny, *Proc. Nat. Acad. Sci. USA* **79:**1917–1920.

Dennison, J. M., and W. W. Hay, 1967, Estimating the needed sampling area for subaquatic ecologic studies, *Jour. Paleontol.* **41:**706–708.

Derstler, K., 1982, Estimating the rate of morphological change in fossil groups, *North American Paleont. Conv., 3rd Proc.* **1:**131–136.

Dingus, L., 1984, Effects of stratigraphic completeness on interpretations of extinction rates across the Cretaceous-Tertiary boundary, *Paleobiology* **10:**420–438.

Dingus, L., and P. M. Sadler, 1982, The effects of stratigraphic completeness on estimates of evolutionary rates, *Syst. Zool.* **31:**400–412.

Dobzhansky, T., 1937, *Genetics and the Origin of Species,* Columbia University Press, New York.

Dobzhansky, T., 1970, *Genetics of the Evolutionary Process,* Columbia University Press, New York.

Dodd, J. R., and R. J. Stanton, Jr., 1981, *Paleoecology, Concepts and Principles,* Wiley, New York.

Doyle, J. A., S. Jardine, and A. Doerenkamp, 1982, *Afropollis,* a new genus of early angiosperm pollen, with notes on the Cretaceous palynostratigraphy and paleoenvironments of northern Gondwana, *Bull. Centres Rech. Explor.-Prod. Elf-Aquitaine* **6:**39–117.

Dullemeijer, P., 1974, *Concepts and Approaches to Animal Morphology,* Van Gorcum, Assen, Netherlands.

Dullemeijer, P., 1980, Functional morphology and evolutionary biology, *Acta Biotheoretica* **29:**151–250.

Dunbar, M. J., 1980, The blunting of Occam's razor, or to hell with parsimony, *Canadian Jour. Zool.* **58:**123–128.

Duncan, T., 1980, Cladistics for the practicing taxonomist—An eclectic view, *Syst. Bot.* **5:**136–148.

Duncan, T., R. B. Phillips, and W. H. Wagner, 1980, A comparison of branching diagrams derived by various phenetic and cladistic methods, *Syst. Bot.* **5:**264–293.

Duncan, T., and T. F. Stuessy, eds., 1984a, *Cladistics: Perspectives on the Reconstruction of Evolutionary History,* Columbia University Press, New York.

Duncan, T., and T. F. Stuessy, 1984b, Introduction [to cladogram construction], in *Cladistics: Perspectives on the Reconstruction of Evolutionary History,* T. Duncan and T. F. Stuessy, eds., Columbia University Press, New York, pp. 91–92.

Dupuis, C., 1979, Permanence et actualité de la systématique: La "systématique phylogénétique" de W. Hennig (historique, discussion, choix de références), *Cahiers des Naturalistes* **34:**1–69.

Eck, R. V., and M. O. Dayhoff, 1966, *Atlas of Protein Sequence and Structure 1966,* National Biomedical Research Foundation, Silver Spring, Md.

Edwards, A. W. F., and L. L. Cavalli-Sforza, 1963, The reconstruction of evolution, *Ann. Human Genet.* **27:**105.

Edwards, A. W. F., and L. L. Cavalli-Sforza, 1964, Reconstruction of evolutionary trees, in *Phenetic and Phylogenetic Classification,* V. H. Heywood and J. McNeill, eds., Systematics Association Publication No. 6, pp. 67–76.

Edwards, L. E., 1978, Range charts and no-space graphs, *Computers and Geoscience* **4:**247–255.

Edwards, L. E., 1982a, Numerical and semi-objective biostratigraphy: Review and predictions, *North America Paleontol. Conv., 3rd Proc.* **1:**147–152.

Edwards, L. E., 1982b, Quantitative biostratigraphy: The methods should suit the data, in *Quantitative Stratigraphic Correlation,* J. M. Cubitt and R. A. Reyment, eds., Wiley, Chichester, England, pp. 45–60.

Edwards, L. E., 1984, Insights on why graphic correlation (Shaw's method) works, *Jour. Geol.* **92:**583–597.

Edwards, L. E., and R. J. Beaver, 1978, The use of a paired comparison model in ordering stratigraphic events, *Mathematical Geology* **10:**261–272.

Efremov, J. A., 1940, Taphonomy: New branch of paleontology, *Pan-American Geologist* **74:**81–93.

Ehrlich, P. R., and R. W. Holm, 1962, Patterns and populations, *Science* **137:**652–657.

Ehrlich, P. R., and P. H. Raven, 1969, Differentiation of populations, *Science* **165:**1228–1231.

Eldredge, N., 1979a, Cladism and common sense, in *Phylogenetic Analysis and Paleontology,* J. Cracraft and N. Eldredge, eds., Columbia University Press, New York, pp. 165–198.

Eldredge, N., 1979b, Alternative approaches to evolutionary theory, *Bull. Carnegie Mus. Nat. Hist.* **13:**7–19.

Eldredge, N., and J. Cracraft, 1980, *Phylogenetic Patterns and the Evolutionary Process,* Columbia University Press, New York.

Eldredge, N., and S. J. Gould, 1972, Punctuated equilibria: An alternative to phyletic gradualism, in *Models in Paleobiology,* T. J. M. Schopf, ed., Freeman, Cooper, San Francisco, pp. 82–115.

Eldredge, N., and S. J. Gould, 1977, Evolutionary models and biostratigraphic strategies, in *Concepts and Methods in Biostratigraphy,* E. G. Kauffman and J. E. Hazel, eds., Dowden, Hutchinson and Ross, Stroudsburg, Pa., pp. 25–40.

Eldredge, N., and M. J. Novacek, 1985, Systematics and paleobiology, *Paleobiology* **11**:65–74.

Eldredge, N., and S. M. Stanley, eds., 1984, *Living Fossils*, Springer-Verlag, New York and Berlin.

Eldredge, N., and I. Tattersall, 1975, Evolutionary models, phylogenetic reconstruction, and another look at hominid phylogeny, *Contrib. to Primatol.* **5**:218–242.

Endler, J. A., 1977, *Geographic Variations, Speciation, and Clines*, Princeton University Press, Princeton, N.J.

Engelmann, G. F., and E. O. Wiley, 1977, The place of ancestor-descendant relationships in phylogeny reconstruction, *Syst. Zool.* **26**:1–11.

Estabrook, G. F., 1968, A general solution in partial orders for the Camin-Sokal model in phylogeny, *Jour. Theor. Biol.* **21**:421–438.

Estabrook, G. F., 1972, Cladistic methodology: A discussion of the theoretical basis for the induction of evolutionary history, *Ann. Rev. Ecol. Syst.* **3**:427–456.

Estabrook, G. F., 1977, Does common equal primitive? *Syst. Bot.* **2**:36–42.

Estabrook, G. F., 1979, Some concepts for the estimation of evolutionary relationships in systematic botany, *Syst. Bot.* **3**:146–158.

Estabrook, G. F., 1984, Phylogenetic trees and character-state trees, in *Cladistics: Perspectives on the Reconstruction of Evolutionary History*, T. Duncan and T. F. Stuessy, eds., Columbia University Press, New York, pp. 135–151.

Estabrook, G. F., C. S. Johnson, Jr., and F. R. McMorris, 1975, An idealized concept of the true cladistic character, *Math. Biosci.* **23**:263–272.

Estabrook, G. F., C. S. Johnson, Jr., and F. R. McMorris, 1976, A mathematical foundation for the analysis of cladistic character compatibility, *Math. Biosci.* **29**:181–187.

Estabrook, G. F., and F. R. McMorris, 1980, When is one estimate of evolutionary relationships a refinement of another? *Jour. Math. Biol.* **10**:367–373.

Estabrook, G. F., J. G. Strauch, Jr., and K. L. Fiala, 1977, An application of compatibility analysis to the Blackiths' data on orthopteroid insects, *Syst. Zool.* **26**:269–276.

Falconer, D. S., 1981, *Introduction to Quantitative Genetics*, Longman Group Ltd., London.

Farris, J. S., 1969, A successive approximation to character weighting, *Syst. Zool.* **18**:374–385.

Farris, J. S., 1970, Methods for computing Wagner trees, *Syst. Zool.* **19**:83–92.

Farris, J. S., 1973, On the use of the parsimony criterion for inferring evolutionary trees, *Syst. Zool.* **22**:250–256.

Farris, J. S., 1974, Formal definitions of paraphyly and polyphyly, *Syst. Zool.* **23**:548–554.

Farris, J. S., 1976, Phylogenetic classification of fossils with recent species, *Syst. Zool.* **25**:271–282.

Farris, J. S., 1977a, Phylogenetic analysis under Dollo's law, *Syst. Zool.* **26**:77–88.

Farris, J. S., 1977b, On the phenetic approach to vertebrate classification, in *Major Patterns in Vertebrate Evolution*, M. K. Hecht, P. C. Goody, and B. M. Hecht, eds., Plenum Press, New York, pp. 823–850.

Farris, J. S., 1978, Inferring phylogenetic trees from chromosome inversion data, *Syst. Zool.* **27**:275–284.

Farris, J. S., 1979, The information content of the phylogenetic system, *Syst. Zool.* **28**:483–519.

Farris, J. S., 1980, The efficient diagnoses of the phylogenetic system, *Syst. Zool.* **29**:386–401.

Farris, J. S., 1981, Distance data in phylogenetic analysis, in *Advances in Cladistics: Proceedings of the First Meeting of the Willi Hennig Society*, V. A. Funk and D. R. Brooks, eds., New York Botanical Garden, Bronx, N.Y., pp. 3–23.

Farris, J. S., 1982, Simplicity and informativeness in systematics and phylogeny, *Syst. Zool.* **31**:413–444.

Farris, J. S., 1983, The logical basis of phylogenetic analysis, in *Advances in Cladistics, Volume 2: Proceedings of the Second Meeting of the Willi Hennig Society*, N. I. Platnick and V. A. Funk, eds., Columbia University Press, New York, pp. 7–36.

Farris, J. S., and A. G. Kluge, 1979, A botanical clique, *Syst. Zool.* **28**:400–411.

Farris, J. S., A. G. Kluge, and M. J. Eckardt, 1970, A numerical approach to phylogenetic systematics, *Syst. Zool.* **19**:172–189.

Farris, J. A., A. G. Kluge, and M. F. Mickevich, 1982, Phylogenetic analysis, the monothetic group method, and myobatrachid frogs, *Syst. Zool.* **31**:317–327.

Felsenstein, J., 1978, Cases in which parsimony or compatibility methods will be positively misleading, *Syst. Zool.* **27**:401–410.

Felsenstein, J., 1979, Alternative methods of phylogenetic inference and their interrelationship, *Syst. Zool.* **28**:49–62.

Felsenstein, J., 1982, Numerical methods for inferring evolutionary trees, *Quart. Rev. Biol.* **57**:379–404.

Felsenstein, J., 1984, The statistical approach to inferring evolutionary trees and what it tells us about parsimony and compatibility, in *Cladistics: Perspectives on the Reconstruction of Evolutionary History,* T. Duncan and T. F. Stuessy, eds., Columbia University Press, New York, pp. 169–191.

Fisher, D. C., 1980, The role of stratigraphic data in phylogenetic inference, *Geol. Soc. Amer., Abstracts with Programs* **12**:426.

Fisher, D. C., 1981, The role of functional analysis in phylogenetic inference: Examples from the history of the Xiphosura, *Amer. Zool.* **21**:47–62.

Fisher, D. C., 1982, Phylogenetic and macroevolutionary patterns within the Xiphosurida, *North American Paleontol. Conv., 3rd Proc.* **2**:175–180.

Fisher, D. C., 1985, Evolutionary morphology: Beyond the analogous, the anecdotal, and the ad hoc, *Paleobiology* **11**:120–138.

Fitch, W. M., 1971, Toward defining the course of evolution: Minimum change for a specific tree topology, *Syst. Zool.* **20**:406–416.

Fitch, W. M., 1977a, The phyletic interpretation of macromolecular sequence information: Simple methods, in *Major Patterns in Vertebrate Evolution,* M. K. Hecht, P. C. Goody, and B. M. Hecht, eds., Plenum Press, New York, pp. 169–204.

Fitch, W. M., 1977b, The phyletic interpretation of macromolecular sequence information: Sample cases, in *Major Patterns in Vertebrate Evolution,* M. K. Hecht, P. C. Goody, and B. M. Hecht, eds., Plenum Press, New York, pp. 211–248.

Fitch, W. M., 1977c, On the problem of discovering the most parsimonious tree, *Amer. Natur.* **111**:223–257.

Fitch, W. M., 1982, A non-sequential method for constructing a hierarchical classification, *Jour. Mol. Evol.* **18**:30–37.

Fitch, W. M., 1984, Cladistic and other methods: Problems, pitfalls, and potentials, in *Cladistics: Perspectives on the Reconstruction of Evolutionary History,* T. Duncan and T. F. Stuessy, eds., Columbia University Press, New York, pp. 221–252.

Fitch, W. M., and E. Margoliash, 1967, The construction of phylogenetic trees, *Science* **155**:279–284.

Flynn, J. J., B. J. MacFadden, and M. C. McKenna, 1984, Land-mammal "ages" faunal heterochrony, and temporal resolution in Cenozoic terrestrial sequences, *Jour. Geol.* **92**:687–705.

Forey, P. L., 1982, Neontological analysis versus palaeontological stories, *Syst. Assoc. Spec. Vol.* **21**:119–234.

Fortey, R. A., and R. P. S. Jefferies, 1982, Fossils and phylogeny—A compromise approach, *Syst. Assoc. Spec. Vol.* **21**:197–234.

Frazzetta, T. H., 1970, From hopeful monsters to bolyerine snakes?, *Amer. Nat.* **104**:55–72.

Frazzetta, T. H., 1975, *Complex Adaptations in Evolving Populations,* Sinauer Associates, Sunderland, Mass.

Füchtbauer, H., 1974, *Sediments and Sedimentary Rocks,* Wiley, New York.

Fuller, W. A., M. T. Nietfeld, and M. A. Harris, eds., 1985, *Abstracts of Papers and Posters: Fourth International Theriological Congress, Edmonton, 13–20 August, 1985, and Fourth*

International Reindeer/Caribou Symposium, Whitehorse, 22–25 August, 1985, University of Alberta, Edmonton.

Funk, V. A., and D. R. Brooks, eds., 1981, *Advances in Cladistics: Proceedings of the First Meeting of the Willi Hennig Society*, The New York Botanical Garden, Bronx, N.Y.

Funk, V. A., and T. F. Stuessy, 1978, Cladistics for the practicing plant taxonomist, *Syst. Bot.* **3:**159–178.

Futuyma, D. J., 1979, *Evolutionary Biology*, Sinauer Associates, Sunderland, Mass.

Futuyma, D. J., 1983, *Science on Trial: The Case for Evolution*, Pantheon Books, New York.

Gaffney, E. S., 1979a, An introduction to the logic of phylogeny reconstruction, in *Phylogenetic Analysis and Paleontology*, J. Cracraft and N. Eldredge, eds., Columbia University Press, New York, pp. 79–111.

Gaffney, 1979b, Tetrapod monophyly: A phylogenetic analysis, *Bull. Carnegie Mus. Nat. Hist.* **13:**92–105.

Gardiner, B., 1982, Tetrapod classification, *Linnean Soc. London Jour. Zoology* **74:**207–232.

Gardner, R. C., and J. C. La Duke, 1979, An estimate of phylogenetic relationships within the genus *Crusea* (Rubiaceae) using character compatibility analysis, *Syst. Bot.* **3:**179–196.

George, W., 1964, *Biologist Philosopher*, Abelard-Schuman, London.

Ghiselin, M. T., 1966a, The adaptive significance of gastropod torsion, *Evolution* **20:**337–348.

Ghiselin, M. T., 1966b, On psychologism in the logic of taxonomic controversies, *Syst. Zool.* **15:**207–215.

Ghiselin, M. T., 1969, *The Triumph of the Darwinian Method*, University of California Press, Berkeley.

Ghiselin, M. T., 1972, Models in phylogeny, in *Models in Paleobiology*, T. J. M. Schopf, ed., Freeman, Cooper, San Francisco, pp. 130–145.

Ghiselin, M. T., 1974, A radical solution to the species problem, *Syst. Zool.* **23:**536–544.

Ghiselin, M. T., 1976, The nomenclature of correspondence: A new look at "homology" and "analogy," in *Evolution, Brain and Behavior: Persistent Problems*, R. B. Masterton, W. Hodos, and H. Jerison, eds., Lawrence Erlbaum, Hillsdale, N.J., pp. 129–142.

Gilmour, J. S. L., 1940, Taxonomy and philosophy, in *The New Systematics*, J. Huxley, ed., Clarendon Press, London, pp. 461–468.

Gilmour, J. S. L., 1961, Taxonomy, in *Contemporary Botanical Thought*, A. M. MacLeod and L. S. Cobley, eds., Quadrangle Books, Chicago, pp. 27–45.

Gingerich, P. D., 1976a, Paleontology and phylogeny: Patterns of evolution at the species level in Early Tertiary mammals, *Amer. Jour. Sci.* **276:**1–28.

Gingerich, P. D., 1976b, Cranial anatomy and evolution of Early Tertiary Plesiadapidae (Mammalia, Primates), *Univ. Michigan, Pap. Paleontol.* **15:**1–141.

Gingerich, P. D., 1979, The stratigraphic approach to phylogeny reconstruction in vertebrate paleontology, in *Phylogenetic Analysis and Paleontology*, J. Cracraft and N. Eldredge, eds., Columbia University Press, New York, pp. 41–77.

Gingerich, P. D., 1980, Evolutionary patterns in Early Cenozoic mammals, *Ann. Rev. Earth Planet. Sci.* **8:**407–424.

Gingerich, P. D., 1982, Time resolution in mammalian evolution: Sampling, lineages, and faunal turnover, *North American Paleontol. Conv., 3rd Proc.* **1:**205–210.

Gingerich, P. D., 1983, Rates of evolution: Effects of time and temporal scaling, *Science* **222:**159–161.

Gingerich, P. D., 1985, Species in the fossil record: Concepts, trends, and transitions, *Paleobiology* **11:**27–41.

Gingerich, P. D., and M. Schoeninger, 1977, The fossil record and primate phylogeny, *Jour. Human Evol.* **6:**484–505.

Goldschmidt, R., 1940, *The Material Basis of Evolution*, Yale University Press, New Haven.

Goodman, M., ed., 1982, *Macromolecular Sequences in Systematic and Evolutionary Biology*, Plenum Press, New York.

Goodman, M., M. L. Weiss, and J. Czelusniak, 1982, Molecular evolution above the species level: Branching patterns, rates, and mechanisms, *Syst. Zool.* **31**:376–399.

Goudge, T. A., 1961, *The Ascent of Life,* University of Toronto Press, Toronto.

Gould, S. J., 1966, Allometry and size in ontogeny and phylogeny, *Biol. Rev.* **41**:587–640.

Gould, S. J., 1971, Geometric similarity in allometric growth: A contribution to the problem of scaling in the evolution of size, *Amer. Nat.* **105**:113–136.

Gould, S. J., 1972, Allometric fallacies and the evolution of *Gryphaea:* A new interpretation based on White's criterion of geometric similarity, *Evol. Biol.* **6**:91–119.

Gould, S. J., 1974, The evolutionary significance of "bizarre" structures: Antler size and skull size in the "Irish elk," *Megaloceros giganteus, Evolution* **28**:191–220.

Gould, S. J., 1977, *Ontogeny and Phylogeny,* Harvard University Press, Cambridge, Mass.

Gould, S. J., 1980a, The promise of paleobiology as a nomothetic, evolutionary discipline, *Paleobiology* 6:96–118.

Gould, S. J., 1980b, Is a new and general theory of evolution emerging? *Paleobiology* **6**:119–130.

Gould, S. J., 1982, Darwinism and the expansion of evolutionary theory, *Science* **216**:380–387.

Gould, S. J., 1983a, The hardening of the modern synthesis, in *Dimensions of Darwinism: Themes and Counterthemes in Twentieth-Century Evolutionary Theory,* M. Grene, ed., Cambridge University Press, Cambridge, pp. 71–93.

Gould, S. J., 1983b, Punctuated equilibrium and the fossil record, *Science* **219**:439–440.

Gould, S. J., 1984, Gingerich's smooth curve of evolutionary rate: A psychological and mathematical artifact, *Science* **226**:994–995.

Gould, S. J., 1985, The paradox of the first tier: An agenda for paleobiology, *Paleobiology* **11**:2–12.

Gould, S. J., and C. B. Calloway, 1980, Clams and brachiopods—Ships that pass in the night, *Paleobiology* **6**:383–396.

Gould, S. J., and N. Eldredge, 1977, Punctuated equilibria: The tempo and mode of evolution reconsidered, *Paleobiology* **3**:115–151.

Gould, S. J., and R. C. Lewontin, 1979, The spandrels of San Marco and the Panglossian paradigm: A critique of the adaptationist programme, *Proc. Roy. Soc. Lond.* **B205**:581–598.

Gould, S. J., D. M. Raup, J. J. Sepkoski, Jr., T. J. M. Schopf, and D. S. Simberloff, 1977, The shape of evolution: A comparison of real and random clades, *Paleobiology* **3**:23–40.

Gould, S. J., and E. S. Vrba, 1982, Exaptation—A missing term in the science of form, *Paleobiology* **8**:4–15.

Grant, M., 1958, *Roman History from Coins,* Cambridge University Press, Cambridge.

Grant, V., 1963, *The Origin of Adaptations,* Columbia University Press, New York.

Grant, V., 1971, *Plant Speciation,* Columbia University Press, New York.

Grant, V., 1977, *Organismic Evolution,* W. H. Freeman, San Francisco.

Greene, E. L., 1912, *Carolus Linnaeus,* Christopher Sower, Philadelphia.

Grene, M., ed., 1983, *Dimensions of Darwinism,* Cambridge University Press, Cambridge.

Griffiths, G. C. D., 1974, On the foundations of biological systematics, *Acta Biotheoretica* **23**:85–131.

Gutmann, W. F., 1972, Die Hydroskelett-Theorie, *Aufsatze u. Reden Senckenb. Naturf. Ges.* **21**:1–91.

Gutmann, W. F., 1975, Konstruktive Vorbedingung und Konsequenz in der phylogenetischen Entwicklund des Korperstammes der Cranioten, *Aufsatze u. Reden Senckenb. Naturf. Ges.* **27**:40–56.

Gutmann, W. F., 1977, Phylogenetic reconstruction: Theory, methodology, and application to chordate evolution, in *Major Patterns in Vertebrate Evolution,* M. K. Hecht, P. C. Goody, and B. M. Hecht, eds., Plenum Press, New York, pp. 645–669.

Haeckel, E., 1866, *Generelle Morphologie der Organismen,* G. Reimer, Berlin.

Haeckel, E., 1868, *Naturliche Schopfungsgeschichte,* G. Reimer, Berlin.

Haldane, J. B. S., 1949, Suggestions as to quantitative measurement of rates of evolution, *Evolution* 3:51–56.

Haldane, J. B. S., and J. Huxley, 1927, *Animal Biology*, Clarendon Press, Oxford.

Hallam, A., 1968, Morphology, palaeoecology and evolution of the genus *Gryphaea* in the British Lias, *Phil. Trans. Roy. Soc. Lond.* B254:91–128.

Hallam, A., 1972, Models involving population dynamics, in *Models in Paleobiology*, T. J. M. Schopf, ed., Freeman, Cooper, San Francisco, pp. 62–80.

Hallam, A., 1975, Evolutionary size increase and longevity in Jurassic bivalves and ammonites, *Nature* 258:493–496.

Hallam, A., ed., 1977, *Patterns of Evolution as Illustrated by the Fossil Record*, Elsevier, Amsterdam.

Hallam, A., 1978, How rare is phyletic gradualism and what is its evolutionary significance? Evidence from Jurassic bivalves, *Paleobiology* 4:16–25.

Hallam, A., 1982, Patterns of speciation in Jurassic *Gryphaea*, *Paleobiology* 8:354–366.

Hallam, A., 1983, Plate tectonics and evolution, in *Evolution from Molecules to Men*, D. S. Bendall, ed., Cambridge University Press, Cambridge, pp. 367–386.

Hanson, E. D., 1977, *The Origin and Early Evolution of Animals*, Wesleyan University Press, Middletown, Conn.

Harland, W. B., A. V. Cox, P. G. Llewellyn, C. A. G. Pickton, A. G. Smith, and R. Walters, 1982, *A Geologic Time Scale*, Cambridge University Press, Cambridge.

Harper, C. W., Jr., 1976, Phylogenetic inference in paleontology, *Jour. Paleontol.* 50:180–193.

Harper, C. W., Jr., 1979, A Bayesian probability view of phylogenetic systematics, *Syst. Zool.* 28:547–553.

Harper, C. W., Jr., 1980, Relative age inference in paleontology, *Lethaia* 13:239–248.

Harper, C. W., Jr., 1981, Inferring succession of fossils in time: The need for a quantitative and statistical approach, *Jour. Paleontol.* 55:442–452.

Harper, C. W., Jr., and N. I. Platnick, 1978, Phylogenetic and cladistic hypotheses: A debate, *Syst. Zool.* 27:354–362.

Harré, R., 1972, *The Philosophies of Science, an Introductory Survey*, Oxford University Press, London.

Hay, J. D., 1970, Stratigraphy and evolutionary trends of radiolaria in north Pacific deep-sea sediments, *Geol. Soc. Amer. Mem.* 126:185–218.

Hay, W. W., 1972, Probabilistic stratigraphy, *Eclogae Geol. Helv.* 65/2:255–266.

Hecht, M. K., 1976, Phylogenetic inference and methodology as applied to the vertebrate record, *Evol. Biol.* 9:335–363.

Hecht, M. K., and J. L. Edwards, 1976, The determination of parallel or monophyletic relationships: The proteid salamanders—a test case, *Amer. Natur.* 110:653–677.

Hecht, M. K., and J. L. Edwards, 1977, The methodology of phylogenetic inference at the species level, in *Major Patterns in Vertebrate Evolution*, M. K. Hecht, P. C. Goody, and B. M. Hecht, eds., Plenum Press, New York, pp. 3–51.

Hecht, M. K., P. C. Goody, and B. M. Hecht, eds., 1977, *Major Patterns in Vertebrate Evolution*, Plenum Press, New York.

Hempel, C. G., 1965, *Aspects of Scientific Explanation, and Other Essays in the Philosophy of Science*, Free Press, New York.

Hempel, C. G., 1966, *Philosophy of Natural Science*, Prentice-Hall Inc., Englewood Cliffs, N.J.

Hennig, W., 1950, *Grundzüge einer Theorie der phylogenetischen Systematik*, Deutscher Zentralverlag, Berlin.

Hennig, W., 1965, Phylogenetic Systematics, *Ann. Rev. Entomol.* 10:97–116.

Hennig, W., 1966a, *Phylogenetic Systematics*, University of Illinois, Urbana.

Hennig, W., 1966b, The Diptera fauna of New Zealand as a problem in systematics and zoogeography, *Pacific Insects Monograph* 9:1–81.

Hennig, W., 1975, "Cladistic Analysis or Cladistic Classification?": A reply to Ernst Mayr, *Syst. Zool.* **24**:244–256.

Hermelin, J. O. R., and B. A. Malmgren, 1980, Multivariate analysis of environmentally controlled variation in *Lagena:* Late Maastrichtian, Sweden, *Cretaceous Research* **1**:193–206.

Hickey, L. J., R. M. West, M. R. Dawson, and D. K. Choi, 1983, Arctic terrestrial biota: Paleomagnetic evidence of age disparity with mid-northern latitudes during the Late Cretaceous and Early Tertiary, *Science* **221**:1153–1156.

Hill, C. R., and P. R. Crane, 1982, Evolutionary cladistics and the origin of angiosperms, *Syst. Assoc. Spec. Vol.* **21**:269–361.

Hitching, F., 1982, *The Neck of the Giraffe,* New American Library, New York.

Ho, M.-W., and P. Y. Saunders, 1979, Beyond neo-Darwinism—An epigenetic approach to evolution, *Jour. Theor. Biol.* **78**:573–591.

Ho, M.-W., and P. Y. Saunders, eds., 1984, *Beyond Neo-Darwinism: An Introduction to the New Evolutionary Paradigm,* Academic Press, London.

Hoffman, A., 1979, Community paleoecology as an epiphenomenal science, *Paleobiology* **5**:357–379.

Hoffman, A., 1980, Ecostratigraphy: The limits of applicability, *Acta Geol. Polonica* **30**:97–108.

Hoffman, A., 1985, Patterns of family extinction depend on definition and geological timescale, *Nature* **315**:659–662.

Honacki, J. H., K. E. Kinman, and J. W. Koeppl, eds., 1982, *Mammal Species of the World: A Taxonomic and Geographic Reference,* Allen Press, Inc., and The Association of Systematics Collections, Lawrence, Kan.

Hubbs, C. L., ed., 1958, Zoogeography, *Publ. Amer. Assoc. Adv. Sci.* **51**:1–509.

Hull, D. L., 1970, Contemporary systematic philosophies, *Ann. Rev. Ecol. Syst.* **1**:19–54.

Hull, D. L., 1974, *Philosophy of Biological Science,* Prentice-Hall, Englewood Cliffs, N.J.

Hull, D. L., 1975, Central subjects and historical narratives, *History and Theory* **14**:253–274.

Hull, D. L., 1976a, Are species really individuals? *Syst. Zool.* **25**:174–191.

Hull, D. L., 1976b, The ontological staus of biological species, in *Boston Studies in the Philosophy of Science,* R. Butts and J. Hintikka, eds., D. Reidel, Dordrecht, vol. 32, pp. 347–358.

Hull, D. L., 1978, A matter of individuality, *Philosophy of Science* **45**:335–360.

Hull, D. L., 1979, The limits of cladism, *Syst. Zool.* **28**:416–440.

Hull, D. L., 1980, Individuality and selection, *Ann. Rev. Ecol. Syst.* **11**:311–332.

Hull, D. L., 1981, The principles of biological classification: The use and abuse of philosophy, *Philosophy of Science Association 1978* **2**:130–153.

Hull, D. L., 1983, Karl Popper and Plato's metaphor, in *Advances in Cladistics,* N. Platnick and V. Funk, eds., Columbia University Press, New York, pp. 177–189.

Hull, D. L., 1984, Cladistic theory: Hypotheses that blur and grow, in *Cladistics: Perspectives on the Reconstruction of Evolutionary History,* T. Duncan and T. F. Stuessy, eds., Columbia University Press, New York, pp. 5–23.

Humphries, C. J., and V. A. Funk, 1984, Cladistic methodology, *Syst. Assoc. Spec. Vol.* **25**:323–362.

Hunter, I. J., 1964, Paralogy, a concept complementary to homology and analogy, *Nature* **204**:604.

Hutchinson, G. E., 1978, *An Introduction to Population Ecology,* Yale University Press, New Haven.

Huxley, J., 1932 [reprinted 1972], *Problems of Relative Growth,* Dover Publications, New York.

Huxley, J., ed., 1940, *The New Systematics,* Clarendon Press, London.

Huxley, J., 1942 [reprinted 1964], *Evolution: The Modern Synthesis,* Wiley, New York.

Huxley, J., 1958, Evolutionary processes and taxonomy with special reference to grades, *Uppsala Univ. Arssks.* **1958**:21–38.

Huxley, T. H., 1862, The anniversary address, *Quart. Jour.Geol. Soc. London* **18**:xl–liv.

International Subcommission on Stratigraphic Classification (ISSC), 1976, *International Stratigraphic Guide,* H. D. Hedberg, ed., Wiley, Chichester, England.

Jablonski, D., K. W. Flessa, and J. W. Valentine, 1985, Biogeography and paleobiology, *Paleobiology* **11:**75–90.

Jablonski, D., J. J. Sepkoski, Jr., D. J. Bottjer, and P. M. Sheehan, 1983, Onshore-offshore patterns in the evolution of Phanerozoic shelf communities, *Science* **222:**1123–1125.

Janvier, P., 1984, Cladistics: Theory, purpose, and evolutionary implications, in *Evolutionary Theory: Paths into the Future,* J. W. Pollard, ed., Wiley, Chichester, England, pp. 39–75.

Jardine, N., and R. Sibson, 1971, *Mathematical Taxonomy,* Wiley, London.

Jefferies, R. P. S., 1979, The origin of chordates—A methodological essay, *Syst. Assoc. Spec. Vol.* **12:**443–477.

Jeletzky, J. A., 1956, Paleontology, a basis of practical geochronology, *Amer. Assoc. Petrol. Geol. Bull.* **40:**679–706.

Jensen, R. J., 1981, Wagner networks and Wagner trees: A presentation of methods for estimating most parsimonious solutions, *Taxon* **30:**576–590.

Jepsen, G. L., G. G. Simpson, and E. Mayr, eds., 1949, *Genetics, Paleontology and Evolution,* Princeton University Press, Princeton.

Johnson, A. L. A., 1981, Detection of ecophenotypic variation in fossils and its application to a Jurassic scallop, *Lethaia* **14:**277–285.

Johnson, J. G., 1982, Occurrence of phyletic gradualism and punctuated equilibria through geologic time, *Jour. Paleontol.* **56:**1329–1331.

Johnson, M. E., 1979, Evolutionary brachiopod lineages from the Llandoverly series of eastern Iowa, *Palaeontology* **22:**549–567.

Johnson, M. E., and V. R. Colville, 1982, Regional integration of evidence for evolution in the Silurian *Pentamerus-Pentameroides* lineage, *Lethaia* **15:**41–54.

Joysey, K. A., and A. E. Friday, eds., 1982, Problems of phylogenetic reconstruction, *Syst. Assoc. Spec. Vol.* **21:**1–442.

Kaplan, D. R., 1984, The concept of homology and its central role in the elucidation of plant systematic relationships, in *Cladistics: Perspectives on the Reconstruction of Evolutionary History,* T. Duncan and T. F. Stuessy, eds., Columbia University Press, New York, pp. 51–70.

Kauffman, E. G., 1977, Evolutionary rates and biostratigraphy, in *Concepts and Methods of Biostratigraphy,* E. G. Kauffman and J. E. Hazel, eds., Dowden, Hutchinson and Ross, Stroudsburg, Pa., pp. 109–141.

Kauffman, E. G., and J. E. Hazel, 1977, *Concepts and Methods of Biostratigraphy,* Dowden, Hutchinson and Ross, Stroudsburg, Pa.

Keast, J. A., 1973, Contemporary biotas and the separation sequence of the southern continents, in *Implications of Continental Drift to the Earth Sciences,* D. H. Tarling and S. K. Runcorn, eds., Academic Press, London, vol. 1, pp. 309–343.

Keast, J. A., 1977, Zoogeography and phylogeny: The theoretical background and methodology to the analysis of mammal and bird fauna, in *Major Patterns in Vertebrate Evolution,* M. K. Hecht, P. C. Goody, and B. M. Hecht, eds., Plenum Press, New York, pp. 249–312.

Kellogg, D. E., 1975, The role of phyletic change in the evolution of *Pseudocubus vema* (Radiolaria), *Paleobiology* **1:**359–370.

Kellogg, D. E., 1976, Character displacement in the radiolarian genus *Eucyrtidium, Evolution* **29:**736–749.

Kellogg, D. E., 1980, Character displacement and phyletic change in the evolution of the radiolarian subfamily Artiscinae, *Micropaleontology* **26:**196–210.

Kellogg, D. E., 1982, Phenology of morphologic change in radiolarian lineages from deep-sea cores: Implications for macroevolution, *North America Paleontol. Conv., 3rd Proc.* **1:**281–284.

Kellogg, D. E., 1983, Phenology of morphologic change in radiolarian lineages from deep-sea cores: Implications for macroevolution, *Paleobiology* **9:**355–362.

Kennedy, W. J., 1977, Ammonite evolution, in *Patterns of Evolution as Illustrated by the Fossil Record*, A. Hallam, ed., Elsevier, Amsterdam, pp. 251–304.

Kennedy, W. J., and W. A. Cobban, 1977, The role of ammonites in biostratigraphy, in *Concepts and Methods of Biostratigraphy*, E. G. Kauffman and J. E. Hazel, eds., Dowden, Hutchinson and Ross, Stroudsburg, Pa., pp. 309–320.

Kidwell, S. M., 1982, Time scales of fossil accumulation: Patterns from Miocene benthic assemblages, *North American Paleont. Conv., 3rd Proc.* **1:**295–300.

Kinne, O., 1963, The effects of temperature and salinity on marine and brackish water animals. I. Temperature, *Oceanog. Marine Biol. Ann. Rev.* **1:**301–340.

Kinne, O., 1964, The effects of temperature and salinity on marine and brackish water animals. II. Salinity and temperature salinity combinations, *Oceanog. Marine Biol. Ann. Rev.* **2:**281–339.

Kinne, O., ed., 1971, *Marine Ecology: A Comprehensive, Integrated Treatise on Life in Oceans and Coastal Waters*, Wiley, New York.

Kiriakoff, S. G., 1959, Phylogenetic systematics versus typology, *Syst. Zool.* **8:**117–118.

Kitchell, J. A., and D. Pena, 1984, Periodicity of extinctions in the geologic past: Deterministic versus stochastic explanations, *Science* **226:**689–692.

Kluge, A. G., 1984, The relevance of parsimony to phylogenetic inference, in *Cladistics: Perspectives on the Reconstruction of Evolutionary History*, T. Duncan and T. F. Stuessy, eds., Columbia University Press, New York, pp. 24–38.

Kluge, A. G., 1985, Ontogeny and phylogenetic systematics, *Cladistics* **1:**13–27.

Kluge, A. G., and J. S. Farris, 1969, Qualitative phyletics and the evolution of anurans, *Syst. Zool.* **18:**1–32.

Kneale, W. and M. Kneale, 1962, *The Development of Logic*, Clarendon Press, Oxford.

Kohlberger, W., 1978, *Review of Problems in Vertebrate Evolution*, Ed. by S. M. Andrews, R. S. Miles, and A. D. Walker, *Syst. Zool.* **27:**373–379.

Kohlberger, W., and R. M. Schoch, 1983, The value of paleontology, *Evol. Theory* **6:**210.

Krassilov, V., 1974, Causal biostratigraphy, *Lethaia* **7:**173–179.

Krassilov, V., 1978, Organic evolution and natural stratigraphic classification, *Lethaia* **11:**93–104.

Krishtalka, L., R. J. Emry, J. E. Storer, and J. F. Sutton, 1982, Oligocene multituberculates (Mammalia: Allotheria): Youngest known record, *Jour. Paleontol.* **56:**791–794.

Kurtén, B., 1964, The population dynamic approach to paleoecology, in *Approaches to Paleoecology*, J. Imbrie and N. D. Newell, eds., Wiley, New York, pp. 91–106.

Kurtén, B., 1967, Continental drift and the palaeogeography of reptiles and mammals, *Soc. Sci. Fennica Commentat. Biol.* **31:**1–8.

Lamarck, J. B., 1984, *Zoological Philosophy*, University of Chicago Press, Chicago.

Lankester, E. R., 1870, On the use of the term homology in modern zoology and the distinction between homogenetic and homoplastic agreements, *Ann. Mag. Nat. Hist.*, ser. 4, **6:** 34–43.

Lauder, G. V., 1981, Form and function: Structural analysis in evolutionary morphology, *Paleobiology* **7:**430–442.

Lauder, G. V., 1982, Historical biology and the problem of design, *Jour. Theor. Biol.* **97:**57–67.

Lazarus, D. B., 1983, Speciation in pelagic Protista and its study in the planktonic microfossil record: A review, *Paleobiology* **9:**327–340.

Lazarus, D. B., and D. R. Prothero, 1984, The role of stratigraphic and morphologic data in phylogeny, *Jour. Paleontol.* **58:**163–172.

Lazarus, D. B., R. P. Scherer, and D. R. Prothero, 1985, Evolution of the radiolarian species-complex *Pterocanium*: A preliminary survey, *Jour. Paleontol.* **59:**183–220.

Le Quesne, W. J., 1969, A method of selection of characters in numerical taxonomy, *Syst. Zool.* **18:**201–205.

Le Quesne, W. J., 1972, Further studies based on the uniquely derived character concept, *Syst. Zool.* **21:**281–288.

Le Quesne, W. J., 1974, The uniquely evolved character concept and its cladistic application, *Syst. Zool.* **23:**513–517.

Levinton, J. S., and C. M. Simon, 1980, A critique of the punctuated equilibria model and implications for the detection of speciation in the fossil record, *Syst. Zool.* **29:**130–142.

Lewin, R., 1983, Origin of species in stressed environments, *Science 222:*1112.

Linnaeus, C., 1758, *Systema Naturae per Regna tria Naturae, Secundum Classes, Ordines, Genera, Species cum Characteribus, Differentiis, Synonymis, Locis. Editio decima, reformata,* Laurentii Salvii, Stockholm.

Lohmann, G. P., 1983, Eigenshape analysis of microfossils: A general morphometric procedure for describing changes in shape, *Mathematical Geol.* **15:**659–672.

Lohmann, G. P., and B. A. Malmgren, 1983, Equatorward migration of *Globorotalia truncatulinoides* ecophenotypes through the Late Pleistocene: Gradual evolution or ocean change? *Paleobiology* **9:**414–421.

Löther, R., 1972, *Die Beherrschung der Mannigfaltigkeit,* Gustav Fischer, Jena.

Løvtrup, S., 1973, Classification, convention and logic, *Zool. Scripta* **2:**49–61.

Løvtrup, S., 1974, *Epigenetics: A Treatise on Theoretical Biology,* Wiley, London.

Løvtrup, S., 1975, On phylogenetic classification, *Acta Zool. Cracov.* **20:**499–523.

Løvtrup, S., 1977a, *The Phylogeny of Vertebrata,* Wiley, London.

Løvtrup, S., 1977b, Phylogenetics: Some comments on cladistic theory and method, in *Major Patterns in Vertebrate Evolution,* M. K. Hecht, P. C. Goody, and B. M. Hecht, eds., Plenum Press, New York, pp. 805–822.

Løvtrup, S., 1982a, The four theories of evolution I: The theories on the reality and on the history of evolution, *Revista di Biologia* **75:**53–60.

Løvtrup, S., 1982b, The four theories of evolution II: The epigenetic theory, *Revista di Biologia* **75:**231–255.

Løvtrup, S., 1982c, The four theories of evolution III: The ecological theory, *Revista di Biologia* **75:**385–398.

Løvtrup, S., 1984, Ontogeny and phylogeny, in *Beyond Neo-Darwinism: An Introduction to the New Evolutionary Paradigm,* M.-W. Ho and P. T. Saunders, eds., Academic Press, London, pp. 159–190.

MacArthur, R. H., and E. O. Wilson, 1967, *The Theory of Island Biogeography,* Princeton University Press, Princeton, N.J.

MacGinitie, 1969, The Eocene Green River flora of northwestern Colorado and northeastern Utah, *Univ. Calif. Publ. Geol. Sci.* **83:**1–140.

McKenna, M. C., 1973, Sweepstakes, filters, corridors, Noah's arks, and beached Viking funeral ships in palaeogeography, in *Implications of Continental Drift to the Earth Sciences,* vol. 1, D. H. Tarling and S. K. Runcorn, eds., Academic Press, London, pp. 295–308.

McKenna, M. C., 1975, Toward a phylogenetic classification of the Mammalia, in *Phylogeny of the Primates,* W. P. Luckett and F. S. Szalay, eds., Plenum Press, New York and London, pp. 21–46.

McKenna, M. C., 1981, Book review of *Comparative Biology and Evolutionary Relationships of Tree Shrews* edited by W. P. Luckett, *Int. Jour. Primatol.* **2:**97–101.

McKenna, M. C., 1984, Holarctic landmass rearrangement, cosmic events, and Cenozoic terrestrial organisms, *Ann. Missouri Bot. Gard.* **70:**459–489.

McKenna, M. C, G. F. Engelmann, and S. F. Barghoorn, 1977, Review of "Cranial anatomy and evolution of Early Tertiary Plesiadapidae (Mammalia, Primates)" by P. D. Gingerich, *Syst. Zool.* **26:**233–238.

McKinney, H. L., ed., 1971, *Lamarck to Darwin: Contributions to Evolutionary Biology 1809–1859,* Coronado Press, Lawrence, Kan.

McKinney, M. L., 1984, Allometry and heterochrony in an Eocene echinoid lineage: Morphological change as a by-product of size selection, *Paleobiology* **10:**407–419.

McKinney, M. L., 1985, Distinguishing patterns of evolution from patterns of deposition, *Jour. Paleontol.* **59**:561–567.

McKinney, M. L., and R. M. Schoch, 1983, A composite terrestrial Paleocene section with completeness estimates, based upon magnetostratigraphy, *Amer. Jour. Sci.* **283**:801–814.

McNamara, K. J., 1980, Evolutionary trends and their functional significance in chasmopine trilobites, *Lethaia* **13**:61–78.

McNamara, K. J., 1982, Heterochrony and phylogenetic trends, *Paleobiology* **8**:130–142.

McNeill, J., 1978, Purposeful phenetics, *Syst. Zool.* **28**:465–482.

McNeill, J., 1982, Phylogenetic reconstruction and phenetic taxonomy, *Zool. Jour. Linn. Soc.* **74**:337–344.

Malmgren, B. A., 1984, Analysis of the environmental influence on the morphology of *Ammonia beccarii* (Linné) in southern European Salinas, *Geobios* **17**:737–746.

Malmgren, B. A., W. A. Berggren, and G. P. Lohmann, 1983, Evidence for punctuated gradualism in the Late Neogene *Globoratalia tumida* lineage of planktonic Foraminifera, *Paleobiology* **9**:377–389.

Malmgren, B. A., W. A. Berggren, and G. P. Lohmann, 1984, Species formation through punctuated gradualism in planktonic Foraminifera, *Science* **225**:317–319.

Malmgren, B. A., and J. P. Kennett, 1981, Phyletic gradualism in a Late Cenozoic planktonic foraminiferal lineage; DSDP Site 284, southwest Pacific, *Paleobiology* **7**:230–240.

Malmgren, B. A., and J. P. Kennett, 1982, The potential of morphologically based phylo-zonation: Application of a Late Cenozoic planktonic foraminiferal lineage, *Marine Micropaleontol.* **7**:285–296.

Martinsson, A., 1973, Editor's column: Ecostratigraphy, *Lethaia* **6**:441–443.

Martinsson, A., 1978, Project ecostratigraphy, *Lethaia* **11**:84.

Martinsson, A., 1980a, International Commission on Stratigraphy, *Lethaia* **13**:26.

Martinsson, A., 1980b, Ecostratigraphy: Limits of applicability, *Lethaia* **13**:363.

Masalin, T. P., 1952, Morphological criteria of phyletic relationships, *Syst. Zool.* **1**:49–70.

Mason, B., and L. G. Berry, 1978, *Elements of Mineralogy,* Freeman, San Francisco.

Matthew, W. D., 1915, Climate and evolution, *Ann. New York Acad. Sci.* **24**:171–318.

Maynard Smith, J., 1975, *The Theory of Evolution,* Penguin Books, Middlesex.

Maynard Smith, J., 1978, *The Evolution of Sex,* Cambridge University Press, Cambridge.

Maynard Smith, J., ed., 1982, *Evolution Now: A Century after Darwin,* Freeman, San Francisco.

Mayr, E., 1940, Speciation phenomena in birds, *Amer. Nat.* **74**:249–278.

Mayr, E., ed., 1952, The problem of land connections across the South Atlantic, with special reference to the Mesozoic, *Bull. Amer. Mus. Nat. Hist.* **99**:79–258.

Mayr, E., ed., 1957, *The Species Problem,* Publication No. 50 of the American Association for the Advancement of Science, Washington, D.C.

Mayr, E., 1963, *Animal Species and Evolution,* Harvard University Press, Cambridge, Mass.

Mayr, E., 1969, *Principles of Systematic Zoology,* McGraw-Hill, New York.

Mayr, E., 1970, *Populations, Species, and Evolution* (an abridgment of *Animal Species and Evolution*), Harvard University Press, Cambridge, Mass.

Mayr, E., 1974, Cladistic analysis or cladistic classification? *Zeit. Zool. Syst. Evolut.-forsch.* **12**:94–128.

Mayr, E., 1976a, The role of systematics in biology, in *Evolution and the Diversity of Life: Selected Essays,* E. Mayr, ed., Harvard University Press, Cambridge, Mass., pp. 416–424.

Mayr, E., 1976b, Theory of biological classification, in *Evolution and the Diversity of Life: Selected Essays,* E. Mayr, ed., Harvard University Press, Cambridge, Mass., pp. 425–432.

Mayr, E., 1976c, Species concepts and definitions, in *Evolution and the Diversity of Life: Selected Essays,* E. Mayr, ed., Harvard University Press, Cambridge, Mass., pp. 493–508.

Mayr, E., 1976d, Is the species a class or an individual? *Syst. Zool.* **25**:192.

Mayr, E., 1981, Biological classification: Toward a synthesis of opposing views, *Science* **214**:510–516.

Mayr, E., 1982, *The Growth of Biological Thought: Diversity, Evolution, and Inheritance,* Harvard University Press, Cambridge, Mass.

Mayr, E., and W. B. Provine, eds., 1980, *The Evolutionary Synthesis: Perspectives on the Unification of Biology.* Harvard University Press, Cambridge, Mass.

Meacham, C. A., 1980, Phylogeny of the Berberidaceae with an evaluation of classifications, *Syst. Bot.* **5**:149–172.

Meacham, C. A., 1984, Evaluating characters by character compatibility analysis, in *Cladistics: Perspectives on the Reconstruction of Evolutionary History,* T. Duncan and T. F. Stuessy, eds., Columbia University Press, New York, pp. 152–165.

Medioli, F. S., and D. B. Scott, 1983, Holocene Arcellacea (Thecamoebians) from eastern Canada, *Cushman Foundation Spec. Publ. No.* **21**:5–63.

Meglitsch, P. A., 1954, On the nature of species, *Syst. Zool.* **3**:49–65.

Melville, R. V., 1984, International Code of Zoological Nomenclature, 3rd ed., *Bull. Zool. Nom.* **41**:196–197.

Meredith, A., 1982, Devolution, *Jour. Theor. Biol.* **96**:49–65.

Meredith, A., and R. M. Schoch, 1984, Devolution and punctuated equilibrium, *Evol. Theory* **7**:68.

Mickevich, M. F., 1978, Taxonomic congruence, *Syst. Zool.* **27**:143–158.

Mickevich, M. F. and M. S. Johnson, 1976, Congruence between morphological and allozyme data in evolutionary inference and character evolution, *Syst. Zool.* **25**:260–270.

Milkman, R., ed., 1982, *Perspectives on Evolution,* Sinauer Associates, Sunderland, Mass.

Mill, J. S., 1879, *A System of Logic,* Longmans, Green, London.

Miller, F. X., 1977, The Graphic correlation method in biostratigraphy, in *Concepts and Methods of Biostratigraphy,* E. G. Kauffman and J. E. Hazel, eds., Dowden, Hutchinson and Ross, Stroudsburg, Pa., pp. 165–186.

Miller, M. F., A. A. Ekdale, and M. D. Picard, eds., 1984, Trace fossils and paleoenvironments: Marine carbonate, marginal marine terrigenous and continental terrigenous settings, *Jour. Paleontol.* **58**:283–587.

Moorhouse, W. W., 1959, *The Study of Rocks in Thin Section,* Harper & Row, New York.

Moore, P. D., 1983, Ecological diversity and stress, *Nature* **306**:17.

Nagel, E., 1961, *The Structure of Science,* Harcourt, Brace and World, New York.

Nelson, C. H., and G. S. Van Horn, 1975, A new simplified method for constructing Wagner networks and the cladistics of *Pentachaeta* (Compositae, Astereae), *Brittonia* **27**:363–373.

Nelson, G., 1969, Origin and diversification of teleostean fishes, *Ann. New York Acad. Sci.* **167**:18–30.

Nelson, G., 1970, Outline of a theory of comparative biology, *Syst. Zool.* **19**:373–384.

Nelson, G., 1971, "Cladism" as a philosophy of classification, *Syst. Zool.* **20**:373–376.

Nelson, G., 1972a, Phylogenetic relationships and classification, *Syst. Zool.* **21**:227–231.

Nelson, G., 1972b, Comments on Hennig's "Phylogenetic Systematics" and its influence on ichthyology, *Syst. Zool.* **21**:364–374.

Nelson, G., 1973, Classification as an expression of phylogenetic relationships, *Syst. Zool.* **22**:344–359.

Nelson, G., 1978a, From Candolle to Croizat: Comments on the history of biogeography, *Jour. Hist. Biol.* **11**:269–305.

Nelson, G., 1978b, Ontogeny, phylogeny, paleontology, and the biogenetic law, *Syst. Zool.* **27**:324–345.

Nelson, G., 1979, Cladistic analysis and synthesis: Principles and definitions, with a historical note on Adanson's *Familles des Plantes* (1763–1764), *Syst. Zool.* **28**:1–21.

Nelson, G., 1981, Summary, in *Vicariance Biogeography: A Critique*, G. Nelson and D. E. Rosen, eds., Columbia University Press, New York, pp. 524–537.

Nelson, G., 1983a, Reticulation in cladograms, in *Advances in Cladistics, vol. 2: Proceedings of the Second Meeting of the Willi Hennig Society*, N. I. Platnick and V. A. Funk, eds., Columbia University Press, New York, pp. 105–111.

Nelson, G., 1983b, Vicariance and cladistics: Historical perspectives with implications for the future, *Syst. Assoc. Spec. Vol.* **23:**469–492.

Nelson, G., 1984, Cladistics and biogeography, in *Cladistics: Perspectives on the Reconstruction of Evolutionary History*, T. Duncan and T. F. Stuessy, eds., Columbia University Press, New York, pp. 273–293.

Nelson, G., 1985, Outgroups and ontogeny, *Cladistics* **1:**29–45.

Nelson, G., and N. I. Platnick, 1978, The perils of plesiomorphy: Widespread taxa, dispersal, and phenetic biogeography, *Syst. Zool.* **27:**474–477.

Nelson, G., and N. I. Platnick, 1980, A vicariance approach to historical biogeography, *Bioscience* **30:**339–343.

Nelson, G., and N. I. Platnick, 1981, *Systematics and Biogeography: Cladistics and Vicariance*, Columbia University Press, New York.

Nelson, G., and D. E. Rosen, eds., 1981, *Vicariance Biogeography: A Critique*, Columbia University Press, New York.

Newell, N. D., 1959, The nature of the fossil record, *Proc. Amer. Phil. Soc.* **103:**264–285.

Newell, N. D., and D. W. Boyd, 1975, Parallel evolution in Early Trigoniacean bivalves, *Bull. Amer. Mus. Nat. Hist.* **154:**53–162.

North American Commission on Stratigraphic Nomenclature (NACSN), 1983, North American Stratigraphic Code, *Amer. Assoc. Petrol. Geol. Bull.* **67:**841–875.

Novacek, M. J., 1982a, Information for molecular studies from anatomical and fossil evidence on higher eutherian phylogeny, in *Macromolecular Sequences in Systematic and Evolutionary Biology*, M. Goodman, ed., Plenum Press, New York, pp. 3–41.

Novacek, M. J., convener, 1982b, Symposium: Phylogeny and rates of evolution, *Syst. Zool.* **31:**337–412.

Novacek, M. J., and M. A. Norell, 1982, Fossils, phylogeny, and taxonomic rates of evolution, *Syst. Zool.* **31:**366–375.

Odin, G. S., ed., 1982, *Numerical Dating in Stratigraphy*, Wiley, Chichester, England.

Odum, E. P., 1971, *Fundamentals of Ecology*, W. B. Saunders, Philadelphia.

O'Grady, R. T., 1984, Evolutionary theory and teleology, *Jour. Theor. Biol.* **107:**563–578.

Osborn, H. F., 1929, The titanotheres of ancient Wyoming, Dakota, and Nebraska, *U.S. Geol. Surv., Monograph* **55:**1–953.

Oster, G. and P. Alberch, 1982, Evolution and bifurcation of developmental programs, *Evolution* **36:**444–459.

Owen, R., 1843, *Lectures on Comparative Anatomy*, Longman, Brown, Green, and Longmans, London.

Owen, R., 1866, *On the Anatomy of Vertebrates. Vol. 1. Fishes and Reptiles*, Longmans, Green, London.

Panchen, A. L., 1982, The use of parsimony in testing phylogenetic hypotheses, *Zool. Jour. Linn. Soc.* **74:**305–328.

Paterson, H. E. H., 1978, More evidence against speciation by reinforcement, *South Afr. Jour. Sci.* **74:**369–371.

Paterson, H. E. H., 1982, Perspectives on speciation by reinforcement, *South Afr. Jour. Sci.* **78:**53–57.

Patterson, C., 1977, The contribution of paleontology to teleostean phylogeny, in *Major Patterns in Vertebrate Evolution*, M. K. Hecht, P. C. Goody, and B. M. Hecht, eds., Plenum, New York, pp. 579–643.

Patterson, C., 1980, Cladistics, *Biologist* **27**:234–240.

Patterson, C., 1981*a*, Significance of fossils in determining evolutionary relationships, *Ann. Rev. Ecol. Syst.* **12**:195–223.

Patterson, C., 1981*b*, Agassiz, Darwin, Huxley, and the fossil record of teleost fishes, *Bull. Brit. Mus. Nat. Hist. (Geol.)* **36**:213–224.

Patterson, C., 1981*c*, Methods of paleobiogeography, in *Vicariance Biogeography: A Critique*, G. Nelson and D. E. Rosen, eds., Columbia University Press, New York, pp. 447–490.

Patterson, C., 1982*a*, Morphological characters and homology, *Syst. Assoc. Spec. Vol.* **21**:21–74.

Patterson, C., ed., 1982*b*, Methods of phylogenetic reconstruction, *Zool. Jour. Linn. Soc., London* **74**:197–344.

Patterson, C., 1982*c*, Classes and cladists or individuals and evolution, *Syst. Zool.* **31**:284–286.

Patterson, C., 1983*a*, Aims and methods in biogeography, *Syst. Assoc. Spec. Vol.* **23**:1–28.

Patterson, C., 1983*b*, How does phylogeny differ from ontogeny?, in *Development and Evolution*, B. C. Goodwin, H. Holder, and C. C. Wylie, eds., Cambridge University Press, Cambridge, pp. 1–31.

Patterson, C., and D. E. Rosen, 1977, Review of ichthyodectiform and other Mesozic teleost fishes and the theory and practice of classifying fossils, *Bull. Amer. Mus. Nat. Hist.* **158**:81–172.

Paul, C. R. C., 1982, The adequacy of the fossil record, *Syst. Assoc. Spec. Vol.* **21**·75–117.

Phillips, R. B., 1984, Considerations in formalizing a classification, in *Cladistics: Perspectives on the Reconstruction of Evolutionary History*, T. Duncan and T. F. Stuessy, eds., Columbia University Press, New York, pp. 257–272.

Planka, E. R., 1974, *Evolutionary Ecology*, Harper & Row, New York.

Platnick, N. I., 1977*a*, Parallelism in phylogeny reconstruction, *Syst. Zool.* **26**:93–86.

Platnick, N. I., 1977*b*, Cladograms, phylogenetic trees, and hypothesis testing, *Syst. Zool.* **26**:438–442.

Platnick, N. I., 1979, Philosophy and the transformation of cladistics, *Syst. Zool.* **28**:537–546.

Platnick, N. I., and H. D. Cameron, 1977, Cladistic methods in textual, linguistic, and phylogenetic analysis, *Syst. Zool.* **26**:380–385.

Platnick, N. I., and V. A. Funk, eds., 1983, *Advances in Cladistics, Vol. 2: Proceedings of the Second Meeting of the Willi Hennig Society*, Columbia University Press, New York.

Platnick, N. I., and G. Nelson, 1978, A method of analysis for historical biogeography, *Syst. Zool.* **27**:1–16.

Pollard, J. W., ed., 1984, *Evolutionary Theory: Paths into the Future*, Wiley, Chichester, England.

Pope, J. K., 1976, Comparative morphology and shell histology of the Ordovician Strophomenacea (Brachiopoda), *Palaeontographica Americana* **8**:129–213.

Popper, K., 1968, *The Logic of Scientific Discovery*, Harper & Row, New York.

Popper, K., 1984, Evolutionary epistemology, in *Evolutionary Theory: Paths into the Future*, J. W. Pollard, ed., Wiley, Chichester, England, pp. 239–255.

Prothero, D. R., and D. B. Lazarus, 1980, Planktonic microfossils and the recognition of ancestors, *Syst. Zool.* **29**:119–129.

Rachootin, S. P., and K. S. Thomson, coordinators, 1980, Seminar on *Evolution above the Species Level*, September–December 1980, held by the Department of Biology, Bingham Laboratories, Yale University, New Haven, Conn.

Rachootin, S. P., and K. S. Thomson, 1981, Epigenetics, paleontology and evolution, in *Evolution Today*, G. G. E. Scudder and J. L. Reveal, eds., Proceedings of the Second International Congress on Systematic and Evolutionary Biology, Carnegie-Mellon University, Pittsburgh, Pa., pp. 181–193.

Radinsky, L., 1983, Allometry and reorganization in horse skull proportions, *Science* **221**:1189–1191.

Raff, R. A., and T. C. Kaufman, 1983, *Embryos, Genes, and Evolution: The Developmental-Genetic Basis of Evolutionary Change*, Macmillan, New York.

Ramsbottom, J., 1938, Linnaeus and the species concept, *Proc. Linn. Soc., London* **150**:192–219.

Raup, D. M., and R. E. Crick, 1981, Evolution of single characters in the Jurassic ammonite *Kosmoceras, Paleobiology* **7**:200–215.

Raup, D. M., and S. J. Gould, 1974, Stochastic simulation and evolution of morphology—Towards a nomothetic paleontology, *Syst. Zool.* **23**:305–322.

Raup, D. M., S. J. Gould, T. J. M. Schopf, and D. S. Simberloff, 1973, Stochastic models of phylogeny and the evolution of diversity, *Jour. Geol.* **81**:525–542.

Raup, D. M., and S. M. Stanley, 1978, *Principles of Paleontology*, Freeman, San Francisco.

Reineck, H.-E., 1960, Uber Zeitlucken in rezenten Flachsee-Sedimenten, *Geol. Rundshau.* **49**:149–161.

Remane, A., 1955, Morphologie als Homologienforschung, *Zool. Anz. Suppl.* **18**:159–183.

Remane, A., 1956, *Die Grundlagen des naturlichen Systems der vergleichenden Anatomie und der Phylogenetik, 2,* Aufl. Geest und Portig K. G., Leipzig.

Remane, A., 1971, *Die Grundlagen des naturlichen Systems der vergleichenden Anatomie und der Phylogenetik,* Koeltz, Konigstein-Taunus.

Rensch, B., 1959, *Evolution above the Species Level,* Wiley, New York.

Reyment, R. A., R. E. Blackith, and N. A. Campbell, 1984, *Multivariate Morphometrics,* Academic Press, New York.

Ricklefs, R. E., 1979, *Ecology,* Chiron Press, New York.

Ridley, M., 1984, Selected theory [Review of *Conceptual Issues in Evolutionary Biology: An Anthology* edited by E. Sober], *Nature* **311**:489–490.

Ridley, M., 1985, *The Problems of Evolution,* Oxford University Press, London.

Riedl, R., 1977, A systems-analytical approach to macro-evolutionary phenomena, *Quart. Rev. Biol.* **52**:351–370.

Riedl, R., 1978, *Order in Living Organisms,* Wiley, Chichester, England.

Rightmire, G. P., 1981, Patterns in the evolution of *Homo erectus, Paleobiology* **7**:241–246.

Rohlf, F. J., 1970, Adaptive hierarchical clustering schemes, *Syst. Zool.* **19**:58–82.

Rosen, D. E., 1974a, Review of *Space, Time, Form: The Biological Synthesis* by L. Croizat, *Syst. Zool.* **23**:288–290.

Rosen, D. E., 1974b, Phylogeny and zoogeography of salmoniform fishes and relationships of *Lepidogalaxias salamandroides, Bull. Amer. Mus. Nat. Hist.* **153**:265–326.

Rosen, D. E., 1975, A vicariance model of Caribbean biogeography, *Syst. Zool.* **24**:431–464.

Rosen, D. E., 1978, Vicariant patterns and historical explanation in biogeography, *Syst. Zool.* **27**:159–188.

Rosen, D. E., 1979, Fishes from the uplands and intermontane basins of Guatemala: Revisionary studies and comparative biogeography, *Bull. Amer. Mus. Nat. Hist.* **162**:267–367.

Rosen, D. E., 1984, Hierarchies and history, in *Evolutionary Theory: Paths into the Future,* J. W. Pollard, ed., Wiley, Chichester, England, pp. 77–97.

Roth, V. L., 1984, On homology, *Biol. Jour. Linn. Soc.* **22**:13–29.

Rudwick, M. J. S., 1964, The inference of structure from function in fossils, *Brit. Jour. Phil. Soc.* **15**:27–40.

Ruse, M., 1973, *The Philosophy of Biology,* Addison-Wesley, Don Mills, Ontario.

Russell, B., 1946, *History of Western Philosophy,* George Allen and Unwin, London.

Russell, E. S., 1916 [reprinted 1982], *Form and Function: A Contribution to the History of Animal Morphology* (with a new introduction by George V. Lauder), University of Chicago Press, Chicago.

Sadler, P. M., 1981, Sediment accumulation rates and the completeness of stratigraphic sections, *Jour. Geol.* **89**:569–584.

Sadler, P. M., 1983, Is the present long enough to measure the past? *Nature* **302**:752.

Sadler, P. M., and L. W. Dingus, 1982, Expected completeness of sedimentary sections: Estimating a time-scale dependent, limiting factor in the resolution of the fossil record, *North American Paleontol. Conv., 3rd Proc.* **2**:461–464.

Saether, O. A., 1979a, Underlying synapomorphies and anagenetic analysis, *Zool. Scripta* **8:**305–312.

Saether, O. A., 1979b, Hierarchy of the Chironomidae with special emphasis on the female genitalia (Diptera), *Ent. Scand. Suppl.* **10:**17–26.

Saether, O. A., 1979c, [Underlying synapomorphy and unique inside-parallelism elucidated by examples from Chironomidae and Chaoboridae (Diptera)], *Ent. Tidskr.* **100:**173–180.

Saether, O. A., 1983, The canalized evolutionary potential: Inconsistencies in phylogenetic reasoning, *Syst. Zool.* **32:**343–359.

Sainte-Hilaire, G., 1825, *Annales des Sciences Nat.* **12:**341.

Sarjeant, W. A. S., ed., 1983, *Terrestrial Trace Fossils*, Hutchinson Ross, Stroudsburg, Pa.

Sarjeant, W. A. S., and W. J. Kennedy, 1973, Proposal of a code for the nomenclature of trace-fossils, *Canadian Jour. Earth Sci.* **10:**460–475.

Savage, D. E., 1977, Aspects of vertebrate paleontological stratigraphy and geochronology, in *Concepts and Methods of Biostratigraphy*, E. G. Kauffman and J. E. Hazel, eds., Dowden, Hutchinson, and Ross, Stroudsburg, Pa., pp. 427–442.

Savage, D. E., and D. E. Russell, 1983, *Mammalian Paleofaunas of the World*, Addison-Wesley, Reading, Mass.

Schaeffer, B., M. K. Hecht, and N. Eldredge, 1972, Phylogeny and paleontology, *Evol. Biol.* **6:**31–46.

Schankler, D. M., 1980. Faunal zonation of the Willwood Formation in the Central Bighorn Basin, Wyoming, *Univ. Mich. Pap. Paleontol.* **24:**99–114.

Schankler, D. M., 1981, Local extinction and ecological re-entry of Early Eocene mammals, *Nature* **293:**135–138.

Schindel, D. E., 1980, Microstratigraphic sampling and the limits of paleontological resolution, *Paleobiology* **6:**408–426.

Schindel, D. E., 1982a, The gaps in the fossil record, *Nature* **297:**282–284.

Schindel, D. E., 1982b, Gaps in the fossil record: Deme histories are not species' histories, *Nature* **299:**490.

Schindel, D. E., 1982c, Resolution analysis: A new approach to gaps in the fossil record, *Paleobiology* **8:**340–353.

Schindel, D. E., 1982d, Punctuations in the Pennsylvanian evolutionary history of *Glabrocingulum* (Mollusca: Archaeogastropoda), *Geol. Soc. Amer. Bull.* **93:**400–408.

Schmidt-Nielsen, K., 1984, *Scaling: Why Is Animal Size So Important?*, Cambridge University Press, Cambridge.

Schoch, R. M., 1982, Gaps in the fossil record: Fossils and stratigraphy, *Nature* **299:**490.

Schoch, R. M., 1983, Third North American Palaeontological Convention, *Geoscience Canada* **10:**204–207.

Schoch, R. M., 1984a, Cladism defended, *New Scientist* **101:**47.

Schoch, R. M., ed., 1984b, *Vertebrate Paleontology*, Van Nostrand Reinhold, New York.

Schoch, R. M., in press, Systematics, functional morphology and macroevolution of the extinct mammalian order Taeniodonta, *Bull. Peabody Mus. Nat. Hist., Yale Univ.* (also available from University Microfilms International, Ann Arbor, Mich., Dissertation No. DA8329313).

Schoch, R. M., and A. Meredith, 1984, Punctuated patterns in the fossil record: The devolutionary mechanism, *Mem. III Cong. Latinoamericano Paleont., Oaxtepec, Mexico,* Instituto de Geología, Universidad Nacional Autónoma de México, Oaxtepec, pp. 624–630.

Schopf, T. J. M., 1981, Punctuated equilibrium and evolutionary stasis, *Paleobiology* **7:**156–166.

Schopf, T. J. M., 1984, Rates of evolution and the notion of "living fossils," *Ann. Rev. Earth Planet. Sci.* **12:**245–292.

Schopf, T. J. M., and A. Hoffman, 1983, Punctuated equilibrium and the fossil record, *Science* **219:**438–439.

Schwartz, J. H., and H. B. Rollins, eds., 1979, Models and methodologies in evolutionary theory, *Bull. Carnegie Mus.* **13:**1–105.

Scriven, M., 1959, Explanation and prediction in evolutionary theory, *Science* **130**:477–482.
Scudder, G. G. E., and J. L. Reveal, eds., 1981, *Evolution Today*, Proceedings of the Second International Congress on Systematic and Evolutionary Biology, Carnegie-Mellon University, Pittsburgh, Pa.
Seilacher, A., 1970, Arbeitskonzept zur konstruktions-Morphologie, *Lethaia* **3**:393–396.
Seilacher, A., 1973, Fabricational noise in adaptive morphology, *Syst. Zool.* **22**:451–465.
Seilacher, A., 1977, Evolution of trace fossil communities, in *Patterns of Evolution, as Illustrated by the Fossil Record*, A. Hallam, ed., Elsevier, New York, pp. 359–376.
Seilacher, A., 1979, Constructional morphology of sand dollars, *Paleobiology* **5**:191–221.
Selley, R. C., 1976, *An Introduction to Sedimentology*, Academic Press, London.
Sepkoski, J. J., Jr., 1978, A kinetic model of phanerozoic taxonomic diversity. I. Analysis of marine orders, *Paleobiology* **4**:223–251.
Sepkoski, J. J., Jr., 1981, A factor analytic description of the phanerozoic marine fossil record, *Paleobiology* **7**:36–53.
Sepkoski, J. J., Jr., and P. R. Crane, eds., 1985, Tenth anniversary issue, *Paleobiology* **11**:1–138.
Shaw, A. B., 1964, *Time in Stratigraphy*, McGraw-Hill, New York.
Shaw, A. B., 1969, Adam and Eve, paleontology, and the non-objective arts, *Jour. Paleontol.* **43**:1085–1098.
Sibley, C. G., chairman, 1969, *Systematic Biology: Proceedings of an International Congress*, Publication 1692, National Academy of Sciences, Washington, D.C.
Siegel, A. F., and R. H. Benson, 1982, A robust comparison of biological shapes, *Biometrics* **38**:341–350.
Silver, L. T., and P. H. Schultze, eds., 1982, *Geological Implications of Impacts of Large Asteroids and Comets on the Earth*, Geol. Soc. Amer. Spec. Pap. 190, Geological Society of America, Boulder, Colo.
Simpson, G. G., 1940, Mammals and land bridges, *Jour. Washington Acad. Sci.* **30**:137–163.
Simpson, G. G., 1943, Criteria for genera, species, and subspecies in zoology and paleozoology, *Ann. New York Acad. Sci.* **44**:145–178.
Simpson, G. G., 1944, *Tempo and Mode in Evolution*, Columbia University Press, New York.
Simpson, G. G., 1945, The principles of classification and a classification of mammals, *Bull. Amer. Mus. Nat. Hist.* **85**:1–350.
Simpson, G. G., 1951, The species concept, *Evolution* **5**:285–293.
Simpson, G. G., 1953a, *The Major Features of Evolution*, Columbia University Press, New York.
Simpson, G. G., 1953b, *Evolution and Geography*, Condon Lectures, Oregon State System of Higher Education, Eugene.
Simpson, G. G., 1959, Mesozoic mammals and the polyphyletic origin of mammals, *Evolution* **13**:405–414.
Simpson, G. G., 1961, *Principles of Animal Taxonomy*, Columbia University Press, New York.
Simpson, G. G., 1965, *The Geography of Evolution: Collected Essays*, Capricorn Books, New York.
Simpson, G. G., 1975, Recent advances in methods of phylogenetic inference, in *Phylogeny of the Primates*, W. P. Luckett and F. S. Szalay, eds., Plenum Press, New York, pp. 3–19.
Simpson, G. G., 1980, *Why and How: Some Problems and Methods in Historical Biology*, Pergamon Press, Oxford, England.
Simpson, G. G., A. Roe, and R. C. Lewontin, 1960, *Quantitative Zoology*, Harcourt, Brace and World, New York.
Slobodchikoff, C. N., ed., 1976, *Concepts of Species*, Dowden, Hutchinson and Ross, Stroudsburg, Pa.
Sneath, P. H. A., 1982, Review of *Systematics and Biogeography: Cladistics and Vicariance* by G. Nelson and N. Platnick, *Syst. Zool.* **31**:208–217.
Sneath, P. H. A., and R. R. Sokal, 1973, *Numerical Taxonomy*, Freeman, San Francisco.

Sober, E., 1975, *Simplicity*, Oxford University Press, London.

Sober, E., 1983*a*, Parsimony in systematics: Philosophical issues, *Ann. Rev. Ecol. Syst.* **14**:335–357.

Sober, E., 1983*b*, Parsimony Methods in Systematics, in *Advances in Cladistics, Vol. 2*, N. I. Platnick and V. A. Funk, eds., Columbia University Press, New York, pp. 37–47.

Sober, E., 1984*a*, Common cause explanation, *Philosophy of Science* **51**:212–241.

Sober, E., ed., 1984*b*, *Conceptual Issues in Evolutionary Biology*, MIT Press, Cambridge, Mass.

Sober, E., 1984*c*, *The Nature of Selection: Evolutionary Theory in Philosophical Focus*, MIT Press, Cambridge, Mass.

Sokal, R. R., 1983*a*, A phylogenetic analysis of the caminalcules. I. The data base, *Syst. Zool.* **32**:159–184.

Sokal, R. R., 1983*b*, A phylogenetic analysis of the caminalcules. II. Estimating the true cladogram, *Syst. Zool.* **32**:185–201.

Sokal, R. R., 1983*c*, A phylogenetic analysis of the caminalcules. III. Fossils and classification, *Syst. Zool.* **32**:248–258.

Sokal, R. R., 1983*d*, A phylogenetic analysis of the caminalcules. IV. Congruence and character stability, *Syst. Zool.* **32**:259–275.

Sokal, R. R., and T. J. Crovello, 1970, The biological species concept: A critical evaluation, *Amer. Nat.* **104**:127–153.

Sokal, R. R., and P. H. A. Sneath, 1963, *Principles of Numerical Taxonomy*, Freeman, San Francisco.

Southam, J. R., W. W. Hay, and T. R. Worsley, 1975, Quantitative formulation of reliability in stratigraphy, *Science* **188**:357–359.

Stanley, S. M., 1975, A theory of evolution above the species level, *Proc. Nat. Acad. Sci. USA* **72**:646–650.

Stanley, S. M., 1979, *Macroevolution: Pattern and Process*, Freeman, San Francisco.

Stanley, S. M., 1985, Rates of evolution, *Paleobiology* **11**:13–26.

Stanley, S. M., P. W. Signor III, S. Lidgard, and A. F. Karr, 1981, Natural clades differ from "random" clades: Simulations and analyses, *Paleobiology* **7**:115–127.

Stebbins, G. L., 1966, *Processes of Organic Evolution*, Prentice-Hall, Englewood Cliffs, N.J.

Steineck, P. L., and R. L. Fleisher, 1978, Towards the classical evolutionary reclassification of Cenozoic Globigerinacea (Foraminiferida), *Jour. Paleontol.* **52**:618–635.

Stevens, P. F., 1980, Evolutionary polarity of character states, *Ann. Rev. Ecol. Syst.* **11**:333–358.

Stoll, N. R., R. P. Dollfus, J. Forest, N. D. Riley, C. W. Sabrosky, C. W. Wright, and R. V. Melville, 1964, *International Code of Zoological Nomenclature*, International Trust for Zoological Nomenclature, London.

Stuessy, T. F., and J. V. Crisci, 1984, Problems in the determination of evolutionary directionality of character-state change for phylogenetic reconstruction, in *Cladistics: Perspectives on the Reconstruction of Evolutionary History*, T. Duncan and T. F. Stuessy, eds., Columbia University Press, New York, pp. 71–92.

Sylvester-Bradley, P. C., ed., 1956, *The Species Concept in Palaeontology*, Systematics Association Publication No. 2, London.

Sylvester-Bradley, P. C., 1977, Biostratigraphical tests of evolutionary theory, in *Concepts and Methods in Biostratigraphy*, E. G. Kauffman and J. E. Hazel, eds., Dowden, Hutchinson and Ross, Stroudsburg, Pa., pp. 41–63.

Szalay, F. S., 1977*a*, Ancestors, descendants, sister groups and testing of phylogenetic hypotheses, *Syst. Zool.* **26**:12–18.

Szalay, F. S., 1977*b*, Phylogenetic relationships and a classification of the eutherian Mammalia, in *Major Patterns in Vertebrate Evolution*, M. K. Hecht, P. C. Goody, and B. M. Hecht, eds., Plenum Press, New York, pp. 315–374.

Szalay, F. S., 1977*c*, Constructing primate phylogenies: A search for testable hypotheses with maximum empirical content, *Jour. Human Evol.* **6**:3–18.

Szalay, F. S., 1981a, Functional analysis and the practice of the phylogenetic method as reflected by some mammalian studies, *Amer. Zool.* **21**:37–45.

Szalay, F. S., 1981b, Phylogeny and the problem of adaptive significance: The case of the earliest primates, *Folia Primatol.* **36**:157–182.

Szalay, F. S., and E. Delson, 1979, *Evolutionary History of the Primates*, Academic Press, New York.

Szalay, F. S., and G. Drawhorn, 1980, Evolution and diversification of the Archonta in an arboreal milieu, in *Comparative Biology and Evolutionary Relationships of Tree Shrews*, W. P. Luckett, ed., Plenum Press, New York, pp. 133–169.

Tattersall, I., and N. Eldredge, 1977, Fact, theory, and fantasy in human paleontology, *Amer. Sci.* **65**:204–211.

Tedford, R. H., 1970, Principles and practices of mammalian geochronology in North America, *North American Paleontol. Conv., 1st Proc.* **1**:666–703.

Telford, S. R., III, 1984, Goldschmidt reaffirmed: A positive step towards a new synthesis? *Evol. Theory* **7**:52.

Tipper, J. C., 1983, Rates of sedimentation, and stratigraphical completeness, *Nature* **302**:696–698.

Thompson, D. W., 1961, *On Growth and Form* (abridged edition, J. T. Bonner, ed.), Cambridge University Press, Cambridge.

Torbett, M. V., and R. Smoluchowski, 1984, Orbital stability of the unseen solar companion linked to periodic extinction events, *Nature* **311**:641–642.

Turner, J. R. G., 1983, "The hypothesis that explains mimetic resemblance explains evolution": The gradualist—saltationist schism, in *Dimensions of Darwinism: Themes and Counterthemes in Twentieth-Century Evolutionary Theory*, M. Grene, ed., Cambridge University Press, Cambridge, pp. 129–169.

Tway, L. E., and J. Zidek, 1982, Catalog of Late Pennsylvanian ichthyoliths, Part I, *Jour. Vert. Paleontol.* **2**:328–361.

Tway, L. E., and J. Zidek, 1983, Catalog of Late Pennsylvanian ichthyoliths, Part II, *Jour. Vert. Paleontol.* **2**:414–438.

Udvardy, M. D. F., 1969, *Dynamic Zoogeography with Special Reference to Land Animals*, Van Nostrand Reinhold, New York.

Valentine, J. W., 1973, *Evolutionary Paleoecology of the Marine Biosphere*, Prentice-Hall, Englewood Cliffs, N.J.

Valentine, J. W., 1975, Adaptive strategy and the origin of grades and ground plans, *Amer. Zool.* **15**:391–404.

Van Andel, T. H., 1981, Consider the incompleteness of the geological record, *Nature* **294**:397–398.

Van Valen, L., 1973, A new evolutionary law, *Evol. Theory* **1**:1–30.

Van Valen, L., 1974, A natural model for the origin of some higher taxa, *Jour. Herpetol.* **8**:109–121.

Van Valen, L., 1978, Why not to be a cladist, *Evol. Theory* **3**:285–299.

Van Valen, L., 1984, A resetting of phanerozoic community evolution, *Nature* **307**:50–52.

Van Valen, L., and R. E. Sloan, 1966, The extinction of the multituberculates, *Syst. Zool.* **15**:261–278.

Vialov, O. S., 1972, The classification of the fossil traces of life, *24th Internat. Geol. Congr., Montreal 1972, Proc. Sect. 7 (Paleontology)*, pp. 639–644.

Von Baer, K. E., 1828, *Ueber Entwickelungsgeschichte der Thiere. Beobachtung und Reflexion*, Borntrager, Konigsberg.

Vrba, E. S., 1980, Evolution, species and fossils: How does life evolve? *South Afr. Jour. Sci.* **76**:61–84.

Vrba, E. S., 1982a, Darwinism in 1982: The triumph and the challenges, *South Afr. Jour. Sci.* **78**:275–278.

Vrba, E. S., 1982b, The evolution of trends, in *Rythmes et Mecanisms de l'Evolution Biologique,* J. Chaline, ed., Colloques Internationaux du Centre National de la Recherche Scientifique, no. 330, Dijon, pp. 239–246.

Vrba, E. S., 1983, Macroevolutionary trends: New perspectives on the roles of adaptation and incidental effect, *Science* **221**:387–389.

Vrba, E. S., 1984a, Patterns in the fossil record and evolutionary processes, in *Beyond Neo-Darwinism: An Introduction to the New Evolutionary Paradigm,* M.-W. Ho and P. T. Saunders, eds., Academic Press, London, pp. 115–142.

Vrba, E. S., 1984b, What is species selection?, *Syst. Zool.* **33**:318–328.

Vrba, E. S., and N. Eldredge, 1984, Individuals, hierarchies and processes: Towards a more complete evolutionary theory, *Paleobiology* **10**:139–164.

Vuilleumeir, F., 1978, Qu'est-ce que la biogeographie?, *C. R. Somm. Seanc. Soc. Biogeogr.* **54**:41–66.

Wagner, W. H., Jr., 1969, The construction of a classification, in *Systematic Biology: Proceedings of an International Conference,* C. G. Sibley, chairman, Publication 1692, National Academy of Sciences, Washington, D.C., pp. 67–90.

Wagner, W. H., Jr., 1980, Origin and philosophy of the groundplan-divergence method of cladistics, *Syst. Bot.* **5**:173–193.

Wagner, W. H., Jr., 1984, Applications of the concepts of groundplan-divergence, in *Cladistics: Perspectives on the Reconstruction of Evolutionary History,* T. Duncan and T. F. Stuessy, eds., Columbia University Press, New York, pp. 95–118.

Walker, E. P., 1975, *Mammals of the World,* The Johns Hopkins University Press, Baltimore.

Wallace, A. R., 1876, *The Geographical Distribution of Animals,* 2 vols., Macmillan, London.

Wallace, A. R., 1880, *Island Life,* Macmillan, London.

Watrous, L. E., and Q. D. Wheeler, 1981, The out-group method of character analysis, *Syst. Zool.* **30**:1–11.

Werdelin, L., and J. O. R. Hermelin, 1983, Testing for ecophenotypic variation in a benthic foraminifer, *Lethaia* **16**:303–307.

West-Eberhard, M. J., 1983, Sexual selection, social competition, and speciation, *Quart. Rev. Biol.* **58**:155–183.

Westoll, T. S., 1949, On the evolution of the Dipnoi, in *Genetics, Paleontology and Evolution,* G. L. Jepsen, G. G. Simpson, and E. Mayr, eds., Princeton University Press, Princeton, N.J., pp. 121–184.

Whewell, W., 1847, *The Philosophy of the Inductive Sciences,* John Parker, London.

Whiffin, T., and M. W. Bierner, 1972, A quick method for computing Wagner trees, *Taxon* **21**:83–90.

White, M. J. D., 1978a, *Modes of Speciation,* Freeman, San Francisco.

White, M. J. D., 1978b, Chain processes in chromosomal speciation, *Syst. Zool.* **27**:285–298.

Wiggins, V. D., 1982, *Expressipollis striatus* n. sp. to *Anacolosidites striatus* n. sp. An Upper Cretaceous example of suggested pollen aperture evolution, *Grana* **21**:39–49.

Wiley, E. O., 1977, Are monotypic genera paraphyletic? A reply to Norman Platnick, *Syst. Zool.* **26**:352–355.

Wiley, E. O., 1978, The evolutionary species concept reconsidered, *Syst. Zool.* **27**:17–26.

Wiley, E. O., 1979a, Ancestors, species, and cladograms—Remarks on the symposium, in *Phylogenetic Analysis and Paleontology,* J. Cracraft and N. Eldredge, eds., Columbia University Press, New York, pp. 211–225.

Wiley, E. O., 1979b, An annotated Linnaean hierarchy, with comments on natural taxa and competing systems, *Syst. Zool.* **28**:308–337.

Wiley, E. O., 1980, Is the evolutionary species fiction? *Syst. Zool.* **29**:76–80.

Wiley, E. O., 1981, *Phylogenetics: The Theory and Practice of Phylogenetic Systematics,* Wiley, New York.

Wiley, E. O., and D. R. Brooks, 1982, Victims of history—A nonequilibrium approach to evolution, *Syst. Zool.* **31**:1–24.

Wiley, E. O., and D. R. Brooks, 1983, Nonequilibrium thermodynamics and evolution: A response to Løvtrup, *Syst. Zool.* **32**:209–219.

Williams, H., F. J. Turner, and C. M. Gilbert, 1954, *Petrography: An Introduction to the Study of Rocks in Thin Sections,* Freeman, San Francisco.

Wilson, E. O., 1965, A consistency test for phylogenies based on contemporaneous species, *Syst. Zool.* **14**:214–220.

Wing, S. L., and L. J. Hickey, 1984, The Platycarya perplex and the evolution of the Juglandaceae, *Amer. Jour. Bot.* **71**:388–411.

Wyss, A., and K. de Queiroz, 1984, Phylogenetic methods and the early history of amniotes: A comment on Caroll (1982), *Jour. Vert. Paleontol.* **4**:604–608.

Zangerl, R., 1948, The Methods of comparative anatomy and its contribution to the study of evolution, *Evolution* **2**:351–374.

Zinsmeister, W. J., and R. M. Feldmann, 1984a, Cenozoic high-latitude heterochroneity of Southern Hemisphere marine faunas, *Science* **224**:281–283.

Zinsmeister, W. J., and R. M. Feldmann, 1984b, Role of high-latitude heterochroneity in the evolution of modern marine faunas, *Geol. Soc. Amer., Abstracts with Programs* **16**:705.

Zuckerkandl, E., 1983, Molecular basis for directional evolution, in *Rythmes et Mecanisms de l'Evolution Biologique,* J. Chaline, ed., Colloques Internationaux du Centre National de la Recherche Scientifique, no. 330, Dijon, p. 337.

Glossary

acceleration. Increase in the rate of development within a lineage, such that a particular feature appears earlier in the ontogeny of a descendant than in the ontogeny of the ancestor. *See* heterochrony.

ad hoc hypothesis. An explanation or hypothesis invoked to account for a particular, single observation in the absence of a wider or more general application of the hypothesis. An ad hoc hypothesis of homoplasy is a hypothesis invoked to dismiss a putative synapomorphy that is not congruent or consistent with other apparent synapomorphies in the context of a particular phylogenetic study. *See* parsimony.

affinity. A relatively imprecise concept of relationship between biological groups or entities, usually based on raw morphological similarity that is thought to indicate genealogical closeness or community of origin.

agamospecies. An asexual species—that is, a species that does not undergo sexual reproduction.

allometry. Change in shape of an organism or series of organisms that is correlated with a change in size. Allometry may refer to changes in shape correlated with size in the ontogeny (development) of an individual, changes in shape and size in an evolutionary lineage, or correlated changes in shape and size among the comparable semaphoronts (e.g., adults) of related species of organisms.

allopatric. Referring to populations or species of organisms that occupy separate and mutually exclusive geographic areas.

alpha taxonomy. Species-level taxonomy.

anagenesis. Evolution within a single lineage of organisms without speciation or divergence, which produces an increase in morphological (or other) diversity through time. *See* cladogenesis.

333

analogy. Similar structures in different organisms that are not the same structure—that is, not homologous; they cannot be traced to a single feature in a common ancestor. *See* homology.
apomorphy. Referring to an evolutionarily derived or advanced condition of a character.
autapomorphy. An apomorphy that is unique to, and distinguishes, a particular taxon.

beta taxonomy. Taxonomy concerned with arranging species into higher taxa.
biota. The organisms (flora and fauna) of a particular region.

category, taxonomic. A named category, such as phylum, order, family, tribe, species, subspecies, that specifies the hierarchical rank or level of inclusion of a certain taxon in a Linnaean classification.
character. A peculiarity, trait, or feature that distinguishes a particular semaphoront, organism, or group of organisms. Approximate synonyms of character include: attribute, characteristic, character state, feature, property, trait.
cladism. (= cladistics.) A school of phylogeny reconstruction (and systematics and taxonomy) that utilizes exclusively the parsimonious (ad hoc hypotheses of homoplasy are minimized) distribution of shared and derived characters (synapomorphies) to reconstruct phylogeny and classify organisms. Only monophyletic groups, *sensu stricto*, are admissible in a formal cladistic classification.
cladogenesis. That aspect of the evolution of a lineage(s) involving splitting events, speciation, and the increase in biological diversity.
cladogram. A usually bifurcating diagram, dendrogram, or phylogram that graphically represents the phylogenetic relationships (or the distribution of synapomorphies) among a group of organisms or taxa.
classification, biological. The actual ordering of organisms, either formally or informally, into groups and subgroups based on some particular criteria. A biological classification usually takes the form of a set of words (names) that denote a hierarchical grouping of organisms (e.g., the Linnaean system of classification).
cline. A character gradient; a gradual and continuous change of a character within a species, usually seen in a series of populations that are geographically or stratigraphically contiguous.
completeness. The portion of a given time span that is represented by some net accumulation of rock in a particular stratigraphic section.
convergence. Similarity of characters in independent, or only distantly related, lineages.
cryptic species. *See* sibling species.

deme. A local population of a species, the members of which freely interbreed.
dendrogram. A branching diagram based on the linking of entities, using any criterion.
derived character. An evolutionary novelty or advanced evolutionary state.
dispersal. The movement of a species or group of organisms from one geographic area to another.

ecophenotypic variation. Variation among members of a species that is due to nongenetic, noninherited modifications (such as the ecological conditions in which an organism develops).
endosymbiosis. The combining of two distantly related lineages to form a new lineage, originating in the occupation of a symbiont in a host species; eventually the symbiont and host becoming inseparable. *See* reticulate evolution.
EU. Evolutionary unit; term sometimes used for taxa at infraspecific levels.
evolution. Organic change: specifically, the paths and ways of development of a lineage (sequence of ancestors and descendants) through time; phylogeny plus various related components, such as ancestor-descendant relationships; degrees of morphological, physiological, behavioral,

or ecological divergence; adaptation; and tempos, patterns, and processes of change through time. *See* genealogy, phylogeny.

evolutionary systematics. The traditional school of systematics and phylogeny reconstruction, exemplified by the work of Mayr and Simpson. This school, based on the new synthesis of the 1930s and 1940s, takes an extremely eclectic and syncretic approach to phylogeny and classification. Methodologies of the phenetic and cladistic schools are adopted as they are found to be useful. Paraphyletic groups are routinely accepted as useful and natural.

evolutionary tree. (= phylogenetic tree.) Schematic diagram that purports to illustrate the evolutionary history of a group of organisms by means of ancestors (real or hypothetical) and descendants.

fauna. The animal life of a particular geographic region. *See* biota.

flora. The plant life of a particular geographic region. *See* biota.

gamma taxonomy. Taxonomy concerned with the mechanisms, processes, and tempos of evolution.

genealogy. The pattern of mating, parentage, and descent among organisms.

grade. A group of organisms that is believed to have reached (in the process of their evolution) a particular arbitrary level of organization, adaptation, or complexity.

heterobathmy. The concept of both relatively primitive and derived characters distributed throughout a group of related taxa; given any set of monophyletic groups, any one group will be primitive for certain characters and evolutionarily advanced for other characters.

heterochrony. Change in the timing of development (either in the rate of development or time of appearance of a particular structure) within an evolutionary lineage; that is, the ontogeny of the ancestor differs from the ontogeny of the descendant. *See* acceleration, hypermorphosis, neoteny, paedomorphosis, recapitulation, retardation.

hierarchy. System of ranks and categories, as in the Linnaean system of classification; nested sets within sets (groups within groups, or taxa within taxa).

hologenetic relationships. The totality of ontogenetic, tokogenetic, and phylogenetic relationships between semaphoronts.

holomorph. An individual organism—composed of the sequence of semaphoronts (connected by ontogenetic relationships) that make up its life cycle. *See* semaphoront.

holophyletic. Referring to a group of organisms which is a monophyletic group *sensu stricto*. *See* monophyletic, paraphyletic.

homeomorph. Two organisms that are not particularly closely related but have extremely similar (in many details) overall morphologies. In some cases, it may be erroneously believed that the homeomorphs are more closely related to each other than they really are, and homeomorphs may be confused with one another.

homology. Similar structures in different organisms that are the same structure—that is, the observed similarities are due to inheritance from a common ancestor of the organisms under consideration. *See* analogy.

homoplasy. Referring to nonhomology that is mistaken for homology, specifically parallelism and convergence.

HTU. Hypothetical taxonomic unit.

hybridization. The successful interbreeding of dissimilar organisms, often of (previously) isolated lineages, to form a new lineage. *See* reticulate evolution.

hypermorphosis. The extension of the ontogeny of a descendant beyond that of the ancestor, such that adult stages of the ancestor become juvenile (or at least subadult and nonterminal) stages of the descendant. *See* heterochrony.

imago. The sexually mature adult stage of certain insects.

infraspecific. Below the species level, or within the species. Infraspecific also usually refers to below the subspecies level.

Linnaean system. Internationally recognized system(s) of classification of biological organisms, including formal rules of nomenclature.

macroevolution. Usually considered evolution at and above the species level, including the speciation process, the origin of biological diversity, and the origination of major evolutionary innovations.

microevolution. Usually considered evolution below the species level; minor evolutionary fine-tuning or random drift within a species or population that involves changes in gene frequencies.

molecular phylogeny. Phylogeny reconstructed on the basis of comparative biochemical and molecular data of organisms.

monophyletic. In the strict sense, a group of organisms that includes an ancestor and all of its descendants; as used by some authors, a monophyletic group need not include all of the descendants of the common ancestor (i.e., a paraphyletic group is considered monophyletic; *see* holophyletic, paraphyletic).

morphocline. (= transformation series.) Hypothesis as to the (evolutionary) pathway, but not direction, of change from one character (or character state) to another. *See* polarity.

morphospecies. A species recognized on the basis of morphological differences (as essentially all species must be in paleontology).

morphotype. A list of the synapomorphies (and sometimes also the plesiomorphies) that characterize a particular taxon.

neontology. The study of extant (living) organisms.

neoteny. The presence of formerly juvenile characters in adult descendants that is brought about by the retardation of the development of adult somatic features relative to the maturation of the reproductive organs. *See* heterochrony.

nomen dubium. A species-level scientific (Linnaean) name that is not applicable with certainty to any particular known species-level taxon, perhaps because the species was originally described on the basis of an inadequate type specimen, or because the type specimen has been lost since the original description.

nomen nudum. A scientific (Linnaean) name that has not been adequately published so as to be made available or legitimate.

nomen oblitum. A forgotten scientific (Linnaean) name; a name that has not been used in the literature for longer than a designated amount of time (specified by the applicable code: zoological, botanical, or bacteriological) and thus has lost its validity.

nomenclature. System of names, particularly as used in a biological classification, and the rules governing their proper application.

numerical taxonomy. The use of numerical algorithms, usually computer aided, in the analysis of data upon which biological classifications are based. *See* phenetics.

ontogeny. The developmental or life history of an organism, from conception to death. Ontogenetic (developmental) relationships link semaphoronts into individuals (holomorphs).

OTU. Operational taxonomic unit.

out-group. A group of organisms relatively closely related to the group of organisms under consideration in a particular study. The distribution of certain characters among the members of an out-group is often used to help determine the polarity of a morphocline within the group under study.

overall similarity. A broad measure of similarity between two organisms or biological taxa calculated on the basis of the similarities of numerous individual characters. *See* phenetics, numerical taxonomy.

paedomorphosis. The presence of formerly juvenile characters in adult descendants, which is brought about by the acceleration of the maturation of the reproductive organs relative to the development of the somatic tissues. The term is also sometimes used to encompass paedomorphosis as here defined and neoteny. *See* heterochrony, neoteny.

paleontology. The study of ancient life.

parallelism. Similar, nonhomologous characters in independent, but supposedly relatively closely related lineages.

parapatric. Populations or species whose geographic ranges are nonoverlapping, but in contact at a border.

paraphyletic. A group of organisms composed of a common ancestor and some, but not all, of its descendants. *See* holophyletic, monophyletic.

parsimony. The principle that ad hoc hypotheses should be minimized (the simplest explanation or solution is to be preferred). *See* ad hoc hypothesis.

patristic similarity. Similarity among organisms due to a common ancestry, without distinguishing between primitive and derived features (usually refers to relatively primitive features).

phenetics. School of taxonomy, also known as numerical taxonomy, that bases classifications of organisms on estimates of taxonomic affinity calculated using various mathematical clustering procedures applied to equally weighted coded taxonomic characters. *See* numerical taxonomy.

phenogram. A branching diagram that usually links organisms or higher taxa on the basis of some measure of overall similarity.

phenon. A phenotypically uniform group of similar specimens.

phenotype. The characteristics and appearance of an individual biological organism as a result of the interaction of its genetic inheritance and the environment in which it develops.

phyletic. Referring to a lineage, line of descent, or line of evolution.

phylogenetic relationships. The relative degrees of relatedness among organisms, species, or monophyletic taxa. Given three organisms, species, or monophyletic taxa (A, B, and C), if A and B have a common ancestor that is not also an ancestor of C, then A and B are phylogenetically more closely related to each other than either is to C—that is, A and B share a close phylogenetic relationship to each other relative to C. *See* phylogeny.

phylogenetic tree. *See* evolutionary tree.

phylogeny. Usually, the pattern of descent, genealogy, or evolutionary history of a group of organisms. Phylogeny *sensu stricto*, as used in this book, is a simple account of the relative degrees of relatedness (phylogenetic relationships) of certain organisms, species, or monophyletic taxa. *See* phylogenetic relationships.

phylograms. Hierarchical diagrams that are intended to depict phylogenetic relationships (e.g., cladograms) *and* evolutionary (phylogenetic) trees and networks.

plesiomorphy. An evolutionarily primitive or ancestral condition of a character.

plesion. In a formal classification utilizing the concept of plesions, a plesion is a fossil species or monophyletic group (equal to any categorical rank) that is the sister group of the group (or groups) that succeeds it.

polarity. Directionality of a morphocline along a spectrum of characters ranging from those that are evolutionarily primitive (early) to those that are evolutionarily advanced or derived (late). *See* morphocline.

polyphyletic. A group of organisms that does not include the common ancestor of all members of the group. *See* monophyletic.

polyploid. An organism that contains three or more times the haploid number of chromosomes.

recapitulation. The appearance of ancestral adult traits or stages as juvenile traits or stages of descendant organisms. *See* heterochrony.

retardation. Decrease in the rate of development within a lineage, such that a particular feature appears later in the ontogeny of a descendant than in the ontogeny of the ancestor. *See* heterochrony.

reticulate evolution. Evolution that occurs as the result of crossing and interbreeding among two or more formerly separate evolutionary lines. *See* hybridization, endosymbiosis.

scenario. A narrative explanation of how (and why) certain organisms evolved.

semaphoront. A single individual organism at a particular point in time (at a particular point in its ontogeny or life cycle). *See* holomorph.

sexual dimorphism. Consistent morphological or phenotypic differences between the males and females of a sexual species.

sibling species. Species that are nearly identical to one another, especially morphologically, and thus extremely difficult to distinguish from one another. Sibling species are usually considered to be more closely related to each other than they are to any other species.

sister groups. Two taxa that are most closely related to each other within the context of a certain study; two taxa that are each other's closest relatives (both taxa originated from the same speciation event that split an ancestral species into two daughter species) are referred to as sister groups, sister taxa, or twin taxa.

speciation. The splitting of a lineage causing an increase in organismal diversity (an increase in species).

species. Distinct groups of organisms that form the basis of Linnaean classification and many low-level phylogenetic analyses.

stem species. The ancestral species or common species of a monophyletic group.

subspecies. A group within a recognized Linnaean species that differs (is diagnosable) from other subspecies of the species; however, the subspecies of a particular species may grade into one another (often subspecies are distinguished on the basis of geographic or temporal [stratigraphic] data).

sympatric. Populations or species of organisms that occupy the same geographic area.

symplesiomorphy. Shared and primitive character; shared plesiomorphy.

synapomorphy. Shared and derived character; shared apomorphy.

systematics. The study of the diversity of organisms; generally includes taxonomy and classification as integral parts of this study.

taphonomy. The study of the processes involved in the transition of living organisms to the fossil state, and also the biases involved in the recovery and interpretation of fossil entities.

taxon. Any grouping of organisms that is potentially namable within a particular classificatory scheme and philosophy of biological classification.

taxonomy. The theoretical study of the ordering of the diversity of organisms, including the criteria and bases used in a classification of the organisms.

tokogenetic relationships. The genetic relationships among the individuals of a biparental species brought about through reproduction involving copulation between members of opposite sexes.

transformation series. *See* morphocline.

type-specimen. For a particular species, the standard of reference for the application of the particular species name. The type-specimen is the name bearer and by definition belongs to the species whose name it bears (of course, the limits of the particular species may be judged by some systematists to encompass those of another named species—if so, by the rule of priority, the oldest name takes precedence). Taxa above the species level, if based on a system of types, take as their types lower-level taxa. Thus, in zoology the type of a genus is a species, and the type of a family is a genus (and the type of this genus is a species, and the type of this species is a type-specimen); in zoology ordinal-level and higher taxa are not governed by the *International Code of Zoological Nomenclature* and are not based on a system of types.

vicariance. Any event (geographic, geologic, ecological, climatic, etc.) that subdivides the ancestral range of a population, species, or taxon.

weighting. The placing of more emphasis on particular characters in the analysis of phylogeny or the construction of a classification. Certain characters may be more heavily weighted relative to other characters because they are considered to be more reliable, or of more value, in reconstructing phylogeny or in the development of a stable classification.

Index

About the Author

ROBERT MILTON SCHOCH, Ph.D, is assistant professor in the Division of Science, College of Basic Studies, Boston University; assistant professor in the Department of Geology, College of Liberal Arts, Boston University; curatorial affiliate in vertebrate paleontology, Peabody Museum of Natural History, Yale University; and research associate at Schiele Museum of Natural History, Gastonia, North Carolina. He is a member of the Paleontological Society, the Society of Vertebrate Paleontology, the Geological Society of America, and Sigma Xi Scientific Research Society. Dr. Schoch has participated in several geological and paleontological expeditions, is the author of numerous scientific papers, and is the editor of *Vertebrate Paleontology,* also published by Van Nostrand Reinhold. Among his most recent research interests are vertebrate paleontology, theoretical paleontology, and evolutionary theory.